Studies in Computational Intelligence

Volume 973

The series "Studies in Computational Intelligence" (SCI) publishes new developments and advances in the various areas of computational intelligence—quickly and with a high quality. The intent is to cover the theory, applications, and design methods of computational intelligence, as embedded in the fields of engineering, computer science, physics and life sciences, as well as the methodologies behind them. The series contains monographs, lecture notes and edited volumes in computational intelligence spanning the areas of neural networks, connectionist systems, genetic algorithms, evolutionary computation, artificial intelligence, cellular automata, self-organizing systems, soft computing, fuzzy systems, and hybrid intelligent systems. Of particular value to both the contributors and the readership are the short publication timeframe and the world-wide distribution, which enable both wide and rapid dissemination of research output.

Indexed by SCOPUS, DBLP, WTI Frankfurt eG, zbMATH, SCImago.

All books published in the series are submitted for consideration in Web of Science.

More information about this series at http://www.springer.com/series/7092

Endre Pap
Editor

Artificial Intelligence: Theory and Applications

Springer

Editor
Endre Pap
Singidunum University
Belgrade, Serbia

ISSN 1860-949X ISSN 1860-9503 (electronic)
Studies in Computational Intelligence
ISBN 978-3-030-72710-9 ISBN 978-3-030-72711-6 (eBook)
https://doi.org/10.1007/978-3-030-72711-6

This Springer imprint is published by the registered company Springer Nature Switzerland AG
The registered company address is: Gewerbestrasse 11, 6330 Cham, Switzerland

Preface

Artificial Intelligence (AI) has become an important part of our everyday life, and it will continue to grow in importance in the future. Therefore, managing further developments in AI theory, and its applications in complex systems, is a matter of urgency. The most important research is in the domains of the human brain and powerful computers endowed with strong software. These are studied both separately and together. As Alan Turing said, "What we want is a machine that can learn from experience." Machine Learning (ML) is a branch of artificial intelligence, and is the study of computer algorithms that allow computer programs to automatically improve through experience.

The present book is an up-to-date collection of 18 chapters, written by internationally recognized experts in the fields of AI and environmental research. It is related to the project ATLAS, https://ai.ipb.ac.rs/, and completed in collaboration with leading scientists in the fields of AI and environmental research.

The volume centers on two main subjects, and is therefore organized into two parts, theory and applications, although all chapters contain both approaches.

Part I of this book is devoted to theoretical issues which are mostly mathematically based.

The chapter "Mathematical Foundation of Artificial Intelligence" is devoted to some of the mathematical tools used in AI, which are largely related to decision-making. Among the main tools are general aggregation functions (operators) with some important special cases, such as triangular norms and copulas. A significant extension of classical mathematical analysis is pseudo-analysis, which is illustrated with a short overview of some of its important applications. General fuzzy measures and the corresponding integrals, including a number of many different generalizations of the Choquet integral, are also presented.

The chapter "Collection and Decomposition Integrals in Multicriteria-Decision Support" focuses on recently introduced collection and decomposition integrals, and covers the Choquet, the Shilkret and the PAN integrals, among others. It discusses some properties and contains several examples of these integrals, in particular ones extending the Lebesgue integral. Possible applications in multicriteria-decision support and domains containing imprecision and uncertainty are shown.

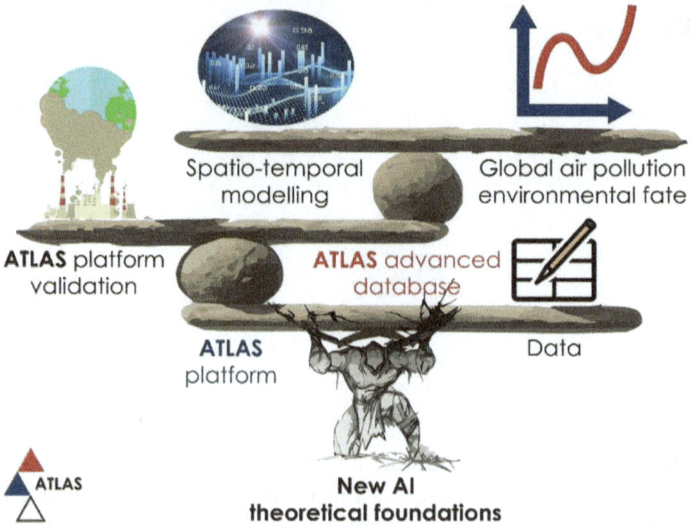

Fig. 1 ATLAS project and completed in collaboration with leading scientists

The chapter "A Refinement of the Jensen Type Inequality for the Pseudo-integral" presents new results related to the Jensen-type inequality for the pseudo-integral, using the Fubini type result for pseudo-integral. The fact that the pseudo-product of two pseudo measures reduces to repeated pseudo-integrals, enables obtaining a refinement of the Jensen-type inequality.

The chapter "Convolutional Neural Networks Hyperparameters Tuning" discusses image classification with convolution neural networks. In this early stage of CNN development, the common method of tuning CNN is by guessing and estimating, also known as the guestimating method. Since this is a hard optimization problem, there is a chance for optimization metaheuristic to be applied. One of the promising approaches is the application of swarm intelligence algorithms.

In the chapter "The Case for Quantifying Artificial General Intelligence with Entropy Semifields" it is supposed that Information Semifields are the underlying calculi that brains operate on, and it is postulated that strong artificial intelligence should try to imitate them. These semifields emerge from pseudo-calculus by the rational choices of generator functions that respect the classical postulates of information theory, and align with the Renyi entropy and divergence. The chapter provides relevant evidence from he fields of Neuroscience, Machine Learning, Cognitive Computation, Information Theory and Complex Systems Theory.

The chapter "Fuzzy Metrics and Its Applications in Image Processing" is devoted to fuzzy metrics and analyzes their application in image processing. Two applications of fuzzy metrics related to image processing are illustrated: Image filtering and Copy-move forgery detection in images. The aim was to improve the sharpness and the quality of the image, as measured by the image quality indices UIQI and CPBD. Fuzzy metric parameters that give the best image quality and sharpness are determined experimentally. Copy-Move Forgery Detection (CMFD) is one of the methods used to detect forgeries.

The chapter "Aggregation Operators and Distributivity Equations" is devoted to the issue of distributivity of aggregation operators, which is crucial in many different areas such as decision-making theory and integration theory. The characterization of all pairs *(F; G)* of special aggregation operators that satisfy the distributivity law are given.

The chapter "The Use of Fuzzy Logic in Various Combinatorial Optimization Problems" presents how fuzzy logic can be used for modeling uncertainties in a number of combinatorial problems and improve their quality. The focus is on the Location Set Covering Problem (LSCP), the Maximal Covering Location Problem (MCLP) and the Minimal Covering Location Problem (MinCLP) as a special modification of MCLP. These problems are applicable in searching optimal places for desired and undesired facilities under given conditions. Firstly, small dimensional instances of problems will be solved with an exact algorithm and using the CPLEX optimizer tool. When the dimensions of the problem become too large for exact solutions, the instances will then be solved with the Particle Swarm Optimization (PSO) metaheuristic.

The chapter "An Improved BAT Algorithm for Solving Job Scheduling Problems in Hotels and Restaurants" is an improvement to the original Bat Algorithm (BA) to speed up the convergence has been made. Each solution consists of many parameters. To conduct a comprehensive comparative analysis between the original BA, the modified BA proposed in this chapter and other state-of-the-art bio143inspired metaheuristics, the performance of both approaches is evaluated on a standard set of 23 (unimodal, multimodal and fixed-dimension multimodal) benchmark functions. Subsequently, the modified BA was applied to solve real-world practical job scheduling problems in hotels and restaurants.

Part II of the book presents a number of important applications of AI.

In the chapter "Patterns of PCB-138 Bioaccumulation in Small Pelagic Fish from the Eastern Mediterranean Sea Using Explainable Machine Learning Prediction" is examined the impact of 17 Fatty Acids (FAs) and 36 toxic organic and inorganic contaminants on the behavior patterns of the indicator congener PCB-138 in marine fish using eXtreme Gradient Boosting (XGBoost), SHapley Additive exPlanations (SHAP) and SHAP value fuzzy clustering. XGBoost indicated non-linear relationships between PCB-138 and the other examined variables that were explained by SHAP values. The ten fuzzy clusters of SHAP values that were obtained revealed that a higher intake of saturated myristic-C14:0 and margaric-C17:0 acids, followed by the intake of nutritionally beneficial eicosadienoic acid (C20:2n-6), mostly doesn't result in higher bio-accumulation of PCB-138.

In the chapter "Patterns of PCB-138 Occurrence in the Breast Milk of Primiparae and Multiparae Using SHapley Additive exPlanations Analysis", a novel SHapley Additive exPlanations (SHAP) method was used to examine the key parameters that govern the distribution of PCB-138 in breast milk. According to the results, PCB-156, PCB-180, HCB, HCH and PCB-118 have a major impact, while PCB-28, PCB-52 and PCB-189 have a minor impact on PCB-138 distribution in breast milk.

Similar contaminant behaviors, which belong to both the indicator congener group (−28, −52, −180) and the toxicologically relevant PCBs (−118, −189), were also noted.

In the chapter "What Information on Volatile Organic Compounds Can Be Obtained from the Data of a Single Measurement Site Through the Use of Artificial Intelligence?", an innovative integrated methodology for the spatio-temporal characterization and concentration forecasts of various pollutants was used. Toxic, mutagenic and carcinogenic representatives of volatile organic species—benzene, toluene, ethylbenzene and xylene, commonly referred to as BTEX—were studied. Increasing air pollutant concentrations over the last few decades have been the focus of contemporary scientific research due to their adverse effects on public health, the environment and climate change. The methodology is based on receptor-oriented air circulation modeling and artificial intelligence, implemented through machine learning and explainable artificial intelligence methods. The correlations and ratios between BTEX compounds were used for estimating their interrelationships and presence in the air, which contributed to the identification of their origin. XGBoost was efficient at forecasting BTEX levels, with low estimated errors (6–15%) significantly below the uncertainty obtained by conventional models for the evaluation of average annual pollutant concentrations.

The chapter "The Linear Fuzzy Space: Theory and Applications" covers the main theoretical results of the linear fuzzy space. It also analyzes its capabilities to model spatial and spatio-temporal systems using representative examples of real-world problems. It contains a brief, yet comprehensive, presentation of linear fuzzy space theory. The third section presents an analysis of fuzzy linear space theory concerning the modeling of spatial and temporal systems using two examples which present its applications in modeling application topology and time series.

The chapter "Image Fuzzy Segmentation Using Aggregated Distance Functions and Pixel Descriptors" is a review of recent research on image fuzzy segmentation using a fuzzy c-means clustering algorithm based on the distance function constructed by applying an aggregation function on the sequence of the initial distance functions and pixel descriptors. In image segmentation algorithms, distance functions compare pixels and represent a decision criterion for classifying the pixels into image segments. The segmentation criterion is determined based on the information fusion process, where the application of the appropriated aggregation function enables the adjustment of the segmentation criteria according to the intuitively expected decision.

The chapter "A Generative Model for the Creation of Large Synthetic Image Datasets Used for Distance Estimation" presents a generative model for the creation of large synthetic image datasets. The model is implemented as a 3D scene, representing an urban environment in Blender. The images are rendered using the Cycles rendering engine which allows for the creation of high fidelity data. The data was used for end-to-end training of a convolutional neural network, designed to estimate distances from stereoscopic images. The evaluation of the neural networks' performance showed that the generative model presented in this chapter is a viable tool for the generation of large image datasets for training predictive models.

The chapter "Appraisal of Apartments in Belgrade Using Hedonic Regression: Model Specification, Predictive Performance, Suitability for Mass Appraisal, and Comparison with Machine Learning Methods" presents a Bayesian hedonic regression model for the appraisal of apartments located within the metropolitan area of Belgrade, Serbia. Nested random effects are used to model the hierarchical structure present in the location identifiers, and thin-plate spline functions are used to capture non-linear effects. Major factors affecting prices include area in m2, floor number, the total number of floors in the building, the availability of an elevator and condition. The model achieves a Mean Average Percent Error (MAPE) of 13.06% in the validation set. Its predictive performance is compared to that of three popular Machine Learning (ML) methods, and its suitability for mass appraisal is examined.

Chapter "The Role of Chatbots in Foreign Language Learning: The Present Situation and the Future Outlook" is devoted to chatbots, computer programs developed to engage in conversations with humans. The most natural and potentially powerful application of chatbots is related to their fundamental nature—language practice. However, their role and outcomes within both formal and informal language learning is currently tangential the best. Existing research in the area has generally focused on chatbots' comprehensibility and the motivation they inspire in their users. In this chapter, an overview of chatbots for learning languages is provided, existing approaches are critically analyzed and major challenges for future research are given.

The chapter "Intelligent Interactive Technologies for Mental Health and Well-Being" presents intelligent technologies used to automate the assessment and evaluation of psychological treatments and mental well-being and functioning. The technologies include different types of robots, video games and conversational agents. The chapter critically analyzes existing solutions with future research implications.

The book is intended for researchers in AI, and researchers in environmental sciences as well as for Ph.D. students.

The editor is grateful to the authors for their excellent contributions. Many thanks go to proofreaders Profs. Jasna Petrović and Luke Deighton Chrisostomides for their effort to help the volume reach its present form. This book was supported by the Science Fund of the Republic of Serbia, #Grant No. 6524105, AI-ATLAS.

Belgrade, Serbia
February 2021

Endre Pap

Contents

Theory

Mathematical Foundation of Artificial Intelligence 3
Endre Pap

Collection and Decomposition Integrals in Multicriteria-Decision Support 31
Radko Mesiar and Adam Šeliga

A Refinement of the Jensen Type Inequality for the Pseudo-integral 47
Mirjana Štrboja, Biljana Mihailović, and Jelena Ivetić

Convolutional Neural Networks Hyperparameters Tuning 65
Eva Tuba, Nebojša Bačanin, Ivana Strumberger, and Milan Tuba

The Case for Quantifying Artificial General Intelligence with Entropy Semifields 85
Francisco J. Valverde-Albacete and Carmen Peláez-Moreno

Fuzzy Metrics and Its Applications in Image Processing 99
Nebojša Ralević

Aggregation Operators and Distributivity Equations 121
Dragan Jočić and Ivana Štajner-Papuga

The Use of Fuzzy Logic in Various Combinatorial Optimization Problems 137
Darko Drakulić, Aleksandar Takači, and Miroslav Marić

An Improved BAT Algorithm for Solving Job Scheduling Problems in Hotels and Restaurants 155
Tarik A. Rashid, Chra I. Shekho Toghramchi, Heja Sindi, Abeer Alsadoon, Nebojša Bačanin, Shahla U. Umar, A. S. Shamsaldin, and Mokhtar Mohammadi

Applications

Patterns of PCB-138 Bioaccumulation in Small Pelagic Fish from the Eastern Mediterranean Sea Using Explainable Machine Learning Prediction 175
Andreja Stojić, Bosiljka Mustać, Gordana Jovanović, Jasna Đinović Stojanović, Mirjana Perišić, Svetlana Stanišić, and Snježana Herceg Romanić

Patterns of PCB-138 Occurrence in the Breast Milk of Primiparae and Multiparae Using SHapley Additive exPlanations Analysis 191
Gordana Jovanović, Marijana Matek Sarić, Snježana Herceg Romanić, Svetlana Stanišić, Marija Mitrović Dankulov, Aleksandar Popović, and Mirjana Perišić

What Information on Volatile Organic Compounds Can Be Obtained from the Data of a Single Measurement Site Through the Use of Artificial Intelligence? 207
Svetlana Stanišić, Mirjana Perišić, Gordana Jovanović, Dimitrije Maletić, Dušan Vudragović, Ana Vranić, and Andreja Stojić

The Linear Fuzzy Space: Theory and Applications 227
Đorđe Obradović, Zora Konjović, Endre Pap, and Andrej Šoštarić

Image Fuzzy Segmentation Using Aggregated Distance Functions and Pixel Descriptors 255
Endre Pap, Ljubo Nedović, and Nebojša Ralević

A Generative Model for the Creation of Large Synthetic Image Datasets Used for Distance Estimation 275
Nebojša Nešić, Mladen Vidović, Ivan Radosavljević, Aleksandra Mitrović, and Đorđe Obradović

Appraisal of Apartments in Belgrade Using Hedonic Regression: Model Specification, Predictive Performance, Suitability for Mass Appraisal, and Comparison with Machine Learning Methods 293
Nemanja Stanišić, Tijana Radojević, and Nenad Stanić

The Role of Chatbots in Foreign Language Learning: The Present Situation and the Future Outlook 313
Jasna Petrović and Mlađan Jovanović

Intelligent Interactive Technologies for Mental Health and Well-Being 331
Mlađan Jovanović, Aleksandar Jevremović, and Milica Pejović-Milovančević

Theory

Mathematical Foundation of Artificial Intelligence

Endre Pap

Abstract In this chapter we present some of the mathematical tools used in AI, mostly related to decision making. One of the main tools are general aggregation functions (operators) with some special important cases as triangular norms and copulas. We present briefly some of their important applications. Important extension of the classical field of real numbers is related to the so called pseudo-operations: pseudo-addition and pseudo-multiplication. Here we present some of the important cases. Then corresponding pseudo additive measures and corresponding integrals are introduced, which make the base for the so called pseudo-analysis. The usefulness of the pseudo-analysis, which is a approach to nonlinearity, uncertainty and optimization, is illustrated with a short overview of some important applications. Very important tools in modeling decision making with overlaped events are general fuzzy measures and corresponding integrals as Choquet integral. We present a fuzzy integral approach to Cumulative Prospect theory. Further, we present many different generalizations of the Choquet integral.

Keywords Artificial intelligence · Aggregation function · Triangular norm · Copula · Fuzzy logic · Fuzzy sets · Pseudo-analysis · Fuzzy measure · Choquet integral · Cumulative Prospect Theory

1 Introduction

Generally there are many different mathematical tools and methods used in AI. Here we present some of the mathematical tools used in AI, mostly related to decision making. Some of them will be used in the next chapters. There are also many other approaches to AI, e.g., with machine learning, see [12], neural networks, with DeepLearning, genetic algorithms. etc. (but here are also used some of the mathemat-

E. Pap (✉)
Singidunum University, Belgrade, Serbia
e-mail: epap@singidunum.ac.rs

E. Pap (ed.), *Artificial Intelligence: Theory and Applications*,
Studies in Computational Intelligence 973,
https://doi.org/10.1007/978-3-030-72711-6_1

ical tools presented here, but also as regression analysis, etc.). We will not consider them here.

First of all one of the main tools are general aggregation functions (operators) with some special important cases as triangular norms and copulas. We present briefly some of their applications as dependence of many dimensional distribution functions, fuzzy logics, fuzzy sets. Important extension of the classical field of real numbers is related to the so called pseudo-operations: pseudo-addition and pseudo-multiplication. Here we present some of the important cases. Then corresponding pseudo additive measures and corresponding integrals are introduced, which make the base for the so called pseudo-analysis. The usefulness of the pseudo-analysis, which is a approach to nonlinearity, uncertainty and optimization, is illustrated with a short overview of some important applications: utility theory, nonlinear equations, fuzzy numbers, information theory, system theory, option pricing, large deviation principle, physics of the universe.

Very important tools in modeling decision making with overlaped events are general fuzzy measures and corresponding integrals as Choquet integral and Sugeno integral. These integrals play important roles in decision making. We present a fuzzy integral approach to Cumulative Prospect theory. Further, we give many new different generalizations of the Choquet integral, as nonadditive integrals with new operations, integral based on sublinear means, superdecomposition integral, integrals defined by two fuzzy measures.

2 Aggregation Functions

One of the basic tools in AI is the process of combining several numerical values into a single representative, one is called *aggregation*, and the numerical function performing this process is called *aggregation function (aggregation operator)*, requiring some natural conditions as monotonicity and boundary conditions. There is large field of application of aggregation: applied mathematics (e.g., probability, statistics, decision theory), computer sciences (e.g., artificial intelligence, operations research), as well as many applied fields (economy and finance, pattern recognition and image processing, data fusion, multicriteria decision aid, automated reasoning, etc.), see [21, 28, 60]. We shall give here only the basic definition of the aggregagation function with some basic examples, as triangular norms and copulas.

2.1 General Definition

We assume that variables are from a common domain $\mathbb{I}$, which is a nonempty real interval, bounded or not of the extended real number system $\overline{\mathbb{R}} = [-\infty, \infty]$, where we use the convention that $0 \cdot \infty = 0$ and $\infty + (-\infty) = -\infty$. We denote by $\mathbb{N}$ the set of natural numbers and $|V|$ is the cardinal number of the set V. We have by [21].

Definition 1 An *aggregation function* in $\mathbb{I}^n$ is a function $\mathsf{A}^{(n)} : \mathbb{I}^n \to \mathbb{I}$ which satisfies the following conditions

(i) is nondecreasing (in each variable)
(ii) the boundary conditions

$$\inf_{x_1,\ldots,x_n \in \mathbb{I}^n} \mathsf{A}^{(n)}(x_1, \ldots, x_n) = \inf \mathbb{I} \quad \text{and} \quad \sup_{(x_1,\ldots,x_n) \in \mathbb{I}^n} \mathsf{A}^{(n)}(x_1, \ldots, x_n) = \sup \mathbb{I}.$$

Specialy, for $\mathbb{I} = [0, 1]$ the condition (ii) reduces on $\mathsf{A}(0, \ldots, 0) = 0$ and $\mathsf{A}(1, \ldots, 1) = 1$.

Example 1 (i) One of the mostly used aggregation function is the *arithmetic mean*, given by

$$\mathsf{AM}^{(n)}(x_1, \ldots, x_n) = \frac{1}{n} \sum_{i=1}^{n} x_i$$

is an aggregation function in any domain $\mathbb{I}^n$.
(ii) The *geometric mean* function $\mathsf{GM} : \mathbb{I}^n \to \mathbb{I}$ is given by

$$\mathsf{GM}(x_1, \ldots, x_n) = \left(\prod_{i=1}^{n} x_i \right)^{1/n}.$$

The aggregation of a singleton is not an aggregation, therefore we use the following convention: $\mathsf{A}^{(1)}(x) = x$ $(x \in \mathbb{I})$. We now introduce the concept of extended aggregation function.

Definition 2 An *extended aggregation function* in $\cup_{n \in \mathbb{N}} \mathbb{I}^n$ is a mapping

$$\mathsf{A} : \bigcup_{n \in \mathbb{N}} \mathbb{I}^n \to \overline{\mathbb{R}}.$$

whose restriction $\mathsf{A}^{(n)} = \mathsf{A}|_{\mathbb{I}^n}$ to $\mathbb{I}^n$ is an aggregation function in $\mathbb{I}^n$ for any $n \in \mathbb{N}$.

For example, the arithmetic mean as an extended aggregation function is the sequence $(\mathsf{AM}^{(n)})_{n \in \mathbb{N}}$, where $\mathsf{AM}^{(n)}$ is defined for all $n \in \mathbb{N}$. Usually in applications important property of the aggregation function is the continuity, which means that any small changes in the arguments (possible minor errors) should not make a big change in the aggregated value (output error). There are many different types of continuity, see [21]. We remark that in many applications aggregation functions are considered for special interval $\mathbb{I} = [0, 1]$. Many applications can be found in [21, 28, 60]. Interesting application of the extended aggregation function on images segmentation can be found in [13].

One very important application is measuring the air quality, which is characterized by air quality index (AQI). Namely, first there is calculated sub-index (AQI_i) for each pollutant (SO_2, PM_{10}, $PM_{2.5}$, NO_2,...) by the following linear interpolation formula

$$ACI_i = \frac{I_{high} - I_{low}}{C_{high} - C_{inv}}(C - C_{low}) + I_{low},$$

where C is the monitored ambient average concentration of pollutant i; C_{low} is the breakpoint lower than or equal to C; C_{high} is the breakpoint higher than or equal to C; and I_{low} and I_{high} are the sub-index values corresponding to C_{low} and C_{high}, respectively. Under the influence of the USA for the overall AQI is then taken the simplest aggregation function

$$AQI = \max(AQI_1, \ldots, AQI_n).$$

There are further investigations for new aggregation functions, which involve the influence of all pollutants, see a review in [71].

2.2 *Triangular Norm*

One of the important special aggregation function in AI is given in the following definition, see [21, 28].

Definition 3 A *triangular norm* T (t-norm) is a function $\mathsf{T} : [0, 1]^2 \to [0, 1]$ such that
(T1) $\mathsf{T}(x, y) = \mathsf{T}(y, x)$ (commutativity)
(T2) $\mathsf{T}(x, \mathsf{T}(y, z)) = \mathsf{T}(\mathsf{T}(x, y), z)$ (associativity)
(T3) $\mathsf{T}(x, y) \leqslant \mathsf{T}(x, z) \text{for} y \leqslant z$ (monotonicity)
(T4) $\mathsf{T}(x, 1) = x$ (boundary condition).

The following are the most important t-norms

$$\mathsf{T}_{\mathbf{M}}(x, y) = \min(x, y), \quad \mathsf{T}_{\mathbf{P}}(x, y) = xy$$

$$\mathsf{T}_{\mathbf{L}}(x, y) = \max(0, x + y - 1), \quad \mathsf{T}_{\mathbf{D}}(x, y) = \begin{cases} \min(x, y) & \text{if } \max(x, y) = 1 \\ 0 & \text{otherwise.} \end{cases}$$

The corresponding dual operation S, t-conorm, is given by

$$\mathsf{S}(x, y) = 1 - \mathsf{T}(1 - x, 1 - y).$$

Many other important t-norms and t-conorms can be found in [28], as well as many important families of them.

We mention here an important application in decision making which is related to the so called conditional distributivity of t-norm T over t-conorm S, i.e.,

$$\mathsf{T}(x, \mathsf{S}(y, z)) = \mathsf{S}(\mathsf{T}(x, y), \mathsf{T}(x, z)) \text{ for } \mathsf{S}(y, z) < 1.$$

Based on the result [28] of the characterization of the pair (T, S) which satisfies the previous equality, there was obtained in [16, 21, 58] a generalization of von Neumann–Morgenstern utility theory.

2.3 Copulas

Important functions that links a multi-dimensional probability distribution function to its one-dimensional margins is given in the following definition, see [25, 28, 52].

Definition 4 A (two-dimensional) copula is a function $\mathsf{C} : [0, 1]^2 \to [0, 1]$ such that:

(i) $\mathsf{C}(x, 0) = \mathsf{C}(0, x) = 0$ and $\mathsf{C}(x, 1) = \mathsf{C}(1, x) = x$ for any $x \in [0, 1]$;
(ii) $\mathsf{C}(x_2, y_2) + \mathsf{C}(x_1, y_1) \geqslant \mathsf{C}(x_1, y_2) + \mathsf{C}(x_2, y_1)$ for $x_1 \leqslant x_2$ and $y_1 \leqslant y_2$.

If C is a copula, then C is non-decreasing in each place and continuous. $\mathsf{T}_{\mathbf{M}}$, $\mathsf{T}_{\mathbf{P}}$ and $\mathsf{T}_{\mathbf{L}}$ are associative and commutative copulas. We remark that $\mathbf{S}_{\mathbf{M}} = \max$ does not satisfy (ii) in Definition 4, and therefore it is not a copula. We have the boundaries for arbitrary copula C :

$$\mathsf{T}_{\mathbf{L}} \leqslant \mathsf{C} \leqslant \mathsf{T}_{\mathbf{M}}.$$

Example 2 The family of Cuadras–Augé copulas $(\mathsf{C}_\lambda^{\mathbf{CA}})_{\lambda \in [0,1]}$ given by

$$\mathsf{C}_\lambda^{\mathbf{CA}} = {\mathsf{T}_{\mathbf{M}}}^{\lambda} \cdot {\mathsf{T}_{\mathbf{P}}}^{1-\lambda}$$

is not associative for $\lambda \neq 0, 1$, and therefore it is not a t-norm, but we have $\mathsf{C}_0^{\mathbf{CA}} = \mathsf{T}_{\mathbf{P}}$ and $\mathsf{C}_1^{\mathbf{CA}} = \mathsf{T}_{\mathbf{M}}$.

Definition 5 A distribution function is a function $F : [-\infty, \infty] \to [0, 1]$ which is left-continuous on $\mathbb{R}$, non-decreasing and $F(-\infty) = 0, \ F(\infty) = 1$.

Let $(X, \mathcal{A}, P)$ be a probability measure space and $U : \Omega \to \mathbb{R}$ an $\mathcal{A}$-measurable function (random variable). Then the function $F_U : [-\infty, \infty] \to [0, 1]$ defined by

$$F_U(x) = P(\{\omega \mid \omega \in X, U(\omega) < x\}) \quad (= P(U < x))$$

is a distribution function of the random variable U. A *joint distribution function* is a function $J : [-\infty, \infty]^2 \to [0, 1]$ which is left-continuous on $\mathbb{R}$ with $J(-\infty, \infty) = 1$ and such that for every $x, y, u, v \in [-\infty, \infty]$ $J(x, -\infty) = J(-\infty, x) = 0$ and

$$J(x, y) + J(u, v) \geqslant J(x, v) + J(u, y).$$

For a joint distribution J the marginal distribution functions F and G are defined by $F(x) = J(x, \infty)$ and $G(x) = J(\infty, x)$.

The following main theorem represents by a copula the joint distribution of a random vector by corresponding marginal distributions.

Theorem 1 (Sklar's theorem) *If J is a two-dimensional distribution function with one-dimensional marginal distribution functions F and G then there exists a copula* C *such that for all $x, y \in \mathbb{R}$,*

$$J(x, y) = \mathsf{C}(F(x), G(y)).$$

If F and G are continuous then C *is unique; otherwise* C *is uniquely determined on* (Range F) $\times$ (Range G).

We remark that the random variables U, V are independent if and only if it is possible to take $\mathsf{C} = \mathsf{T_P}$. Sklar's theorem shows that much of the study of joint distribution functions can be reduced to the investigation of copulas. Therefore copulas can be used for investigation of the dependence, see [25, 52]. The notion of a copula can be extended to n dimensions. An n-copula is an n-dimensional distribution function whose support is in the unit n-cube and whose one-dimensional margins are uniform, see [25, 52].

3 Fuzzy Systems

3.1 Fuzzy Logics

We shall show very briefly where in the theory of fuzzy logics and fuzzy sets occur the preceding operations of t-norms and t-conorms. Taking a t-norm T, the Zadeh strong negation n given by $n(x) = 1 - x$ and, implicitly, with the t-conorm S dual to T given by $\mathsf{S}(x, y) = n\big(\mathsf{T}(n(x), n(y))\big)$, we can introduce the basic connectives in a [0, 1]–valued logic: *conjunction*: $x \wedge_{\mathsf{T}} y = \mathsf{T}(x, y)$, and *disjunction*: $x \vee_{\mathsf{T}} y = \mathsf{S}(x, y)$. Restricting to Boolean (i.e., two–valued) logic with truth values 0 and 1 only, then we obtain the classical logical connectives. As in the classical logic, it is possible to construct implication, bi–implication with negation, conjunction and disjunction. Taking into account that in Boolean logic 'NOT A OR B' is equivalent to 'IF A THEN B', one possibility of modelling the implication in a [0, 1]–valued logic (based on T, n and S) is to define the function $I_{\mathsf{T}} : [0, 1]^2 \to [0, 1]$ by

$$I_{\mathsf{T}}(x, y) = \mathsf{S}(n(x), y) = n(\mathsf{T}(x, n(y)))).$$

It is clear that in this case the *law of contraposition* $I_{\mathsf{T}}(x, y) = I_{\mathsf{T}}(n(y), n(x))$ is always valid. For the two basic t-norms $\mathsf{T_M}, \mathsf{T_L}$ we obtain the following implications:

$$I_{\mathsf{T_M}}(x, y) = \begin{cases} y & \text{if } x + y \geqslant 1, \\ 1 - x & \text{otherwise;} \end{cases}$$

$$I_{\mathsf{T_L}}(x, y) = \begin{cases} 1 & \text{if } x \leqslant y, \\ 1 - x + y & \text{otherwise;} \end{cases}$$

Another way of extending the classical binary implication operator (acting on $\{0, 1\}$) to the unit interval $[0, 1]$ uses the *residuation* R_{T}, see [28]. In general, I_{T} and R_{T} are different (although both are extensions of the Boolean implication), but we have $I_{\mathsf{T_L}} = R_{\mathsf{T_L}}$.

3.2 *Fuzzy Sets*

As the classical set theory is based on the classical logic, so the theory of fuzzy sets is based on fuzzy logics. The theory of fuzzy sets was introduced by Lotfi A. Zadeh in 1965. Given a (crisp) universe X, a *fuzzy subset* A of X (see, e.g., [28]) is characterized by its *membership function* $\mu_A : X \to [0, 1]$, where for $x \in X$ the number $\mu_A(x)$ is interpreted as the *degree of membership* of x in the fuzzy set A. The membership function μ_A of a fuzzy subset A of X is a generalization of the *characteristic function* $\mathbf{1}_B : X \to \{0, 1\}$ of a crisp subset B of X, assigning the value 1 to all elements of X which belong to B, and the value 0 to all remaining elements of X. In order to generalize the Boolean set–theoretical operations like *intersection* and *union*, it is quite natural to use triangular norms and conorms. For a t-norm T and a t-conorm S, for any fuzzy subsets A, B of the universe X, the membership functions of the *intersection* $A \cap B$, the *union* $A \cup B$ and the *complement* A^c are defined by:

$$\mu_{A \cap B}(x) = \mathsf{T}\left(\mu_A(x), \mu_B(x)\right), \quad \mu_{A \cup B}(x) = \mathsf{S}\big(\mu_A(x), \mu_B(x)\big),$$

$$\mu_{A^c}(x) = 1 - \mu_A(x).$$

Let $X_1 \times X_2$ is the crisp Cartesian product of two crisp sets X_1 and X_2. Then a *fuzzy relation* R on $X_1 \times X_2$ is a fuzzy set on $X_1 \times X_2$. The notion of *fuzzy relation equation*, first introduced by Sanchez for $\mathsf{T_M}$, is important in many applications, as for example in automatic control by fuzzy controllers, see [28]. There are many important practical applications of fuzzy systems, but also in many decision making problems, see [60].

There is an important application in assessment and prediction of air quality, see end of Sect. 2.1, using fuzzy systems, see [10, 69].

4 Pseudo-analysis

The first traces of the pseudo-analysis goes to Grossman and Katz [22] and Burgin [6] (see further [7]), then Maslov [38] (see further [36, 39]). These ideas were managed as a complete unified theory later by E. Pap in his pseudo-analysis, and as a special case the g-calculus [53, 54], see [28, 45, 46, 55, 57, 59]. The main idea behind the theory is to replace the usual field of real numbers endowed the usual addition and product, with new operations, so called pseudo-addition and pseudo-multiplication. The advantage of the pseudo-analysis is to enable to treat in a unified way three important problems as nonlinearity, uncertainty and optimization.

4.1 Pseudo-calculus

In this subsection, we recall some notions related to semirings and pseudo-integrals, which are adopted from [55, 57, 59]. Let $[a, b]$ be a closed (in some cases can be considered semiclosed) subinterval of $[-\infty, \infty]$. The full order on $[a, b]$ will be denoted by $\preceq$. A binary operation $\oplus$ on $[a, b]$ is pseudo-addition, if it is commutative, non-decreasing (with respect to $\preceq$) associative and with a zero (neutral) element denoted by $\mathbf{0}$. Let $[a, b]_+ \subseteq [a, b]$ with $\mathbf{0} \preceq x$. A binary operation $\otimes$ on $[a, b]$ is pseudo-multiplication, if it is commutative, positively non-decreasing, i.e., $x \preceq y$ implies $x \otimes z \preceq y \otimes z$ for all $z \in [a, b]_+$, associative and with unit element $\mathbf{1} \in [a, b]$, i.e., for each $x \in [a, b]$, $\mathbf{1} \otimes x = x$. We also assume $\mathbf{0} \otimes x = \mathbf{0}$ and $\otimes$ is distributive over $\oplus$, i.e., $x \otimes (y \oplus z) = (x \otimes y) \oplus (x \otimes z)$. The structure $([a, b], \oplus, \otimes)$ is a (real) semiring. In this paper, all pseudo-operations are supposed to be continuous.

According to considerations of a pair of pan-operations [46, 53, 55, 57], there are three canonical representations on semirings. For many other interesting cases, please refer to [4, 57].

Case I: Idempotent $\oplus$ and not non-idempotent $\otimes$:

$$x \oplus y = \sup(x, y), \; x \otimes y = g^{-1}(g(x) \cdot g(y)),$$

where $g : [a, b] \to [0, \infty]$ is continuous and strictly increasing bijection, called a generator. A full order is induced as follows: $x \preceq y \Leftrightarrow \sup(x, y) = y, \quad \mathbf{0} = a$. A metric is induced as follows: $d(x, y) = |g(x) - g(y)|$.

Case II: g-semiring, i.e.:

$$x \oplus y = g^{-1}(g(x) + g(y)) \quad \text{and} \quad x \otimes y = g^{-1}(g(x) \cdot g(y)),$$

where $g : [a, b] \to [0, \infty]$ is continuous and strictly monotone, called a generator.

(a) If g is a strictly increasing generator, then $\mathbf{0} = a$, the usual order induced by $\oplus$ is as follows: $x \preceq y \Leftrightarrow g(x) \leqslant g(y)$.

(b) If g is a strictly decreasing generator, then $\mathbf{0} = b$, the usual order induced by $\oplus$ as follows: $x \preceq y \Leftrightarrow g(x) \geqslant g(y)$. A metric is induced as follows: $d(x, y) = |g(x) - g(y)|$.

Case III: Idempotent $\oplus$ and $\otimes$:

$$x \oplus y = x \vee y = \sup(x, y),\ x \otimes y = x \wedge y = \inf(x, y).$$

A full order is induced as follows: $x \preceq y \Leftrightarrow \sup(x, y) = y$ and $\mathbf{0} = a, \mathbf{1} = b$. A metric is induced as follows: $d(x, y) = |\arctan(x) - \arctan(y)|$.

Let X be a non-empty set and $\mathcal{A}$ be a $\sigma-$algebra of subsets of X, i.e., $(X, \mathcal{A})$ is a measurable space.

Definition 6 A set-function $m : \mathcal{A} \to [a, b]_+$ is a $\oplus$*-measure* if it satisfies the following conditions

(i) $m(\varnothing) = \mathbf{0}$. (if $\oplus$ is nonidempotent, i.e., $x \oplus x \neq x$ for all $x \in [a, b]$.)

(ii) $m\left(\bigcup_{i=1}^{\infty} A_i\right) = \bigoplus_{i=1}^{\infty} m(A_i) = \lim_{n\to\infty} \bigoplus_{i=1}^{n} m(A_i)$ for any sequence (A_i) of pairwise disjoint sets from $\mathcal{A}$.

The *pseudo-characteristic function* on A ($\emptyset \neq A \subseteq X$) with values in a semiring $([a, b], \oplus, \otimes)$ is defined by $\chi_A(x) = \mathbf{1}$ for $x \in A$ and $\mathbf{0}$ for $x \notin A$. A *step (measurable) function* is the mapping $e : X \to [a, b]$ that has the following representation

$$e = \bigoplus_{i=1}^{n} a_i \otimes \chi_{A_i}$$

for $a_i \in [a, b]$, $a_1 \preceq a_2 \preceq \cdots \preceq a_n$ and sets $A_i \in \mathcal{A}$ are pairwise disjoint if $\oplus$ is nonidempotent. The above step function e can be also represented by $e = \oplus_{i=1}^{n} a_i \otimes \chi_{C_i}$, where $X = C_1, C_2 = \cup_{i=2}^{n} A_i, \ldots, C_n = A_n$.

We have the following definition of pseudo-integral [28, 53, 55, 57, 59, 80]:

Definition 7 Let $m : \mathcal{A} \to [a, b]_+$ be a $\oplus$-measure.

(i) The *pseudo-integral* of a step function $e : X \to [a, b]$ is defined by

$$\int_X^{\oplus} e \otimes dm = \oplus_{i=1}^{n} a_i \otimes m(A_i).$$

(ii) The *pseudo-integral* of a measurable function is defined by

$$\int_X^{\oplus} f \otimes dm = \lim_{n\to\infty} \int_X^{\oplus} e_n \otimes dm,$$

where (e_n) is the sequence of step functions such that $d(e_n(x), f(x)) \to 0$ uniformly as $n \to \infty$.

Remark 1 The pseudo-integral of a measurable function $f : X \to [a, b]_+$ can be equivalently defined by

$$\int_X^{\oplus} f \otimes dm = \sup_{e \preceq f} \int_X^{\oplus} e \otimes dm.$$

Example 3 Let $([a, b], \oplus, \otimes)$ be a g-semiring. Then the pseudo-integral for a function $f : X \to [a, b]$ reduces to the g-integral [54],

$$\int_X^{\oplus} f \otimes dm = g^{-1} \left(\int_X (g \circ f) dg \circ m \right).$$

Example 4 Let $([a, b], \sup, \otimes)$ be a semiring Case I. Then the pseudo-integral for a function $f : X \to [a, b]$ is given by, see [38, 39, 55, 57]

$$\int_X^{\sup} f \otimes dm = \sup_{x \in X}(f(x) \otimes \psi(x)) = \sup_{\alpha \in [a,b]} \alpha \otimes m(\alpha \preceq f),$$

where the function $\psi : X \to [a, b]$ defines a sup-measure m by $m(A) = \sup_{x \in A} \psi(x)$, and $(\alpha \preceq f) = \{x \in X \mid \alpha \preceq f(x)\}$.

It is well known that in the classical mathematical analysis the notion of convolution of functions play important role, as well in many applications. Therefore we have introduced the notion of pseudo-convolution, see [57, 59, 63]. Let G be subset of $\mathbb{R}^n$ and $*$ a commutative binary operation on G such that $(G, *)$ is a cancellative semigroup with unit element e. Let $G_+ = \{x | x \in G, x \geqslant e\}$ be a subsemigroup of G.

Definition 8 The *pseudo-convolution of the first type* of two functions $f : G \to [a, b]$ and $h : G \to [a, b]$ with respect to a $\oplus$-measure m and $x \in G_+$ is given in the following way

$$f \star h(x) = \int_{G_+^x}^{\oplus} f(u) \otimes dm_h(v),$$

for $G_+^x = \{(u, v) | u * v = x, v \in G_+, u \in G_+\}$, and $m_h = m$, where

(i) for the case of sup-measure $m(A) = \sup_{x \in A} h(x)$;
(ii) for the case of inf-measure $m(A) = \inf_{x \in A} h(x)$;
(iii) for the case of $\oplus$- measure m, where $\oplus$ has an additive generator g and $g \circ m$ is the Lebesgue measure, $dm_h = h \odot dm$.

The *second type of pseudo-convolution*, when $(G, *)$ is a group, the pseudo-integral is taken over whole set G:

$$f \star h(x) = \int_G^{\oplus} f(x * (-t)) \otimes dm_h(t).$$

where $(-t)$ is unique inverse element for t and $x \in G$.

4.2 Applications

We shall mention only some of important applications of the pseudo-analysis. For more details see [8, 9, 28, 32, 38, 39, 55, 57, 59, 64].

4.2.1 Nonlinear Equations

Pseudo-analysis is applied for solving nonlinear equations (ODE, PDE, difference equations, etc.) using the pseudo-superposition principle [38, 39, 57, 59], which means that if u_1 and u_2 are solutions of the considered nonlinear equation, then also $a_1 \otimes u_1 \oplus a_2 \otimes u_2$ is a solution for any constants a_1 and a_2 from $[a, b]$. There was obtained an exact solutions of the Burgers equation, as well as for the general Hamilton–Jacobi equations, where the nonlinear Hamiltonian is not smooth, but it is important in control theory, see [38, 39, 57].

4.2.2 Fuzzy Numbers

Generally, a fuzzy number is a fuzzy subset (represented by its membership function $\mu_A : \mathbb{R} \to [0, 1]$) of the real line $\mathbb{R}$, see [28]. More precisely.

Definition 9 Let A be a fuzzy subset of the real line. A is a *fuzzy number* if it is normalized and convex, i.e., if $A(x) = 1$ for some $x \in \mathbb{R}$ and for all $x, y, z \in \mathbb{R}$ such that $x < y < z$, it is $A(y) \geqslant \min(A(x), A(z))$.

Depending on applications there are used different types of fuzzy numbers.

Example 5 An *LR-fuzzy number* $A = (a, \alpha, \beta)_{L,R}$ is a function from reals into the interval $[0, 1]$ satisfying

$$A(x) = \begin{cases} R\left(\frac{x-a}{\beta}\right) & \text{if } a \leqslant x \leqslant a + \beta, \\ L\left(\frac{a-x}{\alpha}\right) & \text{if } a - \alpha \leqslant x \leqslant a, \\ 0 & \text{otherwise,} \end{cases}$$

where $a \in \mathbb{R}$ is the center, $\alpha > 0$ is the left spread, $\beta > 0$ is the right spread of A, L and R are non-increasing and continuous functions from $[0, 1]$ to $[0, 1]$ satisfying $L(0) = R(0) = 1$ and $L(1) = R(1) = 0$.

The arithmetical operations with fuzzy numbers are based on Zadeh's extension principle (see [28]): Let T be an t-norm and $*$ a binary operation on $\mathbb{R}$. Then the operation $*$ is extended to fuzzy numbers A and B by

$$A *_T B(z) = \sup_{x*y=z} \mathsf{T}(A(x), B(y))$$

for $z \in \mathbb{R}$. Some usual operations with fuzzy numbers are following: addition is obtained for $* = +$: $A \oplus_T B(z) = \sup_{x+y=z} T(A(x), B(y))$, multiplication for $* = \cdot$: $A \odot_T B(z) = \sup_{x \cdot y=z} T(A(x), B(y))$. Multiplication and addition are generalized pseudo-convolutions of the second type based on the semiring ([0, 1], max, T).

For the practical computations it is natural to preserve the shape of fuzzy numbers during the arithmetic operations. Preserved shape allows us to control the resulting spread while working with fuzzy numbers. In the sense of generalized pseudo-convolution, preserving the shape means that generalized pseudo-convolution is closed operation on certain family of functions. For some special shapes, there exist some special t-norms preserving that shape (see [28]).

4.2.3 Information Theory

The information theory based on the notion of a probability has been introduced in 1948 by Wiener and Shannon (see for more details the chapter of Sander in [56] and [28]). Let $(X, \mathcal{A}, P)$ be a probability space. The mapping $J : \mathcal{A} \to [0, \infty]$ is called an information measure if for an event $A \in \mathcal{A}$ we have $J(A) = -\log P(A)$. Axiomatization of information measure has been introduced in 1967 by Kampe de Feriet and Forte (see [28]).

Definition 10 Let $(X, \mathcal{A})$ be a probability space. Then the following mapping $J : \mathcal{A} \to [0, \infty]$ is called an *information measure* if it satisfy the following conditions:
(I1) $J(X) = 0$, $J(\emptyset) = \infty$;
(I2) $B \subset A$ implies $J(B) \geqslant J(A)$;
(I3) if A and B are independent in the sense of Banach–Marczewski, then

$$J(A \cap B) = J(A) + J(B);$$

(I4) $J(A \cup B) = F(J(A), J(B))$ whenever $A \cap B = \emptyset$ for a two-place function F;
(I5) $J(\bigcup_{i=1}^{\infty} A_i) = \lim_{n\to\infty} J(\bigcup_{i=1}^{n} A_i)$.

Usually, for F are required: 1. continuity, 2. symmetry ($F(x, y) = F(y, x)$), 3. associativity ($F(F(x, y), z) = F(x, F(y, z))$), 4. monotonicity ($F$ is non-decreasing in both components), 5. neutral element. ($F(x, \infty) = x$). Schweizer and Sklar, in 1969, proposed that information J takes for value a random variable on $[0, \infty]$ instead of a non-negative real number, see [26]. Let S be a continuous t-conorm. Then functions

$$\Pi_{\mathsf{S}}(H, G, x) = \mathsf{S}(H(x), G(x)), \quad \tau_{\mathsf{S}}(H, G, x) = \sup_{u+v=x} \mathsf{S}(H(u), G(v)),$$

where $u, v, x \in [0, \infty]$ and H, G are distributions on $[0, \infty]$, fulfill the required axiom. The second one is an pseudo-convolution of the first type with respect to (max, S).

4.2.4 System Theory

Pseudo-convolutions are also important tools in the system theory, see [3, 39]. There is a numbering mechanism which assigns a number to each event in order of these events taking place. We start with an initial finite value $k_0 \in \mathbb{Z}$ (where $\mathbb{Z}$ is set of all integers). There is a special type of event that corresponds to ticks of an absolute clock. These ticks are numbered in increasing order, starting from some value $t_0 \in \mathbb{Z}$. Mapping from $\mathbb{Z}$ into $\mathbb{Z}$ such that graph is the set of all pairs (event number, clock value) is *dater* associated with a certain type of event.

Definition 11 A *signal* u is a mapping from $(\mathbb{R}, +, \cdot)$ into $([-\infty, \infty], \max, +)$. When a signal is a nondecreasing function it is called a *dater*.

Dater d may not be strictly increasing since events of the same type may occur simultaneously. We have

$$d(k) = \begin{cases} -\infty & \text{when } k < k_0, \\ \infty & \text{event never took place} \\ \text{any finite value} & \text{otherwise.} \end{cases}$$

Domain of d may be extended by setting $d(-\infty) = -\infty$ and $d(\infty) = \infty$. The semiring for the range of d is $([-\infty, \infty], \max, +)$. Now, pseudo-addition is the conventional pointwise maximum: $(d_1 \oplus d_2)(k) = \max(d_1(k), d_2(k))$. Pseudo-multiplication is:

$$(d_1 \otimes d_2)(k) = \sup_{l \in \mathbb{Z}} (d_1(l) + d_2(k - l))$$

and that is a pseudo-convolution of the second type (integral is taken over whole group $\mathbb{Z}$) for $\oplus = \max$, $\otimes = +$ and $* = +$. One of elementary systems is *shift* Γ^g. This system maps inputs to outputs according to the equation $y(k) = u(k - g)$ for all k, where u is a signal.

Definition 12 A linear system S is called *shift-invariant* if it commutes with all shift operators, that is if for all u and for all c, $S\Big(\Gamma^c(u)\Big) = \Gamma^c\Big(S(u)\Big)$.

Important fact is that in the shift-invariant case the input-output relation can be expressed as follows: $y(k) = \sup_{s \in \mathbb{R}} \Big(h(k - s) + u(s)\Big)$, i.e., the pseudo-convolution of the second type (for $\oplus = \max$, $\otimes = +$ and $* = +$), analogously in the same way as the classical convolution in conventional system theory.

4.2.5 Option Pricing

The famous Black–Sholes and Cox–Ross–Rubinstein formulas from 1979 are basic results in the recent theory of option pricing in financial mathematics. They are usually obtained by stochastic analysis. The systematic deterministic approach with pseudo-analysis (more precisaly with idempotent analysis) to the option pricing leads to a different type of generalizations of Black–Sholes and Cox–Ross–Rubinstein formulas characterized by more rough assumptions on common stocks evolution (which are therefore easier to verify) [32], see [57]. This approach reduces the analysis of the option pricing to the study of certain homogeneous nonexpansive maps, which are "strongly" infinite dimensional. It was shown using exclusively deterministic arguments by pseudo-analysis that the hedge exists and is the same for both cases whenever the function f is nondecreasing and convex (possibly not strictly). This approach allows to obtain some generalizations of the standard Black–Sholes formula and reduces the analysis of derivative securities pricing to the study of homogeneous nonexpansive maps, which act in infinite dimensional spaces.

4.2.6 Large Deviation Principle

We shall explain here briefly the connection between probability and idempotent measures, which is given by the large deviation principle. Let X be a topological space and $\mathcal{A}$ be the algebra of its Borel sets.

Definition 13 One says that a family of probabilities (P_ε), $\varepsilon > 0$, on $(X, \mathcal{A})$ has the *large deviation principle* if there exists a *rate function* $I : X \mapsto [0, \infty]$ such that

(i) I is lower semi-continuous and $X_a = \{x \in X \mid I(x) \leqslant a\}$ is a compact set for any $a < \infty$,
(ii) $-\limsup_{\varepsilon\to 0} \varepsilon \log P_\varepsilon(C) \geqslant \inf_{\omega\in C} I(\omega)$ for each closed set $C \subset X$,
(iii) $-\liminf_{\varepsilon\to 0} \varepsilon \log P_\varepsilon(U) \leqslant \inf_{x\in U} I(x)$ for each open set $U \subset X$.

Then $m(A) = \inf_{x\in A} I(x)$ is a positive inf-measure on $\mathcal{A}$. Therefore, it was quite natural to generalize the previous definition to weak large deviation principle [64]. Then by an idempotent version of the Riesz type theorem it was proved that the large deviation principle and its weak version are actually equivalent for some (rather general) "good" spaces X, see [64].

4.2.7 Physics of the Universe

In some situations there are physical quantities whose arithmetic is not the standard one, e.g., velocities in special relativity are added by means of with$g(x) = \tanh^{-1}(x)$, see for g Sect. 4.1, as well as a parallel configuration of resistors is a resistor whose resistance is computed by means of the harmonic addition with $g(x) = \frac{1}{x}$.

Very recent applications of pseudo-analysis was obtained in physics of the universe [8, 9]. We present only the motivations of these papers. At this times, the space-time physics is at crossroads. The experimental value of the cosmological constant is some 10^{120} smaller than its theoretical estimate (Weinberg 1989)—probably the worst disagreement between theory and experiment in history of science. Problems with dark energy and dark matter may indicate that physics is geometry and arithmetic. Can we understand the equality

$$2^{100^{100^{100}}} + 2^{100^{100^{100}}} = 2^{100^{100^{100}}+1} \ ?$$

Using today supercomputers we cannot process numbers greater than the so-called machine infinity, a finite number N_1 which does not increase if we add 1 to it, i.e., $N_1 < \infty$ and $N_1 + 1 = N_1$. This type of arithmetic is either inconsistent, or non-Diophantine (named by [8, 22]). Non-Diophantine arithmetic and the non-Newtonian calculus (pseudo-analysis) implies automatically to two types of "dark universes": the ones where super-small and super-large physical quantities behave differently even though they satisfy the same physical laws, and those identified with zero-measure sets whose physics is equipped with the usual laws, but which are invisible from the point of view of standard quantum measurements. Papers [8, 9] give hints to the arguments that the problems with dark energy and dark matter look like an experimental indication that the arithmetic we all work with is not necessarily the one preferred by Nature.

5 Fuzzy Measures and Corresponding Integrals

The non-additive integrals based on fuzzy measures have numerous applications in many disciplines of mathematics, engineering and economics, especially in the problems that some aggregation procedure, optimization and modeling risk and uncertainty have been required, and the most popular among them is the Choquet integral [1, 5, 11, 14, 19, 20, 55, 65, 66, 72, 74–76]. The monotone non-decreasing set functions, vanishing at the empty set in the literature appear under different names-among others, monotone measures, non-additive measures, fuzzy measures or capacities. The Choquet integral is an important aggregation function, which enable non-probabilistic modeling. One of its advantage is to enable modeling dependent events. Choquet integral [11], see [4, 14, 20, 55], is given by

$$I(\mu, f) = (C) \int f \, d\mu = \int_0^\infty \mu(\{f \geqslant t\}) \, dt,$$

where $f : X \to [0, \infty[$ is a bounded measurable function and $\mu : \mathcal{A} \to [0, \infty]$ is a monotone measure (fuzzy measure) satisfying the following conditions on a σ-algebra $\mathcal{A}$ of a given set X: (i) $\mu(\varnothing) = 0$, (ii) $\mu(A) \leqslant \mu(B)$ whenever $A \subseteq B$ and $A, B \in \mathcal{A}$. The integral on the right-hand side is the improper Riemann integral, the function $h_f(x) = m(\{w \mid f(w) \geqslant x\})$, $x \geqslant 0$, is non-increasing and non-negative, hence, Riemann-integrable. Many of important applications of Choquet integral require further generalizations, see Sect. 7.

In analogy with Choquet integral, but with different operations, Sugeno introduced in 1974 [68] a new type of integral, with old traces by Ky Fan from 1944, see [20, 55, 76].

Definition 14 Let μ be a fuzzy measure on X, and $f : X \to [0, \mu(X)]$. The *Sugeno integral* of f with respect to μ is defined by

$$(S)\int f\,d\mu = \sup_{\alpha\in[0,\mu(X)]} (\alpha \wedge \mu(\{x \mid f(x) > \alpha\}))$$

6 Decision Making Based on Non-additive Measures and Integrals

One of the important application of aggregation functions, and specially useful non-additive measures and corresponding integrals is related to decision theory, see [14, 20, 21, 76]. Making decision often means to aggregate scores or preferences on a given set of alternatives, where the scores or preferences were obtained from several decision makers, voters, experts, etc., or represent different points of view, criteria, objectives, etc. This implies decision under multiple criteria or multiple attributes, multiperson decision making, and multiobjective optimization. Approach with non-additive measures and corresponding integrals enables to model situations with dependent inputs, using many different types of interaction indices, see [20, 76].

In decision theory, further step is the consideration under the bipolar perspective. The integrals which can be viewed as generalization of the asymmetric Choquet integral, based on the concept of symmetric pseudo-operations on the interval $[-a, a] \subseteq [-\infty, \infty]$, have been studied in [24, 47]. The universal integral, recently proposed in [30], generalizes the Choquet, Shilkret and Sugeno integrals, and very recently, in [24] the concept of the discrete universal integral has been proposed in the bipolar formulation. Here we shall present only a one part of decision theory.

Cumulative Prospect Theory (CPT) proposed in [72] is the numerical model for comparing two decisions amounts regarding two pairs of sets, the positive and negative features of the alternatives. In general, CPT is the integral-based model for a representation of the utility functional I, defined on the class of prospects ($\mathcal{A}$-measurable functions, from X to $\mathbb{R}$), where a universal set X is a state of nature (($X, \mathcal{A}$) is a measurable space), based on the difference of the Choquet integrals with respect to fuzzy measures m^+ and m^-. In actuarial science, the integral-based premium principles have been widely used for calculation insurance prices, we will mention some of them: the distortion premium principle, the CPT premium principle, the exponential premium principle, the g-integral based premium principle, we refer the reader to [23, 27, 37, 73, 75, 79]. The bipolar models for decision making were observed in [20]. In actuarial science, the necessity of new premium principles able to model the bipolarity based on the concept of CPT has appeared [27]. Based on apparatus of non-additive measure theory, the main goal of the approach presented here, see [48], is to propose the CPT-like integral-based premium principle which can be viewed as the common frame for the distortion premium principle, the general

CPT premium principle and the mean premium principle. We present here only a small part of [48].

6.1 Integral-Based Premium Principles

Let us consider a class $\mathcal{F}$ of *insurance risks*, $f : X \to \mathbb{R}$ (random variables) on a measurable space $(X, \mathcal{A})$, where $\mathcal{A}$ is a σ-algebra of subsets of X, $P : \mathcal{A} \to [0, 1]$ is a probability measure, and $\mathcal{F}^+$ denotes the class of all non-negative insurance risks. Let $\mathcal{M}$ denotes the class of all fuzzy measures on the measurable space $(X, \mathcal{A})$. We say that a set function $m : \mathcal{A} \to [0, \infty[$, is *a fuzzy measure* if it satisfies $m(\emptyset) = 0$, $m(X) > 0$ and for all $A, B \in \mathcal{A}$ such that $A \subseteq B$ it holds $m(A) \leqslant m(B)$. Let $\mathcal{M}_b$ denotes class of all fuzzy measures such that $m(X) = b$, where $b \in]0, \infty[$. If $m(X) = 1$ we say that a fuzzy measure is *normalized*. Obviously, if m is a σ-additive normalized fuzzy measure then it is a probability measure. If m is a σ-additive fuzzy measure then it is a measure. Recall that *a dual set function* of m, $m \in \mathcal{M}_b$, is a set function $\overline{m} : \mathcal{A} \to [0, \infty[$ defined for all $A \in \mathcal{A}$ by:

$$\overline{m}(A) = m(X) - m(X \setminus A).$$

Obviously, it holds $\overline{m} \in \mathcal{M}_b$. In the case that $X = \{x_1, x_2, \ldots, x_n\}$ the monotone integral reduces to the following form:

$$I(f, m) = \sum_{i=1}^{n} \left(f\left(x_{\alpha(i)}\right) - f\left(x_{\alpha(i-1)}\right)\right) m\left(A_{\alpha(i)}\right),$$

where α denotes a permutation such that $0 \leqslant f\left(x_{\alpha(1)}\right) \leqslant \cdots \leqslant f\left(x_{\alpha(n)}\right) \leqslant 1$, $f\left(x_{\alpha(0)}\right) = 0$ and $A_{\alpha(i)} = \{x_{\alpha(i)}, \ldots, x_{\alpha(n)}\}$. Recall some types of the integral-based premium principles widely studied in the literature: (where $f^+ = f \vee 0$ and $f^- = (-f) \vee 0$, avoiding $\infty - \infty$):

Example 6 (i) *The net premium principle*, for $f \in \mathcal{F}$ and a probability measure $P \in \mathcal{M}_1$:

$$E(f) = \Pi_N(f, P, P) = I(f^+, P) - I(f^-, P).$$

(ii) *The asymmetric Choquet integral-based premium principle*, defined for $f \in \mathcal{F}$ and a fuzzy measure $m \in \mathcal{M}_b$ by:

$$\Pi_{ACH}(f, m, m) = I(f^+, m) - I(f^-, \overline{m}),$$

where $\overline{m}$ is a dual set function of m.

(iii) *The distortion premium principle*, defined for $f \in \mathcal{F}$ and $m \in \mathcal{M}_1$ by:

$$\Pi_{DPP}(f, m, m) = I(f^+, m) - I(f^-, \overline{m}),$$

where $m = g \circ P$ for a non-decreasing distortion function $g : [0, 1] \to [0, 1]$, $g(0) = 0, g(1) = 1$.

(iv) *The general CPT premium principle,* for $f \in \mathcal{F}$ and fuzzy measures $m_1, m_2 \in \mathcal{M}_b$:

$$\Pi_{CPT}(f, m_1, m_2) = I(f^+, m_1) - I(f^-, \overline{m}_2),$$

(as a special case $m_i = g_i \circ P$ for non-decreasing distortion functions $g_i : [0, 1] \to [0, \infty[, g_i(0) = 0, i = 1, 2)$.

Remark 2 (i) The asymmetric Choquet integral-based premium principle is asymmetric, in the sense that for all $f \in \mathcal{F}$ we have $\Pi(-f, m, m) = -\Pi(f, \overline{m}, \overline{m})$, while, e.g., *the symmetric Choquet integral-based premium principle*, defined for $f \in \mathcal{F}$ and a fuzzy measure $m \in \mathcal{M}_b$ by:

$$\Pi_{SCH}(f, m, m) = I(f^+, m) - I(f^-, m)$$

in general, is not the asymmetric premium principle. It is asymmetric iff m is additive.

(ii) If we define functionals $U(f, m_1, m_2) = \Pi_{(*)}(u(f), m_1, m_2)$, for all the above premium principles, where u denotes non-decreasing (strictly increasing) utility function, $u(0) = 0$, (by replacing $I(f, m)$ with $I(u(f), m)$ in all above formulae) we obtain the integral-based premium principles from Von Neumann and Morgenstern utility theory. Notice, that in the case of the sign-dependent premium principles, different utility functions $u_1, u_2, u_i(0) = 0, i = 1, 2$ can be considered. For more details about this integral-based premium principles, appropriate for calculation insurance prices in a competitive market setting, we refer to [65, 72, 74, 78].

The asymmetric premium principles have this form:

$$\Pi(f, m_1, m_2) = I(f^+, m_1) - I(f^-, \overline{m}_2).$$

On the other hand, some premium principle are not in the previous form, although they have representation in the terms of the monotone integral.

Example 7 (i) For $f \in \mathcal{F}$, a probability measure $P \in \mathcal{M}_1$ and an increasing real-valued function ϕ, $\phi(0) = 0$, *the mean premium principle* is defined by:

$$\pi_M(f) = \phi^{-1}(I(\phi(f), P)).$$

(ii) For $f \in \mathcal{F}$, a probability measure $P \in \mathcal{M}_1$ and $\alpha > 0$, *the exponential premium principle* is defined by:

$$\pi_{exp}(f) = \frac{1}{\alpha} \ln(I(e^{\alpha f}, P)).$$

6.2 General CPT-Like Integral-Based Premium Principle

We say that *the integral-based premium principle* is any rule (a real-valued functional) $\Pi : \mathcal{F} \times \mathcal{M} \times \mathcal{M} \to \mathbb{R}$ for assigning a premium (a real-value) $\Pi(f, m_1, m_2)$ to an insurance risk (a random variable) $f : X \to \mathbb{R}$. Let $\mathcal{F}_a$ denotes the class of insurance risks $f : X \to [-a, a]$, $a \in \overline{\mathbb{R}^+}$ and with the notations introduced in the previous section, we have the next definition:

Definition 15 *The CPT-like integral-based premium principle* $\Pi_{CPT_\phi} : \mathcal{F}_a \times \mathcal{M}_b \times \mathcal{M}_b \to \mathbb{R}$ for an insurance risk $f \in \mathcal{F}_a$, and $m_1, m_2 \in \mathcal{M}_b$, is defined by:

$$\Pi_{CPT_\phi}(f, m_1, m_2) = \phi^{-1}\left(I(\phi(f^+), m_1) - I(\phi(f^-), \overline{m}_2)\right),$$

for an odd, increasing, continuous function $\phi : [-a, a] \to [-\infty, \infty]$, $\phi(0) = 0$, $\phi(a) = \infty$, with the convention $\infty - \infty = \infty$.

Remark 3 (i) If $\phi(x) = x$, and $m_1 = m_2 = P$ is a probability measure then $\Pi_{CPT_\phi} = \Pi_N$, i.e. coincides with the mathematical expectation $E(f)$ of a random variable f.
(ii) If $m_1 = m_2 = P$ is a probability measure then $\Pi_{CPT_\phi} = \pi_M$, i.e. it coincides with the mean premium principle of a random variable f.
(iii) If we consider only non-negative risks $f \in \mathcal{F}_1^+$, with $\phi(x) = e^{\alpha x} - 1$, and $m_1 = P$, we have $\Pi_{CPT_\phi} = \pi_{exp}$.

It was shown in [48] that our CPT-like integral-based premium principle from Definition 15 is appropriate for solving the decision problem which was not possible to handle with previous types of CPT integrals.

7 Recent Generalizations of the Choquet Integral

We present some generalizations of the Choquet integral, see also [62]. First generalization is based on pseudo-operations. Second generalization of the Choquet integral is based on sublinear means [15, 61] (Sect. 7.2). The third generalization is based on different collection of finite systems [35, 44] (Sect. 7.3). Fourth generalization is based on two fuzzy measures, one of which is pseudo-additive [81] (Sect. 7.4).

7.1 Fuzzy Integral with Pseudo-operations

Applying pseudo-analysis approach on general monotone set functions (fuzzy measures) we can replace in the Choquet integral the usual addition and multiplication by pseudo-addition and pseudo-multiplication, see [4, 76].

Definition 16 Let $\mu : \mathcal{A} \to [a,b]_+$ be a fuzzy measure.

(i) The generalized fuzzy integral of a step function $e : X \to [a,b]_+$ is defined by

$$(G)\int_X^{\oplus} e \otimes d\mu = \bigoplus_{i=1}^{n} a_i \otimes \mu(C_i).$$

(ii) The generalized fuzzy integral of a measurable function $f : X \to [a;b]_+$ is defined by

$$(G)\int_X^{\oplus} f \otimes d\mu = \sup_{e \preceq f} \int_X^{\oplus} e \otimes d\mu.$$

Remark 4 There are also some other kinds of interesting integrals based on pseudo-operations (or pan-operations), such as the pan-integral [4, 77], pseudo-concave integrals (for a g-addition and g-multiplication) [41] which cover concave integrals [35], and the generalized Lebesgue integral [76] which can cover (S) fuzzy integral, (N)fuzzy integral and Lebesgue integral.

7.2 *Choquet Integral Based on Sublinear Means*

This part is based on the papers [15, 61]. We use the notions from [21, 33]. First, we consider the finite case. Let $N = \{1, 2, \ldots, n\}$.

Definition 17 A monotone measure $\mu : 2^N \to [0, \infty]$ is a submeasure if it satisfies

(i) $\mu(\emptyset) = 0$.
(ii) (Monotonicity) $E, F \in 2^N$ and $E \subset F$ imply $\mu(E) \leqslant \mu(F)$.
(iii) (Subadditivity) $E, F \in 2^N$ and $\mu(E \cup F) \leqslant \mu(E) + \mu(F)$.

Let $M : [0, \infty]^n \to [0, \infty]^{2^N}$ be a mean. We have for $\mathbf{x} \in \mathbb{R}^n_+$ and $E \subseteq N$, where $E = \{i_1, i_2, \ldots, i_s\}$ and $i_1 \leqslant \cdots \leqslant i_s$, that

$$M(\mathbf{x})(E) = M(x_{i_1}, x_{i_2}, \ldots, x_{i_s}).$$

Definition 18 A mean $M : [0, \infty]^n \to [0, \infty]^{2^N}$ is sublinear if for every $E \subseteq N$

(i) (Monotonicity) $\mathbf{x} \leqslant \mathbf{y} \Rightarrow M(\mathbf{x})(E) \leqslant M(\mathbf{y})(E)$.

(ii) (Homogenity) $M(\alpha\mathbf{x})(E) = \alpha M(\mathbf{x})(E)$ for all $\alpha \geqslant 0$.

(iii) (Subadditivity) $M(\mathbf{x} + \mathbf{y})(E) \leqslant M(\mathbf{x})(E) + M(\mathbf{y})(E)$.

We have the following examples of sublinear means: the weighted sum, geometric mean, quasi arithmetical mean (see [21], for f strictly mononotone increasing is a sublinear mean if and only if f is sublinear function), the Choquet integral for μ a submeasure (see [14, 55, 56]), the convex integral [44].

Definition 19 Let μ be a submeasure and let M be a sublinear mean. Then we define

$$(MC)\int \mathbf{x}\,d\mu = \int_0^{\max_{i\in N} M_i(\mathbf{x})} \mu(\{i \in N \mid M_i(\mathbf{x}) \geqslant t\})\,dt,$$

where $M_i(\mathbf{x}) = M(x_1, x_2, \ldots, x_i)$.

The (MC) integral with respect to submeasure μ and $\mathbf{x} \in \mathbb{R}_+^n$ can be written in the following form

$$(MC)\int \mathbf{x}\,d\mu = \sum_{i=1}^{n}(M_{\sigma(i)}(\mathbf{x}) - M_{\sigma(i-1)}(\mathbf{x}))\mu(\{\sigma(i), \ldots, \sigma(n)\}),$$

with a permutation σ such that $M_{\sigma(1)}(\mathbf{x}) \leqslant M_{\sigma(2)}(\mathbf{x}) \leqslant \cdots \leqslant M_{\sigma(n)}(\mathbf{x})$ and the convention $M_{\sigma(0)}(\mathbf{x}) = 0$.

Example 8 (i) Taking M such that $M_i(\mathbf{x}) = x_i$ for $i = 1, 2, \ldots, n$, we obtain the Choquet integral.
(ii) Taking $M(\mathbf{x}) = (C)\int \mathbf{x}d\mu$, for a submeasure μ, we obtain the multilevel Chouquet integral, see [50].
(iii) The geometrical mean can not be represented by Choquet integral but as a sublinear mean it can be used in an (MC) integral construction.

Theorem 2 *The (MC) integral is monotone, homogeneous and subadditive.*

We define for $\mathbf{x} \in \mathbb{R}_+^n$ $\|\mathbf{x}\|_{L^\infty(N,M)} = \max_{E\in 2^N} M(\mathbf{x})(E)$. Let $\mu : 2^N \to [0, \infty]$ be a submeasure and $0 < p < \infty$. Then we define for $\mathbf{x} \in \mathbb{R}_+^n$ the following norms

$$\|\mathbf{x}\|_{L^p(N,M)} = \left(\sum_{i=1}^{n}\left(M_{\sigma(i)}(\mathbf{x})^p - M_{\sigma(i-1)}(\mathbf{x})^p\right)\mu(\{\sigma(i), \ldots, \sigma(n)\}\right)^{1/p},$$

$$\|\mathbf{x}\|_{L^{p,\infty}(N,M)} = \left(\max_{i\in\{1,\ldots,n\}} M_{\sigma(i)}(\mathbf{x})^p \mu(\{\sigma(i), \ldots, \sigma(n)\}\right)^{1/p},$$

with a permutation σ such that $M_{\sigma(1)}(\mathbf{x})^p \leqslant M_{\sigma(2)}(\mathbf{x})^p \leqslant \cdots \leqslant M_{\sigma(n)}(\mathbf{x})^p$ and the convention $M_{\sigma(0)}(\mathbf{x}) = 0$. Choquet-like integral [40] and integrals introduced in [50], see [21], are related to previous definitions of norms. We introduce $L^\infty(N, M)$, $L^p(N, M)$ and $L^{p,\infty}(N, M)$ the spaces of elements $\mathbf{x} \in \mathbb{R}_+^n$ endowed with norms $\|\mathbf{x}\|_{L^\infty(N,M)}$, $\|\mathbf{x}\|_{L^p(N,M)}$ and $\|\mathbf{x}\|_{L^{p,\infty}(N,M)}$, respectively.

Further investigation was extended on the infinite case, which is much more complex, see [1, 15], using outer measures which are monotone and countable subadditive. The L^p theory for outer measure is very useful in the investigations of singular integrals, e.g., Carleson embedding theory and in time frequence analysis, see [1, 15].

7.3 Superdecomposition Integral

This section is based on [17, 34, 35, 44]. For a fixed measurable space $(X, \mathcal{A})$ denote by $\mathbb{X}$ the set of all collection systems $\mathcal{H}$, where $\mathcal{C} \subseteq \mathcal{A} \setminus \{\emptyset\}$ is a *collection* whenever it is finite. Taking $(X, \mathcal{A})$ we denote by $\mathcal{M}$ the set of all monotone measures on $(X, \mathcal{A})$ and with $\mathcal{F}$ the set of all bounded measurable functions $f : X \to [0, \infty[$.

Definition 20 Let $\mathcal{H} \in \mathbb{X}$ be fixed. Then the mapping $I^{\mathcal{H}} : \mathcal{M} \times \mathcal{F} \to [0, \infty]$ given by

$$I^{\mathcal{H}}(\mu, f) = \inf \left\{ \sum_{A \in \mathcal{C}} a_A \cdot \mu(A) \mid \mathcal{C} \in \mathcal{H}, a_A \geqslant 0 \text{ for each } A \in \mathcal{C}, \sum_{A \in \mathcal{C}} a_A \mathbf{1}_A \geqslant f \right\}$$

is called a superdecomposition integral.

It is obvious that each superdecomposition integral $I^{\mathcal{H}}$ is positively homogeneous and increasing in each coordinate.

Example 9 (i) Taking $\mathcal{H}_1 = \{\{A\} \mid A \in \mathcal{A}\}$ we obtain

$$\begin{aligned} I^{\mathcal{H}_1}(\mu, f) &= \inf \{a \cdot \mu(A) \mid A \in \mathcal{A}, a \cdot \mathbf{1}_A \geqslant f\} \\ &= \sup\{f(x) \mid x \in X\} \cdot \mu(\{f > 0\}). \end{aligned}$$

Remark that if μ has no non-trivial null-sets, i.e., $m(A) = 0$ only if $A = \emptyset$, then $I^{\mathcal{H}_1}$ coincide with the greatest universal integral (see [30]) related to the product as underlying multiplication.

(ii) Taking $\mathcal{H}_2 = \{\mathcal{C} \mid \mathcal{C} \text{ is a finite chain in } \mathcal{A}\}$, i.e, $\mathcal{C}$ is a finite chain in $\mathcal{A}$ if and only if there is an integer k and $\mathcal{C} = \{A_1, \ldots, A_k\} \subset \mathcal{A}$ satisfies $A_1 \subset A_2 \subset \cdots \subset A_k$, we obtain the $I^{\mathcal{H}_2}$ is the Choquet integral.

Related the order we have that for $\mathcal{H}, \mathcal{G} \subseteq \mathbb{X}$ such that $\mathcal{H} \subseteq \mathcal{G}$, $I^{\mathcal{H}} \geqslant I^{\mathcal{G}}$, i.e., for each $m \in \mathcal{M}$ and $f \in \mathcal{F}$ it holds $I^{\mathcal{H}}(\mu, f) \geqslant I^{\mathcal{G}}(\mu, f)$. Specially for $\mathcal{H}_3 = \{\mathcal{G} \mid \mathcal{G} \text{ is a finite subset of } \mathcal{A}\}$, it holds $\mathcal{H}_3 \supseteq \mathcal{H}$ for each $\mathcal{H} \in \mathbb{X}$, and therefore $I^{\mathcal{H}_3} \leqslant I^{\mathcal{H}}$, i.e., $I^{\mathcal{H}_3}$ is the smallest superdecomposition integral. For further results and proofs see [44].

We compare in [44] superdecomposition integral with the decomposition integrals [17, 35], which is based on a system $\mathcal{H}$ of finite set systems from $\mathcal{A} \setminus \{\emptyset\}$ (called collections in [17]),

$$I_{\mathcal{H}}(m, f) = \sup \left\{ \sum_{i \in I} a_i m(A_i) \mid (A_i)_{i \in I} \in \mathcal{H}, \sum_{i \in I} a_i \mathbf{1}_{A_i} \leqslant f \right\},$$

where all constants a_i, $i \in I$, are non-negative. Depending on $\mathcal{H}$, several classical integrals can be constructed. For example, for a fixed measurable space $(X, \mathcal{A})$ $I_{\mathcal{H}_1}$ is Shilkret integral [67], $I_{\mathcal{H}_2}$ is a Choquet integral, $I_{\mathcal{H}_3}$ is the concave integral [34]. For more details see [17, 44].

7.4 Choquet Integral Based on Two Fuzzy Measures

We use some notions and notations from [53–55, 57]. For more details see [55, 57]. Now we present some of the results obtained recently in [81], for more results and proofs see there.

Definition 21 Given a semiring $([a,b],\oplus,\otimes)$. Let $\mu:\mathcal{A}\to[a,b]_+$ be a fuzzy measure and $m:Borel([a,b])\to[a,b]$ be an $\oplus$-measure. The generalized Choquet integral of a nonnegative measurable function $f:X\to[a,b]_+$ with respect to μ, m over $A\in\mathcal{A}$ is given by

$$(C_{\mu,m})\int_A^{\oplus} f\otimes d(\mu,m)=\int_{[a,b]_+}^{\oplus}\mu(A\cap(t\prec f))\otimes dm(t).$$

For $X=A$ we use the notation $(C_{\mu,m})\int^{\oplus} f\otimes d(\mu,m)$.

Example 10 For a measurable function $f:X\to[0,\infty]$, a fuzzy measure $\mu:\mathcal{A}\to[0,\infty]$ and $A\in\mathcal{A}$, we denote by $(f>t)=\{x\in X\mid f(x)>t\}$ and Mesiar's Choquet-like integral [40] is given by

$$(\hat{C})\int_A f d\mu=(SM)\int_{[0,\infty]}^{\hat{+}}\mu(A\cap(f>t))\hat{\cdot}d\lambda_{\hat{+}},$$

where the right-hand side integral is Sugeno and Murofushi's pseudo-additive integral based on the structure $([0,\infty],\hat{+},\hat{\cdot})$, see [70]. Taking $([0,\infty],\hat{+},\hat{\cdot})$, and $m:Borel([0,\infty])\to[0,\infty]$ a $\hat{+}$ measure, then the generalized Choquet integral reduces to the integral $(\hat{C})\int_A$.

Proposition 1 *The generalized Choquet integral has the following properties:*

(i) $f=\mathbf{0}$ *or* $\mu(A)=\mathbf{0}$ *implies* $(C_{\mu,m})\int_A^{\oplus} f\otimes d(\mu,m)=\mathbf{0}$;
(ii) $f\preceq g$ *implies* $(C_{\mu,m})\int_A^{\oplus} f\otimes d(\mu,m)\preceq(C_{\mu,m})\int_A^{\oplus} g\otimes d(\mu,m)$;
(iii) $A\subseteq B$ *implies* $(C_{\mu,m})\int_A^{\oplus} f\otimes d(\mu,m)\preceq(C_{\mu,m})\int_B^{\oplus} f\otimes d(\mu,m)$;
(iv) $(C_{\mu,m})\int_A^{\oplus} f\otimes d(\mu,m)=(C_{\mu,m})\int^{\oplus}\chi_A\circ f\otimes d(\mu,m)$;
(v) $(C_{\mu,m})\int_A^{\oplus} c\otimes d\mu=m([\mathbf{0},c])\otimes\mu(A)$, *for* $\mathbf{0}\preceq c$;
(vi) $\mu_1\preceq\mu_2$ *implies* $(C_{\mu,m})\int_A^{\oplus} f\otimes d(\mu_1,m)\preceq(C_{\mu,m})\int_A^{\oplus} f\otimes d(\mu_2,m)$;
(vii) $\mu=\mu_1\oplus\mu_2$, *implies*

$$(C_{\mu,m})\int_A^{\oplus} f\otimes d(\mu,m)=(C_{\mu,m})\int_A^{\oplus} f\otimes d(\mu_1,m)\oplus(C_{\mu,m})\int_A^{\oplus} f\otimes d(\mu_2,m)$$

(for Cases I, II and III).

Example 11 For a nonnegative step function $s=\oplus_{i=1}^{k}a_i\otimes\chi_{A_i}$ with

$$\mathbf{0}=a_0\preceq a_1\preceq a_2\preceq\cdots\preceq a_n=b\ \ (\text{suppose }\ \mathbf{0}\prec b)$$

the corresponding generalized Choquet integral is given by

$$(C_{\mu,m})\int^{\oplus} s\otimes d\mu = \bigoplus_{i=1}^{k} m([a_{i-1},a_i])\otimes\mu(\cup_{i=1}^{k}A_i).$$

We have proved in [81] some important properties of the generalized Choquet integral. Here we present only two of them, without proof.

Theorem 3 (Monotone non-decreasing convergence theorem) *Let* $([a,b],\oplus,\otimes)$ *be a semiring and let* (f_n) *be a sequence of nonnegative measurable functions and* f *be a nonnegative function. If* μ *is lower semicontinuous, then* $f_n\uparrow f$ *implies*

$$(C_{\mu,m})\int^{\oplus} f_n\otimes d(\mu,m)\uparrow (C_{\mu,m})\int^{\oplus} f\otimes d(\mu,m).$$

Theorem 4 (Monotone non-decreasing convergence theorem for fuzzy measures) *Let* $([a,b],\oplus,\otimes)$ *be a semiring and let* (μ_n) *be a sequence of fuzzy measures and* μ *be a set function. Then* $\mu_n\uparrow\mu$ *implies*

$$(C_{\mu,m})\int^{\oplus} f\otimes d(\mu_n,m)\uparrow (C_{\mu,m})\int^{\oplus} f\otimes d(\mu,m).$$

It is interesting that for the generalized Choquet integral, in general, the monotone nonincreasing convergence theorem does not hold. We shall consider some important spacial cases.

Definition 22 Let $([a,b],\oplus,\otimes)$ be a semiring Case II and

$$m([a,t]) = g^{-1}(\alpha(g(t))),$$

where $\alpha:[0,\infty]\to[0,\infty]$ is a non-decreasing function with $\alpha(0)=0$. Then the generalized Choquet integral, denoted by $(CS_\alpha)\int_A^{\oplus} f\otimes d\mu$, is called a pseudo-Choquet-Stieltjes integral of f with respect to μ, α, over $A\in\mathcal{A}$.

Example 12 (i) Let $[a,b]=[0,\infty]$ and $g(x)=x$, $x\in[0,\infty]$. Then $(CS_\alpha)\int_A^{\oplus} f\otimes d\mu$ reduces to the Choquet-Stieltjes integral [18, 50].

(ii) $(CS_\alpha)\int_A^{\oplus} f\otimes d\mu$ reduces for $\alpha(t)=t$, $t\in[0,\infty]$ to the g-Choquet integral [40], denoted by $(C_g)\int_A^{\oplus} f\otimes d\mu$.

We give some representations for generalized Choquet integrals.

Proposition 2 *Let* $([a,b],\oplus,\otimes)$ *be a semiring Case I or III from Sect. 4.1. Then we have*

$$(CS_\alpha)\int_A^{\sup} f\otimes d\mu = \sup_{\alpha\in[a,b]}(\alpha\otimes\mu(A\cap(\alpha\preceq f))$$
$$= \sup\{\mathcal{Q}(s,A)\mid s\preceq f\}.$$

where

$$\mathcal{Q}(s,A)=\bigvee_{i=1}^{n}\alpha_i\otimes\mu(A\cap E_i),\quad s=\bigvee_{i=1}^{n}\alpha_i\otimes\chi_{A\cap E_i},$$

for $\alpha_i\in[a,b]$, $E_i\in\mathcal{A}$, $1\leqslant i\leqslant n$.

We need the following notion, see [49, 55]

Definition 23 Functions f,h are co-monotone on $A\subseteq X$, denoted by $f\sim h|_A$, iff $f(s)\prec f(t)$ implies $h(s)\preceq h(t)$ for all $s,t\in A$.

Theorem 5 *Let* $([a,b],\oplus,\otimes)$ *be a semiring Case II and* $\alpha(t)$ *be a linear increasing function. If* f,h *are co-monotone on A, then the following equality holds:*

$$(CS_\alpha)\int_A^{\oplus}(f\oplus h)\otimes d\mu=(CS_\alpha)\int_A^{\oplus}f\otimes d\mu\oplus(CS_\alpha)\int_A^{\oplus}h\otimes d\mu.$$

Theorem 6 *Let* $([a,b],\oplus,\otimes)$ *be a semiring Case I or III. If* f,h *are co-monotone on A, then the following equality holds:*

$$(CS_\alpha)\int_A^{\sup}(f\vee h)\otimes d\mu=(CS_\alpha)\int_A^{\sup}f\otimes d\mu\vee(CS_\alpha)\int_A^{\sup}h\otimes d\mu.$$

Acknowledgements This research was supported by Science Fund of the Republic of Serbia, #Grant No. 6524105, AI-ATLAS.

References

1. Adams, D.R.: Choquet integrals in potential theory. Publ. Mat. **42**(1), 3–66 (1998)
2. Aumann, R.J., Shapley, L.S.: Values of Non-Atomic Games. Princeton University Press, Princton (1974)
3. Baccelli, F., Cohen, G., Olsder, G.J., Quadra, J.P.T: Synchronization and Linearity: An Algebra for Discrete Event Systems. Wiley, New York (1992)
4. Benvenuti, P., Mesiar, R., Vivona D.: Monotone set functions-based integrals. In: Pap, E. (eds.) Handbook of Measure Theory, vol. II, pp. 1329–1379 . Elsevier, Amsterdam (2002)
5. Bernard, C., Ghossoub, M.: Static portfolio choice under cumulative prospect theory. Math. Financ. Econ. **2**, 277–306 (2010)
6. Burgin, M.S.: Nonclassical models of the natural numbers. Uspekhi Mat. Nauk **32**, 209–210 (1977). (in Russian)
7. Burgin, M., Meissner, G.: $1+1=3$: Synergy arithmetics in economics. Appl. Math. **8**, 133–134 (2017)
8. Czachor, M.: A Loophole of all "Loophole-Free" Bell-type theorems. Found. Sci. **25**, 971–985 (2020). https://doi.org/10.1007/s10699-020-09666-0

9. Czachor, M.: Non-Newtonian mathematics instead of non-Newtonian physics: dark matter and dark energy from a mismatch of arithmetics. Found. Sci. **26**, 75–95 (2021). https://doi.org/10.1007/s10699-020-09687-9(0123456789().,-volV)(0123456789().,-volV)
10. Carbajal-Hernandez, J.J., Sanchez-Fernandez, L.P., Carrasco-Ochoa, J.A.: Assessment and prediction of air quality using fuzzy logic and autoregressive models. Atmos. Environ. **60**, 3–50 (2012)
11. Choquet, G.: Theory of capacities. Ann. Inst. Fourier **5**, 131–295 (1953)
12. Deisenroth, M.P., Faisal, A.A., Ong, C.S.: Mathematics for Machine Learning. Cambridge University Press, Cambridge (2020)
13. Delić, M., Nedović, E., Pap, E.: Extended power-based aggregation of distance functions and application in image segmentation. Inf. Sci. **494**, 155–173 (2019)
14. Denneberg, D.: Non-additive Measure and Integral. Kluwer, Dordrecht (1994)
15. Do, Y., Thiele, C.: L^p theory for outer measures and two themes of Lennart Carlson united. Bull. Amer. Math. Soc. **52**(2), 249–296 (2015)
16. Dubois, D., Pap, E., Prade, H.: Hybrid probabilistic-possibilistic mixtures and utility functions. In: Fodor, J., de Baets, B., Perny, P. (eds.) Preferences and Decisions Under Incomplete Knowledge, volume 51 of Studies in Fuzziness and Soft Computing, pp. 51–73. Springer, Berlin (2000)
17. Even, Y., Lehrer, E.: Decomposition-integral: unifying Choquet and concave integrals. Econom. Theory **56**, 33–58 (2014)
18. Gal, G.: On a Choquet-Stieltjes type integral on intervals. Math. Slovaca **69**, 801–814 (2019)
19. Gilboa, I., Schmeidler, D.: Additive Representations of non-additive measures and the Choquet integral. Ann. Oper. Res. **52**, 43–65 (1994)
20. Grabisch, M.: Set Functions. Games and Capacities in Decision Making. Springer, Cham (2016)
21. Grabisch, M., Marichal, J.L., Mesiar, R., Pap, E.: Aggregation Functions. Encyclopedia of Mathematics and Its Applications, vol. 127, Cambridge University Press, Cambridge (2009)
22. Grossman, M., Katz, R.: Non-Newtonian Calculus. Lee Press, Pigeon Cove (1972)
23. Grbić, T., Medić, S., Perović, A., Mihailović, B., Novković, N., Duraković, N.: A premium principle based on the g-integral. Stoch. Anal. Appl. **35**(3), 465–477 (2017)
24. Greco, S., Mesiar, R., Rindone, F.: Discrete bipolar universal integral. Fuzzy Sets Syst. **252**, 55–65 (2014)
25. Joe, H.: Dependence Modeling with Copulas. Monographs on Statistics and Applied Probability, vol. 134. CRC Press, Boca Raton (2015)
26. Hadžić, O., Pap, E.: Fixed Point Theory in Probabilistic Metric Spaces. Mathematics and Its Applications, vol. 536. Kluwer Academic Publishers, Dordrecht (2001)
27. Kaluszka, M., Krzeszowiec, M.: Pricing insurance contracts under cumulative prospect theory. Insur. Math. Econ. **50**, 159–166 (2012)
28. Klement, E.P., Mesiar, R., Pap, E.: Triangular Norms. Kluwer, Amsterdam (2000)
29. Klement, E.P., Mesiar, R., Pap, E.: Integration with respect to decomposable measures based on a conditionally distributive semiring on the unit interval. Int. J. Uncertain. Fuzziness Knowl.-Based Syst. **8**, 701–717 (2000)
30. Klement, E.P., Mesiar, R., Pap, E.: A universal integral as common frame for Choquet and Sugeno integral. IEEE Trans. Fuzzy Syst. **18**(1), 178–187 (2010)
31. Klement, E.P., Li, J., Mesiar, R., Pap, E.: Integrals based on monotone set functions. Fuzzy Sets Syst. **281**, 88–102 (2015)
32. Kolokoltsov, V.N.: Nonexpansive maps and option pricing theory. Kybernetika **34**(6), 713–724 (1998)
33. Kuczma, M., Gilányi, A. (ed.): An Introduction to the Theory of Functional Equations and Inequalities, Cauchy's Equation and Jensens Inequality, 2nd edn. Birkhäuser, Basel (2009)
34. Lehrer, E.: A new integral for capacities. Econom. Theory **39**, 157–176 (2009)
35. Lehrer, E., Teper, R.: The concave integral over large spaces. Fuzzy Sets Syst. **159**, 2130–2144 (2008)
36. Litvinov, G.L., Maslov, V.P. (eds): Idempotent Mathematics and Mathematical Physics. Contemporary Mathematics, vol. 377. American Mathematical Society, Providence (2003)

37. Mao, T., Yang, F.: Characterizations of risk aversion in cumulative prospect theory. Mathematics and Financial Economics. https://doi.org/10.1007/s11579-018-0229-0
38. Maslov, V.P.: A new superposition principle for optimization problems. Uspekhi Mat Nauk **42**(3), 39–48 (1987)
39. Maslov, V.P., Samborskii, S.N. (eds.): Idempotent Analysis. Advances in Soviet Mathematics, vol. 13. American Mathematical Society, Providence (1992)
40. Mesiar, R.: Choquet-like integral. J. Math. Anal. Appl. **194**, 477–488 (1995)
41. Mesiar, R., Li, J., Pap, E.: Pseudo-concave integrals. In: NLMUA 2011, Advances in Intelligent Systems and Computing, vol. 100, pp. 43–49. Springer, Berlin (2011)
42. Mesiar, R., Li, J., Pap, E.: The Choquet integral as Lebesgue integral and related inequalities. Kybernetika **46**, 1098–1107 (2010)
43. Mesiar, R., Li, J., Pap, E.: Discrete pseudo-integrals. J. Approx. Reason. **54**, 357–364 (2013)
44. Mesiar, R., Li, J., Pap, E.: Superdecomposition integral. Fuzzy Sets Syst. **259**, 3–11 (2015)
45. Mesiar, R., Pap, E.: Idempotent integral as limit of g-integrals. Fuzzy Sets Syst. **102**, 385–392 (1999)
46. Mesiar, R., Rybarik, J.: Pan-operations structure. Fuzzy Sets Syst. **74**, 365–369 (1995)
47. Mihailović, B., Pap, E.: Asymmetric integral as a limit of generated Choquet integrals based on absolutely monotone real set functions. Fuzzy Sets Syst. **181**, 39–49 (2011)
48. Mihailović, B., Pap, E., Štrboja, M., Simićevic A.: A unified approach to the monotone integral-based premium principles under the CPT theory. Fuzzy Sets Syst. **398**, 78–97 (2020)
49. Murofushi, T., Sugeno, M.: A theory of fuzzy measures: representations, the Choquet integral, and null sets. J. Math. Anal. Appl. **159**, 532–549 (1991)
50. Narukawa, Y., Murofushi, T.: The n-step Choquet integral on finite spaces. In: Proceedings of the 9th International Conference on Information Processing and Management of Uncertainty in Knowledge-Based Systems, Annecy, pp. 539–543 (2002)
51. Nedović, Lj., Pap, E.: Aggregation of sequence of fuzzy measures. Iranian J. Fuzzy **17**(2), 39–55 (2020)
52. Nelsen, R.B.: An Introduction to Copulas. Lecture Notes in Statistics. Springer, Berlin (1999)
53. Pap, E.: An integral generated by decomposable measure. Univ. Novom Sadu Zb. Rad. Prirod.-Mat. Fak. Ser. Mat. **20**(1), 135–144 (1990)
54. Pap, E.: g-calculus. Univ. u Novom Sadu, Zb. Rad. Prirod.-Mat. Fak. Ser. Mat. **23**(1), 145–150 (1993)
55. Pap, E.: Null-additive set functions. Mathematics and Its Applications, vol. 337. Kluwer Academic Publishers, Dordrecht (1995)
56. Pap, E. (ed.): Handbook of Measure Theory: I, II. Elsevier, Amsterdam (2002)
57. Pap, E.: Pseudo-additive measures and their applications. In: Pap, E. (ed.) Handbook of Measure Theory, vol. II, pp. 1403–1465. Elsevier, North-Holland (2002)
58. Pap, E.: A generalization of the utility theory using a hybrid idempotent-probabilistic measure. In: Litvinov, G.L., Maslov, V.P. (eds.) Proceedings of the Conference on Idempotent Mathematics and Mathematical Physics, Contemporary Mathematics, vol. 377, pp. 261–274. American Mathematical Society, Providence (2005)
59. Pap, E.: Generalized real analysis and its application. Int. J. Approx. Reason. **47**, 368–386 (2008)
60. Pap, E. (ed.): Intelligent Systems: Models and Applications. Topics in Intelligent Engineering and Informatics, vol. 3. Springer, Berlin (2013)
61. Pap, E.: Sublinear Means. In: 36th Linz Seminar on Fuzzy Sets Theory: Functional Equations and Inequalities, Linz, 2–6 February 2, pp. 75–79 (2016)
62. Pap, E.: Three types of generalized Choquet integral. Bull. Un. Mat. Ital. **13**(4), 545–553 (2020)
63. Pap, E., Štajner, I.: Generalized pseudo-convolution in the theory of probabilistic metric spaces, information, fuzzy numbers, system theory. Fuzzy Sets Syst. **102**, 393–415 (1999)
64. Puhalskii, A.: Large Deviations and Idempotent Probability. CRC Press, Boca Raton (2001)
65. Quiggin, J.: Generalized Expected Utility Theory. The Rank-Dependent Model. Kluwer Academic Publishers, Boston (1993)

66. Schmeidler, D.: Subjective probability and expected utility without additivity. Econometrica **57**, 517–587 (1989)
67. Shilkret, N.: Maxitive measure and integration. Indag. Math. **33**, 109–116 (1971)
68. Sugeno, M.: Theory of fuzzy integrals and its applications. Ph.D. Thesis, Tokyo Institute of Technology (1974)
69. Sowlat, M.H., Gharibi, H., Yunesian, M., Mahmoudi, M.T., Lotfi, S.: A novel, fuzzy-based air quality index (FAQI) for air quality assessment. Atmos. Environ. **45**, 2050–2059 (2011)
70. Sugeno, M., Murofushi, T.: Pseudo-additive measures and integrals. J. Math. Anal. Appl. **122**, 197–222 (1987)
71. Tan, X., Han, L., Zhang, X., Zhou, W., Li, W., Qian, Y.: A review of current quality indexes and improvements under the multi-contaminant air pollution exposure. J. Environ. Manag. **279** (2021). https://doi.org/10.1016/j.jenvaman.2020.111681
72. Tversky, A., Kahneman, D.: Advances in prospect theory. Cumulative representation of uncertainty. J. Risk Uncertain. **5**, 297–323 (1992)
73. Wang, S.S., Young, V.R., Panjer, H.H.: Axiomatic characterization of insurance prices. Insur.: Math. Econ. **21**, 173–183 (1997)
74. Wakker, P.P.: Prospect Theory: For Risk and Ambiguity. Cambridge University Press, New York (2010)
75. Wakker, P., Tversky, A.: An axiomatization of cumulative prospect theory. J. Risk Uncertain. **7**, 147–175 (1993)
76. Wang, Z., Klir, G.J.: Generalized Measure Theory. Springer, Boston (2009)
77. Yang, Q.: The pan-integral on the fuzzy measure space. Fuzzy Math. **3**, 107–114 (1985). (in Chinese)
78. Yaari, M.E.: The dual theory of choice under risk. Econometrica **55**, 95–115 (1987)
79. Young, V.R.: Premium principles. Encyclopedia of Actuarial Science. Wiley, New York (2006)
80. Zhang, D., Pap, E.: Fubini theorem and generalized Minkowski inequality for the pseudo-integral. J. Approx. Reason. **122**, 9–23 (2020)
81. Zhang, D., Mesiar, R., Pap, E.: Pseudo-integral and generalized Choquet integral. Fuzzy Sets Syst. https://doi.org/10.1016/j.fss.2020.12.005

Collection and Decomposition Integrals in Multicriteria-Decision Support

Radko Mesiar and Adam Šeliga

Abstract Integrals generalizing the expected value of random variables play an essential role in multi-criteria decision problems, where they are considered as utility functions. In this chapter, we focus on the recently introduced collection and decomposition integrals, covering, among others, the Choquet, the Shilkret and the PAN integrals. We discuss some properties and give several examples of these integrals, in particular those extending the Lebesgue integral. Later, we discuss this type of integral for interval-valued functions. Finally, possible applications in multi-criteria-decision support and imprecise probability domains are shown.

Keywords Multicriteria-decision · Monotone measure · Decomposition system · Choquet integral · Shilkret integral · Decomposition integral · Interval-valued function

1 Introduction

Two standard approaches to integration can be traced back to Lebesgue [5] and Choquet [1], considering the vertical and horizontal decomposition of simple functions. Formally, vertical decompositions are related to partitions while horizontal decompositions are linked to chains. Recently, Lehrer [6] considered all non-empty subsets of a finite space X as a background of his concave integral (related convex integral was proposed).

A generalization of aforementioned approaches was proposed by Even and Lehrer [3] and was extended in several subsequent works. Note that all proposed integrals can be regarded as the solutions to optimization problems, in particular in economics and

R. Mesiar (✉) · A. Šeliga
Faculty of Civil Engineering, Slovak University of Technology, Radlinského 11, 810 05 Bratislava, Slovakia
e-mail: radko.mesiar@stuba.sk

A. Šeliga
e-mail: adam.seliga@stuba.sk

E. Pap (ed.), *Artificial Intelligence: Theory and Applications*,
Studies in Computational Intelligence 973,
https://doi.org/10.1007/978-3-030-72711-6_2

multi-criteria decision support, where they are applied as specific utility functions. In this chapter, we give an overview of some results concerning decomposition integrals, and include some new ones. After the preliminaries introduced in Sect. 2, we present some examples and properties of decomposition integrals in Sect. 3. Section 4 deals with decomposition integrals which extend the Lebesgue integral, i.e., they coincide with the Lebesgue integral once it is defined (when additive measures are considered). Decomposition integrals for interval-valued functions are studied in Sect. 5. Section 6 shows two applications of decomposition integrals, the first concerning multicriteria decision support, and the second is giving a suggestion for constructing coherent lower and upper previsions. Finally, some concluding remarks are added.

2 Preliminaries

Throughout this chapter we will consider a fixed finite set X referred to as a *space*. Without any loss of generality, we may assume that $X = \{1, 2, \ldots, n\}$ for some natural number $n \in \mathbb{N}$. We will consider only non-negative functions $f: X \to [0, \infty[$ and the class of all such functions will be denoted by $\mathbb{F}$.

By $L([0, \infty[)$ we understand the class of all closed intervals that are a subset of $[0, \infty[$. The class of all non-negative interval-valued functions $v: X \to L([0, \infty[)$ will be denoted by $\mathbb{V}$. We say that a function $f \in \mathbb{F}$ is a selector of an interval-valued function $v \in \mathbb{V}$, denoting it by $f \in v$, if and only if $f(x) \in v(x)$ for all $x \in X$. Also note that for every interval-valued function $v \in \mathbb{V}$ there are two functions $a_v, b_v \in \mathbb{F}$ such that $v(x) = [a_v(x), b_v(x)]$ for all $x \in X$. The function a_v is called a lower boundary of v and b_v is called an upper boundary of v.

Definition 1 A *monotone measure* is any set function $\mu: 2^X \to [0, \infty[$ that is grounded, i.e., $\mu(\emptyset) = 0$, and non-decreasing with respect to set inclusion, i.e., $A \subseteq B \subseteq X$ implies $\mu(A) \leqslant \mu(B)$. A monotone measure μ is called:

- a *capacity*, if $\mu(X) = 1$;
- a *measure*, if $A \cap B = \emptyset$ implies $\mu(A \cup B) = \mu(A) + \mu(B)$ for all $A, B \in 2^X$.

The class of all monotone measures will be denoted by $\mathbb{M}$, the class of all capacities by $\mathbb{M}_1$ and the class of all measures by $\mathbb{M}_+$.

Definition 2 A *collection*, mostly denoted by $\mathcal{D}$, is a non-empty subset of $2^X \setminus \{\emptyset\}$. The class of all collections will be denoted by $\mathbb{D}$. We say that a collection $\mathcal{D} \in \mathbb{D}$ is *logically independent* if and only if

$$\bigcap_{A \in \mathcal{D}} A \neq \emptyset.$$

Example 1 The following sets are examples of collections:

- the maximal collection $\mathcal{D}_{\mathsf{max}} = 2^X \setminus \{\emptyset\}$;

- a sub-partition $\mathcal{D}_{\mathsf{part}} = \{A_i\}_{i=1}^k$, where $A_i \cap A_j = \emptyset$ for $i \neq j$;
- a chain $\mathcal{D}_{\mathsf{ch}} = \{A_i\}_{i=1}^k$, where

$$X \supseteq A_1 \supsetneq A_2 \supsetneq \cdots \supsetneq A_1 \neq \emptyset,$$

  and note that each $\mathcal{D}_{\mathsf{ch}}$ is logically independent;
- a PC collection $\mathcal{D}_{\mathsf{PC}}$ is such that $A, B \in \mathcal{D}_{\mathsf{PC}}$ implies $A \cap B \in \{A, B, \emptyset\}$.

Definition 3 ([3]) A *decomposition system*, mostly denoted by $\mathcal{H}$, is a non-empty set of collections. The class of all decomposition systems will be denoted by $\mathbb{H}$. A decomposition system $\mathcal{H} \in \mathbb{H}$ is called *complete* if and only if for every $A \subseteq 2^X$, $A \neq \emptyset$, there exists a collection $\mathcal{D} \in \mathcal{H}$ such that $A \in \mathcal{D}$.

Example 2 The following sets are examples of decomposition systems:

- the maximal decomposition system $\mathcal{H}_{\mathsf{max}} = \{\mathcal{D}_{\mathsf{max}}\}$;
- the PAN decomposition system $\mathcal{H}_{\mathsf{PAN}} = \{\mathcal{D} \in \mathbb{D} \colon \mathcal{D} \text{ is a sub-partition}\}$;
- the Choquet decomposition system $\mathcal{H}_{\mathsf{Ch}} = \{\mathcal{D} \in \mathbb{D} \colon \mathcal{D} \text{ is a chain}\}$;
- the PC decomposition system $\mathcal{H}_{\mathsf{PC}} = \{\mathcal{D} \in \mathbb{D} \colon \mathcal{D} \text{ is a PC collection}\}$;
- the Shilkret decomposition system $\mathcal{H}_{\mathsf{Sh}} = \{\{A\} \colon \emptyset \neq A \subseteq X\}$.

Definition 4 Let $\mathcal{H} \in \mathbb{H}$ be a decomposition system. A *sub-decomposition integral* with respect to a decomposition system $\mathcal{H}$ is an operator

$$\mathsf{dec}_{\mathcal{H}} \colon \mathbb{F} \times \mathbb{M} \to [0, \infty[$$

defined by

$$\mathsf{dec}_{\mathcal{H}}(f, \mu) = \bigvee_{\mathcal{D} \in \mathcal{H}} \bigvee \left\{ \sum_{A \in \mathcal{D}} \alpha_A \mu(A) \mid \sum_{A \in \mathcal{D}} \alpha_A \mathbb{1}_A \leqslant f, \alpha_A \geqslant 0 \right\}$$

for all $f \in \mathbb{F}$ and $\mu \in \mathbb{M}$.

Definition 5 ([7]) Let $\mathcal{H} \in \mathbb{H}$ be a decomposition system. A *super-decomposition integral* with respect to a decomposition system $\mathcal{H}$ is an operator

$$\mathsf{dec}^*_{\mathcal{H}} \colon \mathbb{F} \times \mathbb{M} \to [0, \infty]$$

defined by

$$\mathsf{dec}^*_{\mathcal{H}}(f, \mu) = \bigwedge_{\mathcal{D} \in \mathcal{H}} \bigwedge \left\{ \sum_{A \in \mathcal{D}} \alpha_A \mu(A) \mid \sum_{A \in \mathcal{D}} \alpha_A \mathbb{1}_A \geqslant f, \alpha_A \geqslant 0 \right\}$$

for all $f \in \mathbb{F}$ and $\mu \in \mathbb{M}$.

Remark 1 Note that, in the definition of super-decomposition integrals, not only the inequality sign and supremum are replaced, the range of values is also extended to include infinity.

Through decomposition integrals we understand both sub-decomposition and super-decomposition integrals. In the original publication [3], sub-decomposition integrals are referred to as decomposition integrals.

If we restrict ourselves to singleton decomposition systems, we obtain the so-called collection integrals introduced in [9].

Definition 6 Let $\mathcal{D} \in \mathbb{D}$ be a collection. A *sub-collection integral* with respect to collection $\mathcal{D}$ is an operator $\mathsf{col}_{\mathcal{D}} : \mathbb{F} \times \mathbb{M} \to [0, \infty[$ given by $\mathsf{col}_{\mathcal{D}} = \mathsf{dec}_{\{\mathcal{D}\}}$ and a *super-collection integral* with respect to collection $\mathcal{D}$ is an operator $\mathsf{col}^*_{\mathcal{D}} : \mathbb{F} \times \mathbb{M} \to [0, \infty]$ given by $\mathsf{col}^*_{\mathcal{D}} = \mathsf{dec}^*_{\{\mathcal{D}\}}$.

Note that for any $\mathcal{H} \in \mathbb{H}$ we have

$$\mathsf{dec}_{\mathcal{H}} = \bigvee_{\mathcal{D} \in \mathcal{H}} \mathsf{col}_{\mathcal{D}} \quad \text{and} \quad \mathsf{dec}^*_{\mathcal{H}} = \bigwedge_{\mathcal{D} \in \mathcal{H}} \mathsf{col}^*_{\mathcal{D}}.$$

3 Examples and Basic Properties of Decomposition Integrals

Many useful non-linear integrals are a particular instance of decomposition integrals. In the following example, we list some of them.

Example 3
- $\mathsf{dec}_{\mathcal{H}_{\mathsf{max}}} = \mathsf{col}_{\mathcal{D}_{\mathsf{max}}}$ is the concave integral [6];
- $\mathsf{dec}_{\mathcal{H}_{\mathsf{PAN}}}$ is the PAN integral [13, 15];
- $\mathsf{dec}_{\mathcal{H}_{\mathsf{Ch}}} = \mathsf{dec}^*_{\mathcal{H}_{\mathsf{Ch}}}$ is the Choquet integral [1];
- $\mathsf{dec}_{\mathcal{H}_{\mathsf{PC}}}$ is the PC integral [12];
- $\mathsf{dec}_{\mathcal{H}_{\mathsf{Sh}}}$ is the Shilkret integral [11];
- $\mathsf{dec}^*_{\mathcal{H}_{\mathsf{max}}} = \mathsf{col}^*_{\mathcal{D}_{\mathsf{max}}}$ is the convex integral [7];

In the following theorem, the basic properties of decomposition integrals are given.

Theorem 1 *Let $\mathcal{H}$ be a decomposition system. Then both decomposition integrals with respect to $\mathcal{H}$, i.e., $\mathsf{dec}_{\mathcal{H}}$ and $\mathsf{dec}^*_{\mathcal{H}}$, are*

- *non-decreasing in both arguments;*
- *positively homogeneous in both arguments.*

To the class of collections $\mathbb{D}$ a relation $\preceq$ can be introduced and on the class of decomposition systems $\mathbb{H}$ a relation $\sqsubseteq$ can be introduced.

Definition 7 Let $\mathcal{D}_1, \mathcal{D}_2 \in \mathbb{D}$ be two collections. We say $\mathcal{D}_1 \preceq \mathcal{D}_2$ if and only if $\mathcal{D}_1 \subseteq \mathcal{D}_2$ holds.

Definition 8 Let $\mathcal{H}_1, \mathcal{H}_2 \in \mathbb{H}$ be two decomposition systems. We say $\mathcal{H}_1 \sqsubseteq \mathcal{H}_2$ if and only if for every $\mathcal{D}_1 \in \mathcal{H}_1$ there exists $\mathcal{D}_2 \in \mathcal{H}_2$ such that $\mathcal{D}_1 \preceq \mathcal{D}_2$.

Note that $\preceq$ is a partial order on $\mathbb{D}$ but $\sqsubseteq$ is not a partial order on $\mathbb{H}$. The relation $\sqsubseteq$ is reflexive, transitive, but is not anti-symmetric, i.e., $\sqsubseteq$ is a preorder.

Theorem 2 *Let $\mathcal{D}_1, \mathcal{D}_2 \in \mathbb{D}$ be collections such that $\mathcal{D}_1 \preceq \mathcal{D}_2$. Then*

$$\mathsf{col}_{\mathcal{D}_1} \leqslant \mathsf{col}_{\mathcal{D}_2} \quad \textit{and} \quad \mathsf{col}^*_{\mathcal{D}_1} \geqslant \mathsf{col}^*_{\mathcal{D}_2}.$$

Moreover, if $\mathcal{H}_1, \mathcal{H}_2 \in \mathbb{H}$ are two decomposition systems such that $\mathcal{H}_1 \sqsubseteq \mathcal{H}_2$, then

$$\mathsf{dec}_{\mathcal{H}_1} \leqslant \mathsf{dec}_{\mathcal{H}_2} \quad \textit{and} \quad \mathsf{dec}^*_{\mathcal{H}_1} \geqslant \mathsf{dec}^*_{\mathcal{H}_2}.$$

Following the results of the publication [8], we obtain the following results.

Theorem 3 *Let $\mathcal{D} \in \mathbb{D}$ be a logically independent collection and let $\mathcal{H} \in \mathbb{H}$ be a complete decomposition system. Then*

$$\mathsf{col}_{\mathcal{D}}(\mathbb{1}_A, \mu) \leqslant \mu(A) \quad \textit{and} \quad \mathsf{col}^*_{\mathcal{D}}(\mathbb{1}_A, \mu) \geqslant \mu(A),$$

$$\mathsf{dec}_{\mathcal{H}}(\mathbb{1}_A, \mu) \geqslant \mu(A) \quad \textit{and} \quad \mathsf{dec}^*_{\mathcal{H}}(\mathbb{1}_A, \mu) \leqslant \mu(A),$$

for all sets $A \subseteq X$.

From the previous theorem, one can easily obtain the following result.

Theorem 4 *Let $\mathcal{H} \in \mathbb{H}$ be a complete decomposition system such that every collection $\mathcal{D} \in \mathcal{H}$ in it is logically independent. Then the sub-decomposition and super-decomposition integrals with respect to the decomposition system $\mathcal{H}$ return the underlying monotone measure, i.e.,*

$$\mathsf{dec}_{\mathcal{H}}(\mathbb{1}_A, \mu) = \mu(A) = \mathsf{dec}^*_{\mathcal{H}}(\mathbb{1}_A, \mu)$$

for all sets $A \subseteq X$.

Another interesting result was obtained in [10], where all sub-collection integrals representable by the Choquet integral were characterized. Naturally, one can also extend this result for the super-collection integrals.

Theorem 5 *A collection integral with respect to a collection $\mathcal{D} \in \mathbb{D}$ can be represented by the Choquet integral if and only if $\mathcal{D}$ is a PC-collection.*

Remark 2 Note that the representability of the collection integral $\mathsf{col}_{\mathcal{D}}$ (or, $\mathsf{col}^*_{\mathcal{D}}$), where $\mathcal{D}$ is a PC-collection, by the Choquet integral is in the following sense: For every monotone measure $\mu \in \mathbb{M}$ there exists a monotone measure $\nu \in \mathbb{M}$ such that

$$\mathsf{col}_{\mathcal{D}}(\cdot, \mu) = \mathsf{Ch}(\cdot, \nu), \qquad \left(\text{or, } \mathsf{col}^*_{\mathcal{D}}(\cdot, \mu) = \mathsf{Ch}(\cdot, \nu)\right).$$

Moreover, the monotone measure ν is given by

$$\nu(A) = \mathsf{col}_{\mathcal{D}}(\mathbb{1}_A, \mu), \qquad \left(\text{or, } \nu(A) = \mathsf{col}^*_{\mathcal{D}}(\mathbb{1}_A, \mu)\right),$$

for all sets $A \subseteq X$.

Remark 3 It seems that the collection integrals based on PC-collections $\mathcal{D}_{\mathsf{PC}}$ have an important place within the collection integrals, because in [10] it was proved that, in general, Chebyshev's, Jensen's, Cauchy's, and Hölder's inequalities do hold for these integrals and only for them.

4 Decomposition Integrals Extending Lebesgue Integral

Restricting ourselves to measures only, the following property of decomposition integrals can be obtained.

Theorem 6 *Let $\mathcal{D} \in \mathbb{D}$ be a collection and let $\mathcal{H} \in \mathbb{H}$ be a decomposition system. Then*

$$\mathsf{col}_{\mathcal{D}}(\cdot, \mu) \leqslant \mathsf{Leb}(\cdot, \mu) \leqslant \mathsf{col}^*_{\mathcal{D}}(\cdot, \mu)$$

and

$$\mathsf{dec}_{\mathcal{H}}(\cdot, \mu) \leqslant \mathsf{Leb}(\cdot, \mu) \leqslant \mathsf{dec}^*_{\mathcal{H}}(\cdot, \mu)$$

for all measures $\mu \in \mathbb{M}_+$.

Remark 4 In general, neither $\mathsf{dec}_{\mathcal{H}} \leqslant \mathsf{dec}^*_{\mathcal{H}}$ nor $\mathsf{dec}_{\mathcal{H}} \geqslant \mathsf{dec}^*_{\mathcal{H}}$ is true.

The following classes of collections and decomposition systems are of interest in this section.

Definition 9 A subclass $\mathbb{D}_+$ of the class of all collections $\mathbb{D}$ consists of those collections $\mathcal{D} \in \mathbb{D}$ for which $\mathsf{col}_{\mathcal{D}}$ extends the Lebesgue integral. Analogously, a subclass $\mathbb{H}_+$ of all decomposition systems consists of the decomposition systems $\mathcal{H} \in \mathbb{H}$ for which $\mathsf{dec}_{\mathcal{H}}$ extends the Lebesgue integral.

Remark 5 Note that in the definition of $\mathbb{D}_+$ we used the sub-collection integral and in the definition $\mathbb{H}_+$ we used the sub-decomposition integral. One can replace the sub-collection integral with the super-collection integral and the sub-decomposition integral with the super-decomposition integral while the classes $\mathbb{D}_+$ and $\mathbb{H}_+$ remain unchanged.

The following characterization of decomposition integrals extending the Lebesgue integral was given in the original publication [3, Proposition 3] of Even and Lehrer.

Theorem 7 *Let $\mathcal{H} \in \mathbb{H}$ be a decomposition system. Then $\mathcal{H} \in \mathbb{H}_+$ if and only if for every function $f \in \mathbb{F}$ there exists a collection $\mathcal{D} \in \mathbb{D}$ and non-negative coefficients $\{\alpha_A\}_{A \in \mathcal{D}}$ such that*

$$f = \sum_{A \in \mathcal{D}} \alpha_A \mathbb{1}_A.$$

Remark 6 The previous characterization is an important theoretical result, but it is inconvenient for practical purposes.

For collections, a full characterization was obtained, see, e.g., [9].

Theorem 8 *Let $\mathcal{D} \in \mathbb{D}$ be a collection. Then $\mathcal{D} \in \mathbb{D}_+$ if and only if*

$$\Big\{\{x\} \mid x \in X\Big\} \subseteq \mathcal{D}.$$

Unfortunately, there is no full characterization of the class $\mathbb{H}_+$. In the following, we give a sufficient condition.

Theorem 9 *Let $\mathcal{H} \in \mathbb{H}$. If for every permutation $\sigma\colon X \to X$ there exists a PC-collection $\mathcal{D}_{\mathsf{PC}} \in \mathcal{H}$ and coefficients $\{\alpha_A\}_{A \in \mathcal{D}_{\mathsf{PC}}}$ such that*

$$\sigma = \sum_{A \in \mathcal{D}_{\mathsf{PC}}} \alpha_A \mathbb{1}_A,$$

then $\mathcal{H} \in \mathbb{H}_+$.

Example 4 Based on the previous theorem, the Choquet integral, the PAN integral, the concave integral, and the PC integral extend the Lebesgue integral, i.e., $\mathcal{H}_{\mathsf{Ch}}, \mathcal{H}_{\mathsf{PAN}}, \mathcal{H}_{\mathsf{max}}, \mathcal{H}_{\mathsf{PC}} \in \mathbb{H}_+$. In general, if $\mathcal{H}_1 \in \mathbb{H}_+$ and $\mathcal{H}_2 \in \mathbb{H}$ are decomposition systems such that $\mathcal{H}_1 \sqsubseteq \mathcal{H}_2$, then also $\mathcal{H}_2 \in \mathbb{H}_+$.

The next result can be regarded as a construction method for decomposition systems $\mathcal{H} \in \mathbb{H}_+$. For any non-empty subset $A \subseteq X$, we denote by $\mathcal{D}^A$ a collection of non-empty subsets of A. Similarly, the notation $\mathcal{H}^A$, $\mathbb{H}^A$, and $\mathbb{H}^A_+$ is considered.

Theorem 10 *Let $\{A_1, A_2, \ldots, A_k\}$ be a partition of X and let $\mathcal{H}^{A_i} \in \mathbb{H}^{A_i}$, for $i = 1, 2, \ldots, k$. Then the decomposition system*

$$\mathcal{H} = \left\{\mathcal{D} \in \mathbb{D} \mid \mathcal{D} \cap A_i \in \mathcal{H}^{A_i}\right\}$$

belongs to $\mathbb{H}_+$*, where* $\mathcal{D} \cap A = \{B \cap A \mid B \in \mathcal{D}\}$*. Therefore*

$$\mathsf{dec}_{\mathcal{H}}(f, \mu) = \sum_{i=1}^{k} \mathsf{dec}_{\mathcal{H}^{A_i}}(f\restriction_{A_i}, \mu\restriction_{2^{A_i}}).$$

5 Decomposition Integrals for Interval-Valued Functions

In this section, we will discuss two different approaches of extending decomposition integrals for interval-valued functions. Interestingly, both of these approaches lead to the same result. The first approach is based on the Aumann extension of the Riemann integral and the second one is a direct modification using interval algebra. In the first part of this section we will concentrate only on sub-collection and sub-decomposition integrals.

Definition 10 Let $\mathcal{H} \in \mathbb{H}$ be a decomposition system. An Aumann-like sub-decomposition integral with respect to the decomposition system $\mathcal{H}$ is an operator

$$\hat{\mathsf{dec}}_{\mathcal{H}} \colon \mathbb{V} \times \mathbb{M} \to 2^{[0,\infty[} \quad \text{such that} \quad \hat{\mathsf{dec}}_{\mathcal{H}}(v, \mu) = \big\{\mathsf{dec}_{\mathcal{H}}(f, \mu) \mid f \in v\big\}$$

for all interval-valued functions $v \in \mathbb{V}$ and all monotone measures $\mu \in \mathbb{M}$.

It is easy to see that $a_v, b_v \in v$ and thus we obtain trivially the following proposition.

Proposition 1 *Let* $\mathcal{H} \in \mathbb{H}$ *be a decomposition system, let* $v \in \mathbb{V}$ *be an interval-valued function, and let* $\mu \in \mathbb{M}$ *be a monotone measure. Then*

$$\big\{\mathsf{dec}_{\mathcal{H}}(a_v, \mu), \mathsf{dec}_{\mathcal{H}}(b_v, \mu)\big\} \subseteq \hat{\mathsf{dec}}_{\mathcal{H}}(v, \mu).$$

In the next proposition we will prove that not only that the sub-decomposition integrals of boundaries of v do belong to $\hat{\mathsf{dec}}_{\mathcal{H}}(v, \mu)$ but also that all values in-between them do.

Proposition 2 *Let* $\mathcal{H} \in \mathbb{H}$ *be a decomposition system, let* $f, g \in \mathbb{F}$ *be two functions such that* $f \leqslant g$*, and let* $\mu \in \mathbb{M}$ *be a monotone measure. Then for every real number* $\gamma \in \big[\mathsf{dec}_{\mathcal{H}}(f, \mu), \mathsf{dec}_{\mathcal{H}}(g, \mu)\big]$ *there exists a function* $h_\gamma \in \mathbb{F}$ *such that* $f \leqslant h_\gamma \leqslant g$ *and* $\mathsf{dec}_{\mathcal{H}}(h_\gamma, \mu) = \gamma$*.*

Proof Let us denote $\alpha = \mathsf{dec}_{\mathcal{H}}(f, \mu)$ and $\beta = \mathsf{dec}_{\mathcal{H}}(g, \mu)$. From the monotonicity we have that $\alpha \leqslant \beta$ and thus $[\alpha, \beta]$ is indeed an interval. Now let $\gamma \in [\alpha, \beta]$ be any real number between α and β. Let us construct two sequences of functions $\{l_n\}_{n \in \mathbb{N}}$ and $\{r_n\}_{n \in \mathbb{N}}$ defined recursively by $l_1 = f$, $r_1 = g$,

$$l_{n+1} = \begin{cases} (l_n + r_n)/2, & \text{if } \mathsf{dec}_{\mathcal{H}}((l_n + r_n)/2, \mu) \leqslant \gamma, \\ l_n, & \text{otherwise,} \end{cases}$$

and

$$r_{n+1} = \begin{cases} r_n, & \text{if } \mathsf{dec}_{\mathcal{H}}((l_n + r_n)/2, \mu) \leqslant \gamma, \\ (l_n + r_n)/2, & \text{otherwise.} \end{cases}$$

Note that

$$r_n - l_n = \frac{g - f}{2^{n-1}}$$

which implies that

$$\lim_{n\to\infty} l_n = \lim_{n\to\infty} r_n.$$

Let us denote this limit function as h_γ. The sequences $\{l_n\}_{n\in\mathbb{N}}$ and $\{r_n\}_{n\in\mathbb{N}}$ were constructed in a such way that $\mathsf{dec}_{\mathcal{H}}(l_n, \mu) \leqslant \gamma$ and $\mathsf{dec}_{\mathcal{H}}(r_n, \mu) \geqslant \gamma$ for all $n \in \mathbb{N}$. This implies that $\mathsf{dec}_{\mathcal{H}}(h_\gamma, \mu) \leqslant \gamma$ and $\mathsf{dec}_{\mathcal{H}}(h_\gamma, \mu) \geqslant \gamma$ and thus a function h_γ such that $\mathsf{dec}_{\mathcal{H}}(h_\gamma, \mu) = \gamma$ and $f \leqslant h_\gamma \leqslant g$ was discovered. □

Combining the previous two propositions and the fact that all $f \in v$ are bounded from below by a_v and from above by b_v, we obtain the main theorem concerning the Aumann-like extension of sub-decomposition integrals.

Theorem 11 *Let $\mathcal{H} \in \mathbb{H}$ be a decomposition system, let $v \in \mathbb{V}$ be an interval-valued function, and let $\mu \in \mathcal{M}$ be a monotone measure. Then*

$$\hat{\mathsf{dec}}_{\mathcal{H}}(v, \mu) = \big[\mathsf{dec}_{\mathcal{H}}(a_v, \mu), \mathsf{dec}_{\mathcal{H}}(b_v, \mu)\big].$$

A direct modification of the original definition of sub-decomposition integrals based on the interval algebra offers another look at interval-valued sub-decomposition integrals.

Definition 11 Let $\mathcal{H} \in \mathbb{H}$ be a decomposition system. An interval-valued sub-decomposition integral is an operator $\bar{\mathsf{dec}}_{\mathcal{H}} : \mathbb{V} \times \mathbb{M} \to [0, \infty[$ given by

$$\bar{\mathsf{dec}}_{\mathcal{H}}(v, \mu) = \bigvee_{\mathcal{D}\in\mathcal{H}} \bar{\mathsf{col}}_{\mathcal{D}}(v, \mu)$$

for all interval-valued functions $v \in \mathbb{V}$ and all monotone measures $\mu \in \mathbb{M}$, where $\bar{\mathsf{col}}_{\mathcal{D}} : \mathbb{V} \times \mathbb{M} \to [0, \infty[$ is an interval-valued sub-collection integral such that $\bar{\mathsf{col}}_{\mathcal{D}}(v, \mu)$ equals

$$\bigvee \left\{ \sum_{A\in\mathcal{D}} [\alpha_A, \beta_A]\mu(A) \mid \sum_{A\in\mathcal{D}} [\alpha_A, \beta_A]\mathbb{1}_A \leqslant v, [\alpha_A, \beta_A] \geqslant [0, 0] \right\}$$

for all interval-valued functions $v \in \mathbb{V}$ and all monotone measures $\mu \in \mathbb{M}$.

Let us start by proving a similar result as in the case of the Aumann-like extension of sub-decomposition integrals for interval-valued sub-collection integrals.

Proposition 3 *Let $\mathcal{D} \in \mathbb{D}$ be a collection, let $v \in \mathbb{V}$ be an interval-valued function and let $\mu \in \mathbb{M}$ be a monotone measure. Then*

$$\bar{\mathsf{col}}_{\mathcal{D}}(v, \mu) = \left[\mathsf{col}_{\mathcal{D}}(a_v, \mu), \mathsf{col}_{\mathcal{D}}(b_v, \mu)\right].$$

Proof To simplify the notation, let us denote $\mathsf{col}_{\mathcal{D}}(a_v, \mu)$ by ρ and $\mathsf{col}_{\mathcal{D}}(b_v, \mu)$ by θ. In the definition of $\bar{\mathsf{col}}_{\mathcal{D}}$, let us loosen the condition $[\alpha_A, \beta_A] \geqslant [0, 0]$ by two conditions $\alpha_A \geqslant 0$ and $\beta_A \geqslant 0$. This would imply that $\bar{\mathsf{col}}_{\mathcal{D}}(v, \mu) \leqslant [\rho, \theta]$. Now let us assume two more strict conditions. The first one is that $\alpha_A = 0$. This one implies that $\bar{\mathsf{col}}_{\mathcal{D}}(v, \mu) \geqslant [0, \theta]$. The other one is that $\alpha_A = \beta_A$ which implies the inequality $\bar{\mathsf{col}}_{\mathcal{D}}(v, \mu) \geqslant [\rho, \rho]$. In other words,

$$\bar{\mathsf{col}}_{\mathcal{D}}(v, \mu) \geqslant [0, \theta] \vee [\rho, \rho] = [\rho, \theta]$$

and thus the proposition follows. □

Theorem 12 *Let $\mathcal{H} \in \mathbb{H}$ be a decomposition system, let $v \in \mathbb{V}$ be an interval-valued function, and let $\mu \in \mathbb{M}$ be a monotone measure. Then*

$$\bar{\mathsf{dec}}_{\mathcal{H}}(v, \mu) = \left[\mathsf{dec}_{\mathcal{H}}(a_v, \mu), \mathsf{dec}_{\mathcal{H}}(b_v, \mu)\right].$$

Proof Using the previous proposition and the properties of interval operations we obtain that

$$\begin{aligned} \bar{\mathsf{dec}}_{\mathcal{H}}(v, \mu) &= \bigvee_{\mathcal{D} \in \mathcal{H}} \bar{\mathsf{col}}_{\mathcal{D}}(v, \mu) = \bigvee_{\mathcal{D} \in \mathcal{H}} \left[\mathsf{col}_{\mathcal{D}}(a_v, \mu), \mathsf{col}_{\mathcal{D}}(b_v, \mu)\right] \\ &= \left[\bigvee_{\mathcal{D} \in \mathcal{H}} \mathsf{col}_{\mathcal{D}}(a_v, \mu), \bigvee_{\mathcal{D} \in \mathcal{H}} \mathsf{col}_{\mathcal{D}}(b_v, \mu)\right] = \left[\mathsf{dec}_{\mathcal{H}}(a_v, \mu), \mathsf{dec}_{\mathcal{H}}(b_v, \mu)\right] \end{aligned}$$

as intended. □

Example 5 Imagine a company which employs three people—two workers and one manager. The considered space is $X = \{1, 2, 3\}$, where the number 1 represents the first worker, 2 represents the second worker, and 3 represents the manager. The company's building has a work space where at most two people can work simultaneously. In addition, if the manager is at work, then he does not leave for the day. This leads to the following decomposition system

$$\mathcal{H} = \Big\{ \underbrace{\{\{1, 3\}, \{2, 3\}, \{3\}\}}_{\mathcal{D}_1}, \underbrace{\{\{1\}, \{2\}, \{1, 2\}\}}_{\mathcal{D}_2} \Big\}.$$

The monotone measure μ encodes the productivity per hour of different groups of people. If the workers work individually, then $\mu(\{1\}) = 5$ and $\mu(\{2\}) = 6$. If they work together, then $\mu(\{1, 2\}) = 12$. On the other hand, the manager's productivity is zero, i.e., $\mu(\{3\}) = 0$, and a combination of the manager and a worker yield an increase in production, e.g., $\mu(\{1, 3\}) = \mu(\{2, 3\}) = 7$. The working hours of the first worker are between 4 and 6 h, the second worker's are between 5 and 7 h, and the manager works exactly 5 h, i.e., $v(1) = [4, 6]$, $v(2) = [5, 7]$ and $v(3) = [5, 5]$. Note that, in this example, $a_v(1) = 4$, $a_v(2) = 5$, and $a_v(3) = 5$. Similarly, $b_v(1) = 6$, $b_v(2) = 7$, and $b_v(3) = 5$. Then

$$\mathsf{dec}_{\mathcal{H}}(a_v, \mu) = 50 \quad \text{and} \quad \mathsf{dec}_{\mathcal{H}}(b_v, \mu) = 72,$$

and thus $\hat{\mathsf{dec}}_{\mathcal{H}}(v, \mu) = [50, 72]$. Now, for the second part, notice that

$$\bar{\mathsf{col}}_{\mathcal{D}_1}(v, \mu) = [35, 35] \quad \text{and} \quad \bar{\mathsf{col}}_{\mathcal{D}_2}(v, \mu) = [50, 72]$$

and thus

$$\bar{\mathsf{dec}}_{\mathcal{H}}(v, \mu) = [35, 35] \vee [50, 72] = [50, 72].$$

In this example we illustrated both mechanisms of the introduced extensions of a sub-decomposition integral for interval-valued functions.

Remark 7 Note that analogous definitions of Aumann-like extension and a direct modification using interval algebra can be given for super-decomposition integrals, whose codomain is the set $2^{[0,\infty]}$. Then analogous proves of the fact

$$\hat{\mathsf{dec}}^*_{\mathcal{H}}(v, \mu) = \bar{\mathsf{dec}}^*_{\mathcal{H}}(v, \mu) = \left[\mathsf{dec}^*_{\mathcal{H}}(a_v, \mu), \mathsf{dec}^*_{\mathcal{H}}(b_v, \mu)\right]$$

can be given as in the case of sub-decomposition integrals.

6 Applications of Decomposition Integrals

In multicriteria-decision making, a rather common approach is based on the evaluation of considered alternatives by means of a set of criteria $\{C_1, C_2, \ldots, C_n\}$ which are evaluated on a common scale, e.g., on $[0, 1]$. Their degrees of satisfaction have the form of a score vector $\mathbf{x} = (x_1, x_2, \ldots, x_n)$, where x_i expresses the degree of satisfaction of criterion C_i by the considered alternative. Then, e.g., $\mathbf{x} \in [0, 1]^n$ and the global evaluation of considered alternatives is then performed by means of a utility function $\mathsf{U}\colon [0, 1]^n \to \mathbb{R}$ (and, to satisfy the Pareto condition that alternative T with score vector $\mathbf{t}$ dominating the score vector $\mathbf{v}$ of the alternative V, $\mathbf{t} \geqslant \mathbf{v}$, forces that V cannot be preferred to T, utility function U should be monotone increasing with respect to the standard partial order on $[0, 1]^n$). Note that we can normalize the utility function U to satisfy boundary conditions $\mathsf{U}(0, 0, \ldots, 0) = 0$ and $\mathsf{U}(1, 1, \ldots, 1) = 1$ with no influence on the resulting recommendation for preferences over the considered set of alternatives. Additive utility

$$\mathsf{U}(x_1, x_2, \ldots, x_n) = \sum_{i=1}^{n} w_i x_i,$$

where $w_1, w_2, \ldots, w_n \geqslant 0$ and $w_i > 0$ for some $i \in \{1, 2, \ldots, n\}$ (in the case of normalized utility,

$$\sum_{i=1}^{n} w_i = 1,$$

where U is a weighted arithmetic mean) is one of the most common utility functions in multicriteria-decision support and is applied in almost all domains dealing with quantitative criteria. Though the applicability of additive utilities is high (and, formally, they can be seen as an application of the Lebesgue integral), their main disadvantage is in the fact that they are not able to capture possible interaction among criteria. Grabisch and Labreuche [4] have presented a rather simple example dealing with evaluations of students based on their performance in Mathematics, Physics and Literature (note that they have used a scale [0, 20] typical for some French schools). In their example, additive utilities were not able to model the desired preferences, but the Choquet integral was able to solve this problem. Note that in the last 20 years, the Choquet integral (one of the most prominent decomposition integrals) became a standard tool in the domain of multicriteria-decision making. However, there are rather simple situations where the Choquet integral cannot represent the considered preferences, but the use of some decomposition integrals is helpful. The following is such an instance: related to the original example of Grabisch and Labreuche while advocating the use of the Choquet integral.

Example 6 For student S, let $\mathbf{s} = (s_1, s_2, s_3)$ express the score vector of S representing his performance in Mathematics, Physics and Literature, respectively, expressed on the scale $[0, 4]$. Consider four students S, T, Y, Z, and their respective normed score vectors $\mathbf{s} = (4, 2, 2)$, $\mathbf{t} = (3, 3, 2)$, $\mathbf{y} = (3, 2, 1)$, and $\mathbf{z} = (4, 1, 1)$. Our preferences are as follows: S is strictly preferred to T, and Y is strictly preferred to Z. This means that we look for a utility function $\mathsf{U} \colon [0, 4]^3 \to \mathbb{R}$ such that

$$\mathsf{U}(\mathbf{s}) > \mathsf{U}(\mathbf{t}) \quad \text{and} \quad \mathsf{U}(\mathbf{y}) > \mathsf{U}(\mathbf{z}).$$

Consider U to be Choquet integral with respect to a monotone measure μ. Then

$$\mathsf{U}(\mathbf{s}) = 2\mu_{123} + 2\mu_1 \quad \text{and} \quad \mathsf{U}(\mathbf{t}) = 2\mu_{123} + \mu_{12}.$$

Due to $\mathsf{U}(\mathbf{s}) > \mathsf{U}(\mathbf{t})$, it should hold $2\mu_1 > \mu_{12}$. Similarly,

$$\mathsf{U}(\mathbf{y}) = \mu_{123} + \mu_{12} + \mu_1 \quad \text{and} \quad \mathsf{U}(\mathbf{z}) = \mu_{123} + 3\mu_1.$$

Therefore due to $\mathsf{U}(\mathbf{y}) > \mathsf{U}(\mathbf{z})$, it should hold that $\mu_{12} > 2\mu_1$, which is a contradiction. Hence, no Choquet integral can express our preferences in this setting. On the other hand, consider a collection $\mathcal{D} = \{\{1, 2\}, \{1, 3\}\}$. Then, for any $\mu \in \mathbb{M}$, it holds that

$$\begin{aligned} \mathsf{col}_{\mathcal{D}}(\mathbf{s}, \mu) &= 2\mu_{12} + 2\mu_{13}, \\ \mathsf{col}_{\mathcal{D}}(\mathbf{t}, \mu) &= \max\{3\mu_{12}, \mu_{12} + 2\mu_{13}\}, \\ \mathsf{col}_{\mathcal{D}}(\mathbf{y}, \mu) &= 2\mu_{12} + \mu_{13}, \text{ and} \\ \mathsf{col}_{\mathcal{D}}(\mathbf{z}, \mu) &= \mu_{12} + \mu_{13}. \end{aligned}$$

Evidently, for any μ such that $\mu_{12} > 0$, it holds

$$\mathsf{col}(\mathbf{s}, \mu) > \mathsf{col}(\mathbf{t}, \mu) \quad \text{and} \quad \mathsf{col}(\mathbf{y}, \mu) > \mathsf{col}(\mathbf{z}, \mu),$$

i.e., the related sub-collection integral may serve as a utility function yielding the desired preferences on the set of students S, T, Y and Z.

Another promising application of sub-collection integrals as a method of constructing coherent lower and coherent upper previsions based on these integrals. In this section, the class of all functions $f\colon X \to \mathbb{R}$ will be denoted by $\mathbb{F}_{\mathbb{R}}$.

Definition 12 A coherent lower prevision is any operator $\mathsf{clp}\colon \mathbb{F}_{\mathbb{R}} \to \mathbb{R}$ such that the conditions

(P1) $\mathsf{clp}(f) \geqslant \inf f(X)$;
(P2) $\mathsf{clp}(\lambda f) = \lambda \mathsf{clp}(f)$; and
(P3) $\mathsf{clp}(f + g) \geqslant \mathsf{clp}(f) + \mathsf{clp}(g)$

hold for all functions $f, g \in \mathbb{F}_{\mathbb{R}}$ and all $\lambda \in [0, \infty[$.

A modification of the definition of the shift-invariant sub-collection integrals yield a new, so-called super-additive, integral that is defined for the class of all functions $\mathbb{F}_{\mathbb{R}}$. The following definitions, results and their proofs can be found in [2].

Definition 13 Let $\mathcal{D} \in \mathbb{D}$ be a collection and $\mu \in \mathbb{M}$ be a monotone measure such that $\mathsf{col}_{\mathcal{D}}(\cdot, \mu)$ is a shift-invariant sub-collection integral, i.e., for every $f \in \mathbb{F}$ and every $\lambda \geqslant 0$ the equality

$$\mathsf{col}_{\mathcal{D}}(f + \lambda \mathbb{1}_X, \mu) = \mathsf{col}_{\mathcal{D}}(f, \mu) + \lambda \mathsf{col}_{\mathcal{D}}(\mathbb{1}_X, \mu)$$

holds. Then, the operator $\mathsf{int}^{\mu}_{\mathcal{D}}\colon \mathbb{F}_{\mathbb{R}} \to \mathbb{R}$ given by

$$\mathsf{int}^{\mu}_{\mathcal{D}}(f) = \mathsf{col}_{\mathcal{D}}\big(f - (\inf f)\mathbb{1}_X, \mu\big) + (\inf f)\mathsf{col}_{\mathcal{D}}(\mathbb{1}_X, \mu)$$

for all functions $f \in \mathbb{F}$, is called a super-additive integral with respect to a collection $\mathcal{D}$ and a monotone measure μ.

In the following theorem, the properties of the super-additive integral are summarized.

Theorem 13 *Let* $\mathsf{int}_{\mathcal{D}}^{\mu}$ *be a super-additive integral. Then*

(1) $\mathsf{int}_{\mathcal{D}}^{\mu}(f) = \mathsf{col}_{\mathcal{D}}(f, \mu)$ *for all* $f \in \mathbb{F}$, *i.e.,* $\mathsf{int}_{\mathcal{D}}^{\mu}$ *extends* $\mathsf{col}_{\mathcal{D}}(\cdot, \mu)$*;*
(2) $\mathsf{int}_{\mathcal{D}}^{\mu}(\lambda f) = \lambda \mathsf{int}_{\mathcal{D}}^{\mu}(f)$ *for all* $f \in \mathbb{F}_{\mathbb{R}}$ *and all* $\lambda \geqslant 0$, *i.e.,* $\mathsf{int}_{\mathcal{D}}^{\mu}$ *is positively homogeneous operator;*
(3) $\mathsf{int}_{\mathcal{D}}^{\mu}(f + \lambda \mathbb{1}_X) = \mathsf{int}_{\mathcal{D}}^{\mu}(f) + \lambda \mathsf{int}_{\mathcal{D}}^{\mu}(\mathbb{1}_X)$ *for all functions* $f \in \mathbb{F}_{\mathbb{R}}$ *and all* $\lambda \in \mathbb{R}$, *i.e.,* $\mathsf{int}_{\mathcal{D}}^{\mu}$ *is shift-invariant;*
(4) $\mathsf{int}_{\mathcal{D}}^{\mu}(f) \geqslant (\inf f)\mathsf{int}_{\mathcal{D}}^{\mu}(\mathbb{1}_X)$ *for all* $f \in \mathbb{F}_{\mathbb{R}}$, *i.e.,* $\mathsf{int}_{\mathcal{D}}^{\mu}$ *is bounded below by a normed infimum; and*
(5) $\mathsf{int}_{\mathcal{D}}^{\mu}(f + g) \geqslant \mathsf{int}_{\mathcal{D}}^{\mu}(f) + \mathsf{int}_{\mathcal{D}}^{\mu}(g)$ *for all* $f, g \in \mathbb{F}_{\mathbb{R}}$, *i.e.,* $\mathsf{int}_{\mathcal{D}}^{\mu}$ *is super-additive.*

Note that properties 2 and 5 of the previous theorem correspond to properties (P2) and (P3) of coherent lower previsions, respectively. Property 4 of the previous theorem almost corresponds to property (P1) and after a normalization, we obtain a coherent lower prevision from a super-additive integral.

Theorem 14 *Let* $\mathcal{D} \in \mathbb{D}$ *be a collection and let* $\mu \in \mathbb{M}$ *be a monotone measure such that* $\mathsf{int}_{\mathcal{D}}^{\mu}$ *is a super-additive integral. Then the operator* $\mathsf{clp}_{\mathcal{D}}^{\mu} \colon \mathbb{F}_{\mathbb{R}} \to \mathbb{R}$ *given by*

$$\mathsf{clp}_{\mathcal{D}}^{\mu}(f) = \frac{\mathsf{int}_{\mathcal{D}}^{\mu}(f)}{\mathsf{int}_{\mathcal{D}}^{\mu}(\mathbb{1}_X)}$$

for all functions $f \in \mathbb{F}_{\mathbb{R}}$, *assuming that* $\mathsf{int}_{\mathcal{D}}^{\mu}(\mathbb{1}_X) \neq 0$, *is a coherent lower prevision.*

Remark 8 Based on this construction method, coherent upper previsions can be constructed using the conjugacy property: If clp is a coherent lower prevision then the mapping $\mathsf{cup} \colon \mathbb{F}_{\mathbb{R}} \to \mathbb{R}$ given by

$$\mathsf{cup}(f) = -\mathsf{clp}(-f)$$

for all functions $f \in \mathbb{F}_{\mathbb{R}}$, is a coherent upper prevision.

Example 7 If $\mathcal{D}$ consists of disjoint sets, i.e., for all $A, B \in \mathcal{D}$ such that $A \neq B$ one has $A \cap B = \emptyset$, then $\mathsf{int}_{\mathcal{D}}^{\mu}$ is a super-additive integral for any monotone measure $\mu \in \mathbb{M}$. Note that $\mathsf{int}_{\mathcal{D}}^{\mu}(\mathbb{1}_X) \neq 0$ if and only if there exists a set $A \in \mathcal{D}$ such that $\mu(A) > 0$. The coherent lower prevision constructed based on this super-additive integral is given by

$$\mathsf{clp}_{\mathcal{D}}^{\mu}(f) = \frac{\sum\limits_{A \in \mathcal{D}} \mu(A) \inf f(A)}{\sum\limits_{A \in \mathcal{D}} \mu(A)}$$

and the corresponding coherent upper prevision is given by

$$\mathsf{cup}^{\mu}_{\mathcal{D}}(f) = \frac{\sum\limits_{A \in \mathcal{D}} \mu(A) \sup f(A)}{\sum\limits_{A \in \mathcal{D}} \mu(A)}.$$

7 Concluding Remarks

In this chapter we have summarized several results concerning decomposition integrals (including sub-decomposition, super-decomposition, sub-collection and super-collection integrals). These integrals were also considered for interval-valued functions. An interesting application was also included.

Note that decomposition integrals form a promising tool for multi-criteria decision problems as well as for several economic problems related to optimization. Due to possible high computational complexity (e.g., concave and convex integrals are both (NP)-hard problems when processing them), there is a need for efficient algorithms to compute decomposition integrals on spaces X with high cardinality. For some of the first attempts we recall [9, 14]. As another possible direction for further research, interval-valued monotone measures could be considered.

Acknowledgements Both authors were supported by the Slovak Research and Development Agency under the contracts no. APVV-17-0066 and no. APVV-18-0052. Also the support of the grant VEGA 1/0006/19 is kindly announced.

References

1. Choquet, G.: Theory of capacities. Annales de l'institut Fourier **5**, 131–295 (1954)
2. Doria, S., Mesiar, R., Šeliga, A.: Construction method of coherent lower and upper previsions based on collection integrals. Bollettino dell'Unione Mathematica Italiana, available online
3. Even, Y., Lehrer, E.: Decomposition-integral: unifying Choquet and the concave integrals. Econ. Theory **56**, 33–58 (2014)
4. Grabisch, M., Labreuche, Ch.: A decade of application of the Choquet and Sugeno integrals in multi-criteria decision aid. Ann. Oper. Res. **175**, 247–286 (2010)
5. Lebesgue, H.: Intégrale, longueur, aire. Annali di Matematica Pura ed Applicata **7**(1), 231–359 (1902)
6. Lehrer, E.: A new integral for capacities. Econ. Theory **39**, 157–176 (2009)
7. Mesiar, R., Li, J., Pap, E.: Superdecomposition integrals. Fuzzy Sets Syst. **259**, 3–11 (2015)
8. Mesiar, R., Stupňanová, A.: Decomposition integrals. Int. J. Approx. Reason. **54**, 1252–1259 (2013)
9. Šeliga, A.: Decomposition integral without alternatives, its equivalence to Lebesgue integral, and computational algorithms. J. Autom. Mobile Robot. & Intell. Syst. **13**, 41–48 (2019)
10. Šeliga, A., Smrek, P.: Collection integral vs. Choquet integral. Fuzzy Sets Syst., available online
11. Shilkret, N.: Maxitive measure and integration. Indagationes Mathematicae **33**, 109–116 (1971)

12. Stupňanová, A.: A note on decomposition integrals. In: IPMU 2012, Communications in Computer and Information Science, vol. 300, pp. 542–548 (2012)
13. Wang, Z., Klir, G.J.: Generalized Measure Theory. Springer, Berlin (2009)
14. Wang, Z., Leung, K.S., Wong, M.L., Fang, J.: A new type of nonlinear integrals and the computational algorithm. Fuzzy Sets Syst. **112**, 223–231 (2000)
15. Yang, Q.: The PAN-integral on the fuzzy measure space. Fuzzy Math. **3**, 107–114 (1985). in Chinese

A Refinement of the Jensen Type Inequality for the Pseudo-integral

Mirjana Štrboja, Biljana Mihailović, and Jelena Ivetić

Abstract This paper presents new results related to the Jensen type inequality for the pseudo-integral. Using the Fubini type result for the pseudo-integral, i.e. the fact that the pseudo-integral with respect to the pseudo-product of two σ-$\oplus$-measures can be reduced to repeated pseudo-integrals, we obtain a refinement of the Jensen type inequality for the pseudo-integral.

Keywords Pseudo-integral · Semiring · Jensen type inequality · Pseudo-weight function

1 Introduction

The Jensen inequality for Lebesque integral and its counterparts for nonadditive integrals, such as the Choquet integral [6] and the Sugeno integral [23], have numerous applications, in probability theory, decision-making, actuarial science and mathematical economics. In the framework of generalized measure theory, the Jensen type inequality related to the Sugeno integral, the Choquet integral, the pseudo-integral and other integrals based on a nonadditive measure, was studied by various authors in [1–5, 8, 9, 12, 13, 18–21, 24]. The Jensen type inequality for bipolar pseudo-integrals was considered in [27], whereas a refinement of Jensen's inequality for the Lebesgue integral was studied in [22].

M. Štrboja
Department of Mathematics and Informatics, Faculty of Sciences, University of Novi Sad, Trg Dositeja Obradovića 4, 21000 Novi Sad, Serbia
e-mail: mirjana.strboja@dmi.uns.ac.rs

B. Mihailović (✉) · J. Ivetić
Faculty of Technical Sciences, University of Novi Sad, Trg Dositeja Obradovića 6, 21000 Novi Sad, Serbia
e-mail: lica@uns.ac.rs

J. Ivetić
e-mail: jelenaivetic@uns.ac.rs

E. Pap (ed.), *Artificial Intelligence: Theory and Applications*,
Studies in Computational Intelligence 973,
https://doi.org/10.1007/978-3-030-72711-6_3

The Jensen type inequality for the pseudo-integral was introduced in [18]. The main aim of this paper is to propose a refinement of the Jensen type inequality for this integral using the fact that the pseudo-integral with respect to the pseudo-product of two σ-$\oplus$-measures can be reduced to repeated pseudo-integrals, i.e. by the Fubini type theorem for the pseudo-integral recently proposed in [26].

The paper is organized as follows. In Sect. 2 some basic notions and the definitions of the semiring, σ-$\oplus$-measures and the pseudo-integral are recalled. An overview of the results concerning the Jensen type inequality for the pseudo-integral is given in Sect. 3. As the main result of this paper, in Sect. 4, a refinement of the Jensen type inequality for the pseudo-integral is obtained. Finally, in Sect. 5 some concluding remarks are given.

2 Preliminaries

In this section, we recall some basic notions and the definitions of semiring and σ-$\oplus$-measures, as well as the definition of the pseudo-integral and some recent results [10, 11, 16, 25, 26].

A triple $([a, b], \oplus, \odot)$ is said to be *a semiring* where $[a, b]$ is a closed (in some cases semiclosed) subinterval of $[-\infty, \infty]$, and $(\oplus, \odot)$ is a pair of binary operations defined on $[a, b]$ with the particular properties listed below. A full order on $[a, b]$ is denoted by $\preceq$.

A binary operation $\oplus$ on interval $[a, b]$ is said to be *a pseudo-addition* if for all $x, y, z \in [a, b]$, the following properties are fulfilled:

(O1) $x \oplus y = y \oplus x$,
(O2) $(x \oplus y) \oplus z = x \oplus (y \oplus z)$,
(O3) $x \preceq y$ implies $x \oplus z \preceq y \oplus z$,
(O4) there exists $\mathbf{0} \in [a, b]$ such that $x \oplus \mathbf{0} = x$, for all $x \in [a, b]$.

A binary operation $\odot$ on interval $[a, b]$ is said to be *a pseudo-multiplication* if for all $x, y, z \in [a, b]$, it holds:

(O5) $x \odot y = y \odot x$,
(O6) $(x \odot y) \odot z = x \odot (y \odot z)$,
(O7) if $x \preceq y$ and $\mathbf{0} \preceq z$, then $x \odot z \preceq y \odot z$,
(O8) there exists a unit element $\mathbf{1} \in [a, b]$, i.e., for each $x \in [a, b]$, $\mathbf{1} \odot x = x$.

Additionally, it is required:

(O9) $x \odot (y \oplus z) = (x \odot y) \oplus (x \odot z)$,
(O10) $\mathbf{0} \odot x = \mathbf{0}$.

We refer the reader to [11, 16, 17] for more details.

Let d be a metric on $[a, b]$ such that if x is the limit inferior and limit superior of a sequence $(x_n)_{n \in \mathbb{N}}$ then the sequence $(d(x_n, x))_{n \in \mathbb{N}}$ tends to zero. The metric d

also satisfies at least one of the following conditions for all $x, y, x', y' \in [a, b]$:

(D1) $d\left(x \oplus y, x' \oplus y'\right) \leqslant \max\left\{d\left(x, x'\right), d\left(y, y'\right)\right\}$,
(D2) $d\left(x \oplus y, x' \oplus y'\right) \leqslant d\left(x, x'\right) + d\left(y, y'\right)$.

As a consequence of the condition (D1) (also (D2)), it follows that if a sequence $(d(x_n, y_n))_{n\in\mathbb{N}}$ converges to zero, then $(d(x_n \oplus z, y_n \oplus z))_{n\in\mathbb{N}}$ also converges to zero.

The monotonicity of a metric d is required, i.e., for all $x, y, z \in [a, b]$ such that $x \leqslant z \leqslant y$, it holds that $d(x, y) \geqslant \sup\{d(y, z), d(x, z)\}$.

From this point, a metric d satisfying the above properties is simply called *a compatible metric* on $[a, b]$.

In this paper we consider semirings equipped with the continuous (except perhaps at points $(b, \mathbf{0})$ and $(\mathbf{0}, b)$, or $(a, \mathbf{0})$ and $(\mathbf{0}, a)$) pseudo-addition and pseudo-multiplication while a is the neutral element of the pseudo-addition. Here, we focus on the following three cases as canonical representatives:

Case I: *Idempotent* $\oplus$ *and non-idempotent* $\odot$

$x \oplus y = \sup(x, y)$, $\odot$ is an arbitrary non-idempotent pseudo-multiplication on the interval $[a, b]$. A full order is induced as follows: $x \preceq y$ if and only if $\sup(x, y) = y$. The pseudo-multiplication $\odot$ is generated by an increasing bijection $g : [a, b] \to [0, \infty]$, $x \odot y = g^{-1}(g(x) \cdot g(y))$. It holds that $\mathbf{0} = a$ and $\mathbf{1} = g^{-1}(1)$.

Case II: *g-semiring*

The operations are defined by

$$x \oplus y = g^{-1}(g(x) + g(y)) \text{ and } x \odot y = g^{-1}(g(x) \cdot g(y)),$$

where $g : [a, b] \to [0, \infty]$ is a continuous and strictly increasing generator. It holds that $\mathbf{0} = a$ and $\mathbf{1} = g^{-1}(1)$. The usual order on the interval $[a, b]$ is induced by $\oplus$ as follows: $x \preceq y$ if and only if $g(x) \leq g(y)$.

Case III: *Idempotent* $\oplus$ *and idempotent* $\odot$

$x \oplus y = \sup(x, y)$, $x \odot y = \inf(x, y)$, on the interval $[a, b]$. Here is $\mathbf{0} = a$, $\mathbf{1} = b$ and the pseudo-addition induces the usual order.

Let X be a non-empty set and $\mathcal{A}$ be a σ-algebra of subsets of X, i.e., $(X, \mathcal{A})$ is a measurable space. The definitions of a σ-$\oplus$-measure and the pseudo-integral are given below [10, 16, 17, 25].

Definition 1 A σ-$\oplus$-measure is a set function $m : \mathcal{A} \to [a, b]$ satisfying the following conditions:

(M1) $m(\emptyset) = \mathbf{0}$ (for non-idempotent $\oplus$),
(M2) $m\left(\bigcup_{i=1}^{\infty} A_i\right) = \bigoplus_{i=1}^{\infty} m(A_i)$ for any sequence $(A_i)_{i\in\mathbb{N}}$ of mutually disjoint sets from $\mathcal{A}$.

Remark 1 (i) In Case II, a set function $m : \mathcal{A} \to [a, b]$ is a σ-$\oplus$-measure if and only if $g \circ m : \mathcal{A} \to [0, \infty]$ is the classical measure.

Table 1 The symbols $\mathcal{D}_{(X,\mathcal{A})}$ and $\mathcal{M}_{(X,\mathcal{A})}$

	$\mathcal{D}_{(X,\mathcal{A})}$	$\mathcal{M}_{(X,\mathcal{A})}$
Case I	$\{f \mid f \in \mathcal{F}^b_{(X,\mathcal{A})}\}$	$\mathcal{H}_{(X,\mathcal{A})}$
Case II	$\{f \mid f \in \mathcal{F}^b_{(X,\mathcal{A})}\}$	$\{m \mid m \in \mathcal{H}_{(X,\mathcal{A})},\ m(X) < b\}$
Case III	$\mathcal{F}_{(X,\mathcal{A})}$	$\mathcal{H}_{(X,\mathcal{A})}$

(ii) A σ-$\oplus$-measure is continuous from below, but, in general, it is not continuous from above, see [16].

The set of all measurable functions $f : X \to [a, b]$ is denoted by $\mathcal{F}_{(X,\mathcal{A})}$. If $f \in \mathcal{F}_{(X,\mathcal{A})}$ and it holds that $f(x) \leqslant M < b$ for all $x \in X$ and some $M \in [a, b[$, then we write $f \in \mathcal{F}^b_{(X,\mathcal{A})}$. Furthermore, the class of all σ-$\oplus$ -measures $m : \mathcal{A} \to [a, b]$ on a measurable space $(X, \mathcal{A})$ is denoted by $\mathcal{H}_{(X,\mathcal{A})}$. Hereinafter, the symbols $\mathcal{D}_{(X,\mathcal{A})}$ and $\mathcal{M}_{(X,\mathcal{A})}$ have the meanings given in Table 1, depending on the type of semiring.

The pseudo-characteristic function of a set A is defined as usual:

$$\chi_A(x) = \begin{cases} \mathbf{0}\ , & x \notin A, \\ \mathbf{1}\ , & x \in A. \end{cases}$$

Definition 2 Let $m \in \mathcal{M}_{(X,\mathcal{A})}$.

(i) If $e : X \to [a, b]$ is a step function given by

$$e = \bigoplus_{i=1}^{n} a_i \odot \chi_{A_i}, \tag{1}$$

where a_i belongs to $[a, b]$ for all $i \in \{1, 2, \ldots, n\}$ and the family $(A_i)_{i\in\{1,2,\ldots,n\}}$ is contained in $\mathcal{A}$, and if $\oplus$ is non-idempotent, every two distinct sets in this family are disjoint, then the pseudo-integral of the function e is expressed as

$$\int_X^{\oplus} e \odot dm = \bigoplus_{i=1}^{n} a_i \odot m\,(A_i)\,.$$

(ii) The pseudo-integral of a function $f \in \mathcal{D}_{(X,\mathcal{A})}$ is expressed in terms of a sequence of step functions $(e_n)_{n\in\mathbb{N}}$ as

$$\int_X^{\oplus} f \odot dm = \lim_{n\to\infty} \int_X^{\oplus} e_n \odot dm,$$

where $(d(e_n(x), f(x)))_{n\in\mathbb{N}}$ converges to zero uniformly.

3 The Jensen Inequality for the Pseudo-integral

This section contains an overview of the results related to the Jensen type inequality for the pseudo-integral.

In Case I and Case III due to [25] for $f \in \mathcal{D}_{(X,\mathcal{A})}$, the pseudo-integral can be presented in the form:

$$\int_X^{\oplus} f \odot dm = \sup_{\alpha \in [a,b]} \left(\alpha \odot m(\{t \in X \mid f(t) \geqslant \alpha\})\right). \tag{2}$$

If a σ-sup-measure m is complete, i.e., there exists a function $\psi : X \to [a, b]$ such that $m(A) = \sup_{x \in A} \psi(x)$, for all $A \in \mathcal{A}$, then the pseudo-integral for a function $f \in \mathcal{D}_{(X,\mathcal{A})}$ can be written in the following form:

$$\int_X^{\oplus} f \odot dm = \sup_{x \in X} (f(x) \odot \psi(x)). \tag{3}$$

In the rest of the paper, the following notation is used:

Definition 3 Let $\varphi : [a, b] \to [a, b]$, $\boldsymbol{g} : [a, b] \to [0, \infty]$, where $\boldsymbol{g}$ is a strictly increasing function and $\boldsymbol{g} \circ \varphi \circ \boldsymbol{g}^{-1} : [N, M] \to [0, \infty]$ is well-defined, where $M = \boldsymbol{g}(b)$ and $N = \boldsymbol{g}(a)$. We will say that φ is $\boldsymbol{g}$*-convex* if $\boldsymbol{g} \circ \varphi \circ \boldsymbol{g}^{-1}$ is a convex function on $[N, M]$.

In [20] it was shown that the Jensen type inequality for the pseudo-integral (3) is valid for function φ satisfying the following condition

$$\varphi(\lambda \odot x) \leqslant \lambda \odot \varphi(x), \tag{4}$$

for all $x \in [a, b]$ and $\mathbf{0} \leqslant \lambda \leqslant \mathbf{1}$. That result requires complete σ-sup-measures. However, this type of inequality is also valid for the pseudo-integral (2) based on a wider class of σ-sup-measures and it is covered by the same type of inequality for the generalized Sugeno integral observed in [8]. According to [8] we have the following theorem related to the Jensen type inequality for the pseudo-integral (3).

Theorem 1 *Let* $f \in \mathcal{D}_{(X,\mathcal{A})}$, $m \in \mathcal{M}_{(X,\mathcal{A})}$ *and* $\varphi \in \mathcal{D}_{([a,b],\mathcal{B}([a,b]))}$ *is a nondecreasing, continuous function such that (4) holds for all* $x \in [a, b]$ *and* $\mathbf{0} \leqslant \lambda \leqslant m(A)$. *Then the inequality*

$$\varphi\left(\int_X^{\sup} f \odot dm\right) \leqslant \int_X^{\sup} (\varphi \circ f) \odot dm \tag{5}$$

holds.

Since the condition (4) is too restrictive, in the following lemma we formulate a sufficient condition for (4).

Lemma 1 *Let $\varphi : [a, b] \to [a, b]$ and $g : [a, b] \to [0, \infty]$. If φ is g-convex and $\varphi(\mathbf{0}) = \mathbf{0}$, then:*

for all $x \in [a, b]$ and $\mathbf{0} \leqslant \lambda \leqslant \mathbf{1}$, it holds

$$\varphi(\lambda \odot x) \leqslant \lambda \odot \varphi(x).$$

Proof Since $\varphi(\mathbf{0}) = \mathbf{0}$ and $g(\mathbf{0}) = 0$, $g \circ \varphi \circ g^{-1}$ is a convex function on $[0, M]$, where $M = g(b)$, and g is strictly increasing, for $x \in [a, b]$ and $\mathbf{0} \leqslant \lambda \leqslant \mathbf{1}$ it holds $0 \leqslant g(\lambda) \cdot g(x) \leqslant M$, and hence

$$\begin{aligned} g(\varphi(\lambda \odot x)) &= g\left(\varphi(g^{-1}(g(\lambda) \cdot g(x)))\right) \\ &= g\left(\varphi(g^{-1}(g(\lambda) \cdot g(x) + (1 - g(\lambda)) \cdot 0))\right) \\ &\leqslant g(\lambda) \cdot g\left(\varphi(g^{-1}(g(x)))\right) + (1 - g(\lambda)) \cdot g\left(\varphi(g^{-1}(0))\right) \\ &= g(\lambda) \cdot g(\varphi(x)). \end{aligned}$$

Therefore, for all $x \in [a, b]$ and $\mathbf{0} \leqslant \lambda \leqslant \mathbf{1}$, it holds

$$\varphi(\lambda \odot x) \leqslant g^{-1}(g(\lambda) \cdot g(\varphi(x))) = \lambda \odot \varphi(x).$$

□

As a corollary of Theorem 1 and Lemma 1 we have the following theorem related to Case I.

Theorem 2 *Let $f \in \mathcal{D}_{(X,\mathcal{A})}$, $m \in \mathcal{M}_{(X,\mathcal{A})}$ and $\varphi \in \mathcal{D}_{([a,b],\mathcal{B}([a,b]))}$ be a g-convex function on the range of f such that $\varphi(\mathbf{0}) = \mathbf{0}$. Then the inequality (5) holds.*

The following example related to this case of semiring is given in [18].

Example 1 Let $g(x) = x^\alpha$ for some $\alpha \in [1, \infty[$. Then we have $x \odot y = x + y$ and Jensen type inequality (5) reduces on

$$\varphi\left(\sup_{x \in X}(f(x) \cdot \psi(x))\right) \leqslant \sup_{x \in X}(\varphi(f(x)) \cdot \psi(x)),$$

where ψ is a density function related to m.

Recall that in Case II for the pseudo-integral of $f \in \mathcal{D}_{(X,\mathcal{A})}$ it holds

$$\int_X^{\oplus} f \odot dm = g^{-1}\left(\int_X (g \circ f)\, d(g \circ m)\right), \tag{6}$$

where the integral on the right side is the Lebesgue integral with respect to the measure $g \circ m$.

According to Definition 3 and observations in [18] we have the following Jensen type inequality for the pseudo-integral related to Case II.

Theorem 3 *Let $m \in \mathcal{M}_{(X,\mathcal{A})}$ such that $m(X) = \mathbf{1}$, a generator g of the pseudo-addition $\oplus$ and the pseudo-multiplication $\odot$ is a strictly increasing function. If $f \in \mathcal{D}_{(X,\mathcal{A})}$ and $\varphi \in \mathcal{D}_{([a,b],\mathcal{B}([a,b]))}$ is g-convex on the range of f, then we have*

$$\varphi\left(\int_X^{\oplus} f \odot dm\right) \leqslant \int_X^{\oplus} (\varphi \circ f) \odot dm.$$

The following examples given in [18] illustrate the previous theorem.

Example 2 (i) Let $g(x) = x^{\alpha}$ for some $\alpha \in [1, \infty[$. The corresponding pseudo-operations are $x \oplus y = \sqrt[\alpha]{x^{\alpha} + y^{\alpha}}$ and $x \odot y = xy$. Then the inequality from Theorem 3 has the following form

$$\varphi\left(\sqrt[\alpha]{\int_{[0,1]} f(x)^{\alpha}\, dx}\right) \leqslant \sqrt[\alpha]{\int_{[0,1]} \varphi(f(x))^{\alpha}\, dx}.$$

(ii) Let $g(x) = e^{x}$. The corresponding pseudo-operations are $x \oplus y = \ln(e^{x} + e^{y})$ and $x \odot y = x + y$. Then we have the following inequality

$$\varphi\left(\ln \int_{[0,1]} e^{f(x)}\, dx\right) \leqslant \ln\left(\int_{[0,1]} e^{\varphi(f(x))\, dx}\right).$$

Now we consider Case III. If function φ satisfies $\varphi(x) \leqslant x$ for every $x \in [a, m(X)]$ and $\odot = \inf$ in Theorem 1, condition (4) holds. Also, if $\odot = \inf$ and condition (4) holds then $\varphi(x) \leqslant x$ for every $x \in [a, m(X)]$. Therefore, inequality (5) reduces to Jensen's inequality for the pseudo-integral related to Case III.

Theorem 4 *If $\varphi : [a, b] \to [a, b]$ is a continuous and nondecreasing function such that $\varphi(x) \leqslant x$ for every $x \in [a, m(X)]$, then the inequation*

$$\varphi\left(\int_X^{\sup} f \odot dm\right) \leqslant \int_X^{\sup} (\varphi \circ f) \odot dm$$

holds for all $f \in \mathcal{D}_{(X,\mathcal{A})}$.

Theorem 4 is also covered by Jensen's inequality for the Sugeno integral which was obtained in [21]. The following example is given in [18].

Example 3 If $g(x) = e^{x}$, we have $x \odot y = x + y$. Then the Jensen type inequality from Theorem 4 reduces to

$$\varphi\left(\sup_{x\in X}(f(x)+\psi(x))\right)\leqslant \sup_{x\in X}\left(\varphi(f(x))+\psi(x)\right),$$

where ψ is a density function related to m.

4 The Main Results

Let $([a,b],\oplus,\odot)$ be a semiring (Case I, Case II or Case III).

Let $(X_1,\mathcal{A}_1)$ and $(X_2,\mathcal{A}_2)$ be two measurable spaces. The collection $\mathcal{Z}$ of the finite disjoint union of rectangles $A\times B$, such that $A\in\mathcal{A}_1$, $B\in\mathcal{A}_2$ is an algebra. The σ-algebra on $X_1\times X_2$ generated by the algebra $\mathcal{Z}$ will be denoted by $\mathcal{A}_1\otimes\mathcal{A}_2$. On $X_1\times X_2$ there exists a σ-$\oplus$-measure m given by

$$m\left(A_1\times A_2\right)=m_1\left(A_1\right)\odot m_2\left(A_2\right), \tag{7}$$

where $m_i\in\mathcal{M}_{(X_i,\mathcal{A}_i)}$ and $A_i\in\mathcal{A}_i$, $i=1,2$.

Here, we recall the Fubini type theorem for the pseudo-integral from [26].

Theorem 5 *Let $m_i\in\mathcal{M}_{(X_i,\mathcal{A}_i)}$ for $i=1,2$. If*

(i) in Case I and Case III: $f\in\mathcal{D}_{(X_1\times X_2,\mathcal{R})}$, where $\mathcal{R}=\left\{A\in\mathcal{P}(X_1\times X_2)|A=\bigcup_{n=1}^{\infty}A_n,\ A_n\in\mathcal{Z}\text{ for all } n\in\mathbb{N}\right\}$,

(ii) in Case II: $f\in\mathcal{D}_{(X_1\times X_2,\mathcal{A}_1\otimes\mathcal{A}_2)}$, then

$$\int_{X_1\times X_2}^{\oplus} f\odot dm=\left(\int_{X_1}^{\oplus}\left(\int_{X_2}^{\oplus} f\odot dm_2\right)\odot dm_1\right)$$
$$=\left(\int_{X_2}^{\oplus}\left(\int_{X_1}^{\oplus} f\odot dm_1\right)\odot dm_2\right). \tag{8}$$

We recall the next theorem from [14].

Theorem 6 *Let $f\in\mathcal{D}_{(X_1\times X_2,\mathcal{A}_1\otimes\mathcal{A}_2)}$. If one of the following conditions is satisfied:*

(i) in Case I and Case III: m_i is a complete σ-sup-measure on $(X_i,\mathcal{A}_i)$ for $i=1,2$,
(ii) in Case II: $m_i\in\mathcal{M}_{(X_i,\mathcal{A}_i)}$ for $i=1,2$, then (8) holds.

Notice that in Theorem 6 the completeness of σ-sup-measures m_1 and m_2 in Case I and Case III is required, whereas in Theorem 5 the equality (8) holds for arbitrary σ-sup-measures m_1 and m_2. Also, Theorem 5 requires that function f is $\mathcal{R}$-measurable, whereas Theorem 6 requires that f is an $\mathcal{A}_1\otimes\mathcal{A}_2$-measurable function.

Based on the monotone convergence theorem for the pseudo-integral from [25] we obtain the following result.

Theorem 7 *Let* $(X, \mathcal{A})$ *be a measurable space,* $f \in \mathcal{D}_{(X,\mathcal{A})}$ *and* $m \in \mathcal{M}_{(X,\mathcal{A})}$*. Then for set function* $\nu : \mathcal{A} \to [a, b]$ *defined by*

$$\nu(A) = \int_A^{\oplus} f \odot dm \tag{9}$$

it holds $\nu \in \mathcal{M}_{(X,\mathcal{A})}$*, whereas for any measurable function* $u \in \mathcal{D}_{(X,\mathcal{A})}$ *it holds*

$$\int_X^{\oplus} u \odot d\nu = \int_X^{\oplus} (u \odot f) \odot dm. \tag{10}$$

Proof The first part of the statement is proven in [25]. Let us prove the second part of the statement. Let u be a step function, i.e., $u = \bigoplus_{i=1}^{k} u_i \odot \chi_{U_i}$ for $u_i \in [a, b]$, where sets $U_1, U_2, \ldots, U_k$ are a partition of set X and $U_i \in \mathcal{A}$, $i = 1, \ldots, k$.

$$\begin{aligned}
\int_X^{\oplus} u \odot d\nu &= \bigoplus_{i=1}^{k} \int_X^{\oplus} u_i \odot \chi_{U_i} \odot d\nu \\
&= \bigoplus_{i=1}^{k} u_i \odot \int_X^{\oplus} \chi_{U_i} \odot d\nu \\
&= \bigoplus_{i=1}^{k} u_i \odot \nu(U_i) \\
&= \bigoplus_{i=1}^{k} u_i \odot \int_{U_i}^{\oplus} f \odot dm \\
&= \bigoplus_{i=1}^{k} u_i \odot \int_X^{\oplus} (\chi_{U_i} \odot f) \odot dm \\
&= \int_X^{\oplus} \left(\left(\bigoplus_{i=1}^{k} u_i \odot \chi_{U_i} \right) \odot f \right) \odot dm \\
&= \int_X^{\oplus} (u \odot f) \odot dm.
\end{aligned}$$

Let u be a measurable function and $(u_n)_{n \in \mathbb{N}}$ a nondecreasing sequence of the step functions such that pointwise convergence to function u occurs. Since $\lim_{n \to \infty} u_n(x) =$

$u(x)$, for $f \in \mathcal{D}_{(X,\mathcal{A})}$ it holds that $u_n(x) \odot f(x) \leqslant u_{n+1}(x) \odot f(x)$ for any $n \in \mathbb{N}$ and $x \in X$, and $\lim_{n\to\infty} (u_n(x) \odot f(x)) = u(x) \odot f(x)$. Therefore, due to the monotone convergence theorem for the pseudo-integral we have

$$\begin{aligned}
\int_X^{\oplus} u \odot dv &= \lim_{n\to\infty} \int_X^{\oplus} u_n \odot dv \\
&= \lim_{n\to\infty} \int_X^{\oplus} (u_n \odot f) \odot dm \\
&= \int_X^{\oplus} (u \odot f) \odot dm.
\end{aligned}$$

□

The main result in this section is related to the notion of a pseudo-weight function defined as follows.

Definition 4 A function $\omega \in \mathcal{D}_{(X_1\times X_2,\mathcal{A}_1\otimes\mathcal{A}_2)}$ is said to be a pseudo-weight function if

$$\int_{X_1}^{\oplus} \omega(x,y) \odot dm_1(x) = \mathbf{1} \text{ and } \int_{X_2}^{\oplus} \omega(x,y) \odot dm_2(y) = \mathbf{1}$$

for any $y \in X_2$ and $x \in X_1$, respectively.

Let $m_1 \in \mathcal{M}_{(X_1,\mathcal{A}_1)}$ and $m_2 \in \mathcal{M}_{(X_2,\mathcal{A}_2)}$ satisfy the conditions of Theorem 6 depending on the type of semiring.

Theorem 8 *Let $\omega \in \mathcal{D}_{(X_1\times X_2,\mathcal{A}_1\otimes\mathcal{A}_2)}$ be a pseudo-weight function, $f \in \mathcal{D}_{(X_1,\mathcal{A}_1)}$ and*

$$I = \int_{X_2}^{\oplus} \varphi\left(\int_{X_1}^{\oplus} (f(x) \odot \omega(x,y)) \odot dm_1(x)\right) \odot dm_2(y).$$

If one of the following conditions is satisfied:

(i) $\oplus = \sup$, $\odot = \odot_g$, *function* $\varphi \in \mathcal{D}_{([a,b],\mathcal{B}([a,b]))}$ *is* g*-convex on the range of* f *such that* $\varphi(\mathbf{0}) = \mathbf{0}$,

(ii) $\oplus = \oplus_g$, $\odot = \odot_g$, *function* $\varphi \in \mathcal{D}_{([a,b],\mathcal{B}([a,b]))}$ *is* g*-convex on the range of* f *and* $m_1(X_1) = m_2(X_2) = \mathbf{1}$,

(iii) $\oplus = \sup$, $\odot = \inf$ *and* $\varphi : [a,b] \to [a,b]$ *is continuous and nondecreasing function such that* $\varphi(x) \leqslant x$ *for every* $x \in [a, \sup(m_1(X_1), m_2(X_2))]$, *then*

$$\varphi\left(\int_{X_1}^{\oplus} f \odot dm_1\right) \leqslant I \leqslant \int_{X_1}^{\oplus} \varphi(f) \odot dm_1. \tag{11}$$

Proof According to Theorem 7, for any fixed $y \in X_2$ the set function defined by

$$\nu_y(A) = \int_{A}^{\oplus} \omega(x, y) \odot dm_1(x),$$

where $A \in \mathcal{A}_1$, is a σ-$\oplus$-measure on $(X_1, \mathcal{A}_1)$ and $\nu_y(X_1) = \mathbf{1}$, and

$$\int_{X_1}^{\oplus} f(x) \odot d\nu_y(x) = \int_{X_1}^{\oplus} (f(x) \odot \omega(x, y)) \odot dm_1(x).$$

Due to $\nu_y(X_1) = \mathbf{1}$ and the previous equality we have that for function given by $u(y) = \int_{X_1}^{\oplus} (f(x) \odot \omega(x, y)) \odot dm_1(x)$, $y \in X_2$, it holds that $u \in \mathcal{D}_{(X_2, \mathcal{A}_2)}$.

We have

$$\begin{aligned}
\varphi\left(\int_{X_1}^{\oplus} f \odot dm_1\right) &= \varphi\left(\int_{X_1}^{\oplus}\left(f(x) \odot \int_{X_2}^{\oplus} \omega(x, y) \odot dm_2(y)\right) \odot dm_1(x)\right) \\
&= \varphi\left(\int_{X_2}^{\oplus}\left(\int_{X_1}^{\oplus} (f(x) \odot \omega(x, y)) \odot dm_1(x)\right) \odot dm_2(y)\right) \\
&= \varphi\left(\int_{X_2}^{\oplus} u(y) \odot dm_2(y)\right).
\end{aligned}$$

We can apply the Jensen type inequality for the pseudo-integral (Theorems 2, 3 or Theorem 4 depending on the assumptions in (i), (ii) and (iii)) and we get

$$\varphi\left(\int_{X_1}^{\oplus} f \odot dm_1\right) \leqslant \int_{X_2}^{\oplus} \varphi\left(\int_{X_1}^{\oplus} (f(x) \odot \omega(x, y)) \odot dm_1(x)\right) \odot dm_2(y),$$

i.e., the left side inequality (11) holds.

Now, we will prove the other inequality in (11). Obviously, if it holds that $\int_{X_1}^{\oplus} \varphi(f) \odot dm_1 = b$, then the other inequality in (11) is valid. Let us suppose that

$\int_{X_1}^{\oplus} \varphi(f) \odot dm_1 < b$. Applying the Jensen type inequality for the pseudo-integral we obtain

$$\begin{aligned}
I &= \int_{X_2}^{\oplus} \varphi\left(\int_{X_1}^{\oplus} f(x) \odot d\nu_y(x)\right) \odot dm_2(y) \\
&\leq \int_{X_2}^{\oplus} \left(\int_{X_1}^{\oplus} \varphi(f(x)) \odot d\nu_y(x)\right) \odot dm_2(y) \\
&= \int_{X_2}^{\oplus} \left(\int_{X_1}^{\oplus} (\varphi(f(x)) \odot \omega(x, y)) \odot dm_1(x)\right) \odot dm_2(y) \\
&= \int_{X_1}^{\oplus} \left(\int_{X_2}^{\oplus} (\varphi(f(x)) \odot \omega(x, y)) \odot dm_2(y)\right) \odot dm_1(x) \\
&= \int_{X_1}^{\oplus} \varphi(f(x)) \odot \left(\int_{X_2}^{\oplus} (\omega(x, y)) \odot dm_2(y)\right) \odot dm_1(x) \\
&= \int_{X_1}^{\oplus} \varphi(f(x)) \odot dm_1(x).
\end{aligned}$$

□

The following examples are related to Theorem 8 (i) and (ii).

Example 4 Let $X_1 = X_2 = [0, 1]$ and $\mathcal{A}_1 = \mathcal{A}_2 = \mathcal{B}([0, 1])$.

(i) Let us observe the $\boldsymbol{g}$-semiring $([0, \infty], \oplus_g, \odot_g)$ where $\boldsymbol{g}(x) = \sqrt{x}, x \in [0, \infty]$, is a generator for $\oplus = \oplus_g, \odot = \odot_g$ and $m_i = \boldsymbol{g}^{-1} \circ \lambda,\ i = 1, 2,$ where λ is the Lebesgue measure on $[0, 1]$. Let $\omega(x, y) = (1 + \sin(2\pi x)\cos(2\pi y))^2$, $(x, y) \in [0, 1]^2$. Due to

$$\begin{aligned}
\int_{[0,1]}^{\oplus} \omega(x, y) \odot dm_1(x) &= \boldsymbol{g}^{-1}\left(\int_0^1 \boldsymbol{g}\left((1 + \sin(2\pi x)\cos(2\pi y))^2\right) dx\right) \\
&= \left(\int_0^1 (1 + \sin(2\pi x)\cos(2\pi y))\, dx\right)^2 \\
&= 1,
\end{aligned}$$

for any $y \in [0, 1]$ and analogously we obtain

$$\int_{[0,1]}^{\oplus} \omega(x, y) \odot dm_2(y) = 1,$$

for any $x \in [0, 1]$, ω is a pseudo-weight function.
For $\varphi(x) = x^3$, $x \in \mathbb{R}$ and $f(x) = x^4$, $x \in [0, 1]$, it holds that

$$\begin{aligned}
I &= \int_{[0,1]}^{\oplus} \varphi\left(\int_{[0,1]}^{\oplus} (f(x) \odot \omega(x, y)) \odot dm_1(x)\right) \odot dm_2(y) \\
&= \int_{[0,1]}^{\oplus} \left(\left(\int_0^1 \left(x^2 (1 + \sin(2\pi x)\cos(2\pi y))\right) dx\right)^2\right)^3 \odot dm_2(y) \\
&= \int_{[0,1]}^{\oplus} \left(\frac{1}{3} - \frac{\cos(2\pi y)}{2\pi}\right)^6 \odot dm_2(y) \\
&= \left(\int_0^1 \left(\frac{1}{3} - \frac{\cos(2\pi y)}{2\pi}\right)^3 dy\right)^2 \\
&= \left(\frac{1}{27} + \frac{1}{8\pi^2}\right)^2 \\
&\approx 0.00247031.
\end{aligned}$$

We have

$$\begin{aligned}
\varphi\left(\int_{[0,1]}^{\oplus} f \odot dm_1\right) &= \varphi\left(\int_{[0,1]}^{\oplus} x^4 \odot dm_1\right) = \varphi\left(\left(\int_0^1 x^2 dx\right)^2\right) \\
&= \varphi\left(\frac{1}{9}\right) = \left(\frac{1}{9}\right)^3 \approx 0.00137174
\end{aligned}$$

and

$$\int_{[0,1]}^{\oplus} (\varphi \circ f) \odot dm_1 = \int_{[0,1]}^{\oplus} x^{12} \odot dm_1 = \left(\int_0^1 x^6 dx\right)^2 = \left(\frac{1}{7}\right)^2 \approx 0.0204082.$$

Therefore, (11) holds.

(ii) Let us observe the semiring $([-\infty, \infty], \sup, \cdot)$ and the σ-sup-measures on $([0, 1], \mathcal{B}([0, 1]))$ defined by $m_1(A) = \sup_{t\in A} \frac{t}{2}$ and $m_2(A) = \sup_{t\in A} \frac{t^2}{2}$ where $A \in \mathcal{B}([0, 1])$. Let $\omega(x, y) = 1 + x + y - xy$, $(x, y) \in [0, 1]^2$. Due to

$$\int_{[0,1]}^{\oplus} \omega(x, y) \odot dm_1(x) = \sup_{x\in[0,1]} \left((1 + x + y - xy) \cdot \frac{x}{2}\right) = 1,$$

for any $y \in [0, 1]$, and

$$\int_{[0,1]}^{\oplus} \omega(x, y) \odot dm_2(y) = \sup_{y\in[0,1]} \left((1 + x + y - xy) \cdot \frac{y^2}{2}\right) = 1,$$

for any $x \in [0, 1]$, ω is a pseudo-weight function.
Let $\varphi(x) = x^2$, $x \in \mathbb{R}$ and $f(x) = 1 - x$, $x \in [0, 1]$. It holds that

$$\begin{aligned} I &= \int_{[0,1]}^{\oplus} \varphi\left(\int_{[0,1]}^{\oplus} (f(x) \odot \omega(x, y)) \odot dm_1(x)\right) \odot dm_2(y) \\ &= \int_{[0,1]}^{\oplus} \left(\int_{[0,1]}^{\oplus} ((1 - x)(1 + x + y - xy)) \odot dm_1(x)\right)^2 \odot dm_2(y) \\ &= \int_{[0,1]}^{\oplus} \left(\sup_{x\in[0,1]} \left((1 - x)(1 + x + y - xy)\frac{x}{2}\right)\right)^2 \odot dm_2(y) \\ &= \int_{[0,1]}^{\oplus} (h(y))^2 \odot dm_2(y), \end{aligned}$$

where

$$h(y) = \sup_{x\in[0,1]} \left((1 - x)(1 + x + y - xy)\frac{x}{2}\right) = \begin{cases} l\left(\frac{2y-\sqrt{y^2+3}}{3y-3}\right), & y \in [0, 1[, \\ l\left(\frac{1}{2}\right), & y = 1, \end{cases}$$

and

$$l(x) = (1 - x)(1 + x + y - xy)\frac{x}{2}.$$

Hence

$$I = \sup_{y\in[0,1]} \left((h(y))^2 \frac{y^2}{2}\right) \approx 0.03125.$$

We have

$$\varphi\left(\int_{[0,1]}^{\oplus} f \odot dm_1\right) = \varphi\left(\int_{[0,1]}^{\oplus} (1-x) \odot dm_1\right) = \varphi\left(\sup_{x\in[0,1]} (1-x)\frac{x}{2}\right)$$
$$= \varphi\left(\frac{1}{8}\right) = \left(\frac{1}{8}\right)^2 = \frac{1}{64} \approx 0.015625,$$

and

$$\int_{[0,1]}^{\oplus} (\varphi \circ f) \odot dm_1 = \sup_{x\in[0,1]} \left((1-x)^2 \frac{x}{2}\right) = \frac{2}{27} \approx 0.0740741.$$

Therefore, (11) holds.

Remark 2 Let us observe the assumptions given in Theorem 8 (iii). Due to $\odot = \inf$, monotonicity of the pseudo-integral and monotonicity of the function φ, we have

$$I = \int_{X_2}^{\oplus} \varphi\left(\int_{X_1}^{\oplus} (f(x) \odot \omega(x,y)) \odot dm_1(x)\right) \odot dm_2(y)$$
$$\leqslant \int_{X_2}^{\oplus} \varphi\left(\int_{X_1}^{\oplus} f(x) \odot dm_1(x)\right) \odot dm_2(y).$$

Due to $\odot$-homogeneity of the pseudo-integral it holds

$$I \leqslant \varphi\left(\int_{X_1}^{\oplus} f(x) \odot dm_1(x)\right) \odot m_2(X_2) \leqslant \varphi\left(\int_{X_1}^{\oplus} f(x) \odot dm_1(x)\right).$$

Hence, in this case as a consequence of inequality (11) and the previous inequality we have

$$I = \varphi\left(\int_{X_1}^{\oplus} f(x)\, dm_1(x)\right).$$

As a consequence of Theorem 8, we obtain the Hermite–Hadamard type inequality for the pseudo-integral in Case II.

Corollary 1 *Let a function $\varphi \in \mathcal{D}_{([a,b],\mathcal{B}([a,b]))}$ be g-convex on $[c,d] \subseteq (a,b)$. Then the following holds:*

$$\varphi\Big(g^{-1}\Big(\frac{g(c)+g(d)}{2}\Big)\Big) \leqslant I \leqslant g^{-1}\Big(\frac{g(\varphi(c))+g(\varphi(d))}{2}\Big), \tag{12}$$

where

$$I = g^{-1}\left(\frac{1}{g(d)-g(c)}\int_{g(c)}^{g(d)} g\circ\varphi\circ g^{-1}(u)du\right).$$

Proof Let $X_1 = \{0, 1\}$ with $\mathcal{A}_1 = 2^{X_1}$ and $m_1(\{0\}) = m_1(\{1\}) = g^{-1}(\frac{1}{2})$, and $X_2 = [0, 1]$ with $g \circ m_2 = \lambda$, a Lebesgue measure. From the previous theorem, by taking $\omega(0, y) = g^{-1}(2(1-y))$, $\omega(1, y) = g^{-1}(2y)$, $0 \leqslant y \leqslant 1$, $f(x) \in [c, d]$, $f(0) = c$, $f(1) = d$, with change of variables $u = g(c)(1-y) + g(d)y$, we obtain the claim. □

5 Conclusions

In this paper the Jensen type inequality for the pseudo-integral, originally introduced in [18], has been presented in accordance with the Jensen type inequality for discrete pseudo-integrals in the bipolar formulation presented in [27], with the corresponding inequality for the generalized Sugeno integral obtained in [8] and the inequality for the Choquet integral studied in [13]. Using the Fubini type result for the pseudo-integral, recently shown in [26], a refinement of the Jensen type inequality for the pseudo-integral has been obtained. The established result for g-integral is in accordance with the refinement of Jensen's inequality for the Lebesgue integral studied in [22]. Consequently, the Hermite–Hadamard type inequality for g-integral has been obtained. Since the pseudo-integral for non-negative functions was considered in the framework of principles for premium calculation in [7] and in [15], as the special type of more general CPT-like integrals, we believe that the obtained refinement of Jensen's inequality for the pseudo-integral will be a useful tool in actuarial science and decision-making problems.

Acknowledgements The authors acknowledge the financial support of the Ministry of Education, Science and Technological Development of the Republic of Serbia (the first author Grant No. 451-03-9/2021-14/200125, the second author and the third author were supported by the project "Innovative scientific and artistic research from the FTS (activity) domain" Grant No. 451-03-68/2020-14/200156.

References

1. Agahi, H., Mesiar, R., Babakhani, A.: Generalized expectation with general kernels on g-semirings and its applications. RACSAM **111**, 863–875 (2017)
2. Agahi, H., Mesiar, R., Ouyang, Y., Pap, E., Štrboja, M.: Berwald type inequality for Sugeno integral. Appl. Math. Comput. **217**, 4100–4108 (2010)

3. Agahi, H., Mesiar, R., Ouyang, Y., Pap, E., Štrboja, M.: Hölder and Minkowski type inequalities for pseudo-integral. Appl. Math. Comput. **217**, 8630–8639 (2011)
4. Agahi, H., Mesiar, R., Ouyang, Y., Pap, E., Štrboja, M.: General Chebyshev type inequalities for universal integral. Inf. Sci. **107**, 171–178 (2012)
5. Agahi, H., Mesiar, R., Ouyang, Y., Pap, E., Štrboja, M.: On Stolarsky inequality for Sugeno and Choquet integrals. Inf. Sci. **266**, 134–139 (2014)
6. Choquet, G.: Theory of capacities. Ann. Inst. Fourier (Grenoble) **5**, 131–292 (1953-1954)
7. Grbić, T., Medić, S., Perović, A., Mihailović, B., Novković, N.: Duraković, N: A Premium Principle Based on the g-integral. Stoch. Anal. Appl. **35**(3), 465–477 (2017)
8. Kaluszka, M., Okolewski, A., Boczek, M.: On the Jensen type inequality for generalized Sugeno integral. Inf. Sci. **266**, 140–147 (2014)
9. Klement, E.P., Li, J., Mesiar, R., Pap, E.: Integrals based on monotone set functions. Fuzzy Sets Syst. **218**, 88–102 (2015)
10. Kolokoltsov, V.N., Maslov, V.P.: Idempotent Analysis and Its Applications. Kluwer Academic Publishers, Dordrecht (1997)
11. Kuich, W.: Semirings, Automata, Languages. Springer, Berlin (1986)
12. Li, J., Mesiar, R., Pap, E., Klement, E.P.: Convergence theorems for monotone measures. Fuzzy Sets Syst. **281**, 103–127 (2015)
13. Mesiar, R., Li, J., E. Pap, E.: The Choquet integral as Lebesgue integral and related inequalities. Kybernetika **46**, 1098-1107 (2010)
14. Mihailović, B., Štrboja, M.: Generalized Minkowski type inequality for pseudo-integral. In: Proceedings of 15th IEEE International Symposium on Intelligent Systems and Informatics, Subotica, Serbia, pp. 99–104 (2017)
15. Mihailović, B., Pap, E., Štrboja, M., Simićević, A.: A unified approach to the monotone integral-based premium principles under the CPT theory. Fuzzy Sets Syst. **398**, 78–97 (2020)
16. Pap, E.: Null-Additive Set Functions. Kluwer Academic Publishers, Dordrecht (1995)
17. Pap, E: Pseudo-Additive Measures and Their Applications. In: Pap, E. (ed.) Handbook of Measure Theory, Vol II, , pp. 1403–1465 . Elsevier, Amsterdam (2002)
18. Pap, E., Štrboja, M: Generalization of the Jensen inequality for pseudo-integral. Inf. Sci. **180**, 543–548 (2010)
19. Pap, E., Štrboja, M: Generalizations of Integral Inequalities for Integrals Based on Nonadditive Measures. In: Pap, E. (ed.) Intelligent Systems: Models and Application, pp. 3–22. Springer, Berlin (2013)
20. Pap, E., Štrboja, M: Jensen type inequality for the pseudo-integral based on the smallest universal integral. In: Proceedings of 12th IEEE International Symposium on Intelligent Systems and Informatics, Subotica, Serbia, pp. 143–147 (2014)
21. Román-Flores, H., Flores-Franulič, A., Chalco-Cano, Y.: A Jensen type inequality for fuzzy integrals. Inf. Sci. **177**, 3192–3201 (2007)
22. Rooin, J.: A refinement of Jensen's inequality. J. Inequal. Pure Appl. Math. **6**(2) (2005)
23. Sugeno, M.: Theory of fuzzy integrals and its applications. Ph.D. Dissertation, Tokyo Institute of Technology (1974)
24. Štrboja, M., Grbić, T., Štajner-Papuga, I., Grujić, G., Medić, S.: Jensen and Chebyshev inequalities for pseudo-integrals of set-valued functions. Fuzzy Sets Syst. **222**, 18–32 (2013)
25. Štrboja, M., Pap, E., Mihailović, B.: Transformation of the pseudo-integral and related convergence theorems. Fuzzy Sets Syst. **355**, 67–82 (2019)
26. Štrboja, M., Mihailović, B.: Generalized Minkowski type inequality for pseudo-integral. Math. Slovaca **71**, 57–74 (2021)
27. Todorov, M., Štrboja, M., Pap, E., Mihailović, B.: Jensen type inequality for the bipolar pseudo-integrals. Fuzzy Sets Syst. **379**, 82–101 (2020)

Convolutional Neural Networks Hyperparameters Tuning

Eva Tuba, Nebojša Bačanin, Ivana Strumberger, and Milan Tuba

Abstract Digital images have revolutionized work in numerous scientific fields such as healthcare, astronomy, biology, agriculture as well as in every day life. One of the frequent tasks in applications with digital images is image classification which is a very challenging task. Major progress was made when convolution neural networks were introduced. The use of CNN has produced significant improvements in applications that require image classification. With today's technology, it is relatively simple to implement and use CNNs but in order to obtain the best possible results it is necessary to find the optimal architecture and hyperparameters for every single task. Due to a large number of hyperparameters, it is difficult to find the optimal configuration and there is no deterministic way to do it. In this early stage of CNN development, the common method of tuning CNN is by guessing and estimating, known as the guestimating method. Since this is a hard optimization problem, there is a chance to apply an optimization metaheuristic. There are several studies that have applied different optimization methods for tuning CNN hyperparameters have applied. One promising approach is the application of swarm intelligence algorithms. In this paper, a brief review of the CNN hyperparameters tuning will be presented and discussed.

Keywords Artificial intelligence · Convolutional neural network · Hyperparameters · Swarm intelligence algorithm · Layer · SI algorithm

---

E. Tuba · N. Bačanin · I. Strumberger · M. Tuba (✉)
Singidunum University, Danijelova 32, Belgrade, Serbia
e-mail: tuba@ieee.org

E. Tuba
e-mail: etuba@ieee.org

N. Bačanin
e-mail: nbacanin@singidunum.ac.rs

I. Strumberger
e-mail: istrumberger@singidunum.ac.rs

E. Pap (ed.), *Artificial Intelligence: Theory and Applications*,
Studies in Computational Intelligence 973,
https://doi.org/10.1007/978-3-030-72711-6_4

1 Introduction

Modern life has been defined by our dependency on technological advances, particularly those in the field of computer science. Nowadays, it is uncommon for a household to be without a PC, Internet, smartphones and other gadgets along with various software solutions for the different problems. They are all used on a daily basis to perform simple tasks such as helping us decide what to cook, where to go shopping, learning how to fix things etc. Younger generations are almost entirely oblivious to the life as it once was. In addition to these manifest advances and qualitative progressions known to a mass audience, some revolutionary progress in the field of computer science and artificial intelligence that indirectly affects our everyday life has made recently, i.e. convolutional neural networks (CNNs). CNN's have produced immense improvements in terms of signal classification, which forms an essential part of numerous applications used both in everyday life and science [11, 13, 32].

The idea of CNNs is three decades old, but just as a theoretical concept. It was presented by LeCun in 1990 for handwritten digit recognition [23]. However, its use and development were limited due to the lack of technical abilities. These limitations have been overcome in the past several years. Mass data storage and the growth of computational power enabled full exploitation of CNN's concept. CNN is a special type of artificial neural network that analyzes and considers the spatial correlation of the input which is important for the classification of various signals, especially digital images. Digital images have a significant role in a wide range of applications used in medicine, agriculture, security, astronomy and more. Since almost any task that involves digital images can be represented as a classification problem, the need for an efficient classifier is of significant importance.

Convolutional neural networks significantly improved the classification accuracy while the available technology and software tools made it easy to use. Previously, digital images had to be prepared for use. This step could include some preprocessing such as contrast adjustment, sharpening or smoothing the image but also the extraction of features that will be used as the input instead of the raw image. Even if the preparatory steps are taken, it is not possible to measure the classification accuracy of other classifiers using a CNN if the input is an unprocessed raw image. A good example is handwritten digit recognition task that is a well studied problem and often used as a benchmark task for testing the quality of different classifiers [46, 47]. The best classification accuracy of other classifiers could not go beyond 99% without preprocessing (non-linear methods such as k-nearest neighbors, random forest 97%, support vector machine 98%) while by adding more complex preprocessing steps the accuracy can go slightly above 99%. Nowadays, by using CNN, an accuracy of above 99% can be achieved with just a few lines of code, without the preprocessing step. By adjusting the network, the accuracy can increase to 99.79% which is currently the highest reported accuracy for the MNIST handwritten digit dataset [26].

Building a CNN that will outperform other classifiers is relatively simple, as there are supporting libraries for almost all programming languages. The problem is how to build the best possible CNN. A convolutional neural network has a large number of hyperparameters that should be fine tuned for each task individually. Finding the best combination of values for hyperparameters is a hard optimization problem, i.e. there is no deterministic method that can be used for solving that problem. The first results achieved by CNNs gave little or no attention to hyperparameters. Usually, the CNN is built and set based on some previous knowledge. Architecture and hyperparameters are determined by guessing and estimating better scenarios. This process was later called guestimating.

With the desire to achieve the best possible results, researchers started to apply different mathematical methods for tuning CNN's hyperparameters. Optimization represents one of the most important mathematical fields which has been studied extensively for decades and produced some exceptional results. Some of the efficient metaheuristics used for solving hard optimization problems are nature inspired algorithms, especially swarm intelligence algorithms (SI). Recent studies include application of the SI algorithms for tuning a CNN used for various classification problems. This paper will summarize different adaptations of the SI algorithm for finding the optimal CNN.

Reminder of the paper is organized as follows. Section 2 gives a detailed description of CNN and its hyperparameters. The general idea of SI algorithms is presented in Sect. 3.1. The adaptation and application of the SI algorithms for tuning CNN hyperparameters are categorized and explained in Sect. 3. The paper is concluded in Sect. 4.

2 Convolutional Neural Networks

A convolutional neural network was initially proposed by Yann LeCun and it was used for handwritten digit recognition [23, 24]. An extension of that research was presented almost 10 years later where a CNN was used for document recognition [25]. The initial network was used to sort mail in post offices by reading zip codes and for automatic bank check reading. Even though this deep neural network was a huge improvement in state-of-the-art methods, further research on convolutional neural networks waited for over two decades. The problem was a lack of data since the CNN requires a large amount of training data, as well as a lot of computational power which were not available at that time.

Regardless of the theory and a successful application of it, convolutional neural networks did not attract major attention from the research community. The rise in the interest and usage was led by the AlexNet. The AlexNet was presented in 2012 and it won an image classification competition [20]. The AlexNet CNN is presented in Fig. 1. At that time, large datasets such ImageNet [34] with millions of images sorted into hundreds of categories became available. The convolutional neural network named AlexNet was the only approach that achieve an error rate lower than 25% and the only CNN approach at that time. Starting in the year 2012, the CNN method began

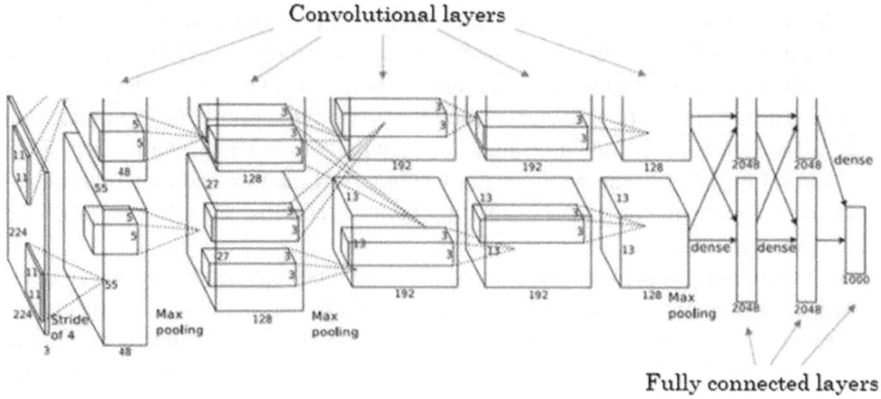

Fig. 1 AlexNet convolutional neural network

being used extensively, which marked a major turning point in artificial intelligence. It is now widely used in many scientific, as well as everyday, applications

Nowadays, CNNs are successfully used in numerous applications for medical image analysis [60], security [59], plant disease detection [45], and more.

2.1 Architecture of Convolutional Neural Networks

One of the main reasons why CNNs are so efficient and widely used is because of their architecture. The CNN extracts features by itself, using convolution of the image and filters it extract features which are then passed on to the next layer. These features are further convoluted with different filters. Finally, these features are used as the input for neural network whose output is a probability of belonging to each of the classes.

One convolutional neural network is a deep neural network that consists of several types of layers. The input is an image (or other signal) while the output layer contains one node for each class. Between these two layers there can be convolutional layers, pooling layers, and fully connected or dense layers. An example of a convolutional neural network, LeNet network [23] is shown in Fig. 2.

Since CNNs are neural networks, they have the same list of hyperparameters. This list contains the learning rate, the number of epochs, the batch size, and the activation function.

Another set of CNN's hyperparameters is related to the network architecture or structure. Before using a CNN it is necessary to determine the number of convolutional, pooling and fully connected layers as well as their order. By adding more layers, usually the results can be improved; however there is a risk of overfitting. There are various method that are used to prevent overfitting and one of them is the

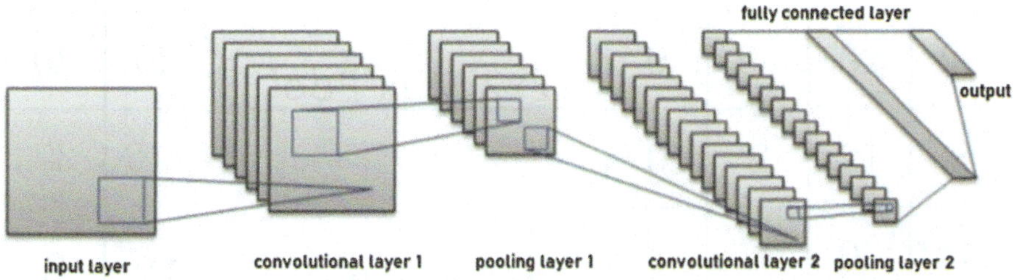

Fig. 2 LeNet—example of convolutional neural network

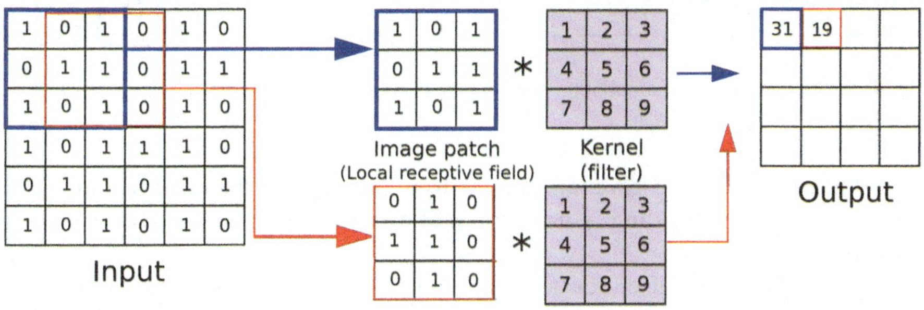

Fig. 3 A convolutional layer

dropout rate, i.e. the probability of discarding some units from the learning process. This is also one of the hyperparameters of the CNN and artificial neural networks.

The convolution layer was named after the operation that it performs. The input of the convolutional layer is convolved with a filter or kernel and the result of this operation is a set of feature maps. An example of a convolutional layer is presented in Fig. 3.

Convolutional layers closer to the input layer extract low level features such as lines, while the later convolutional layers are able to identify more abstract features, like the shapes of specific object (e.g. the eyes, the nose, etc.). The problem with the extracted features is that they are highly sensitive to small changes in the position of certain objects in the input image. These differences in the input image can be the result of rotation, shifting, cropping, and similar changes. One of the methods used to prevent this is changing the stride of the filter convolution across the image. Another, more robust method, is adding a pooling layer.

The hyperparameters that can be set for each convolutional layer are:

- the size of the filter (in Fig. 3 is 3×3),
- the stride—a number of places that the kernel is moved by (in Fig. 3 the stride is 1, the kernel is moved by one place over the input),
- the padding—how many rows and columns will be added around the input matrix,
- the number of output feature maps.

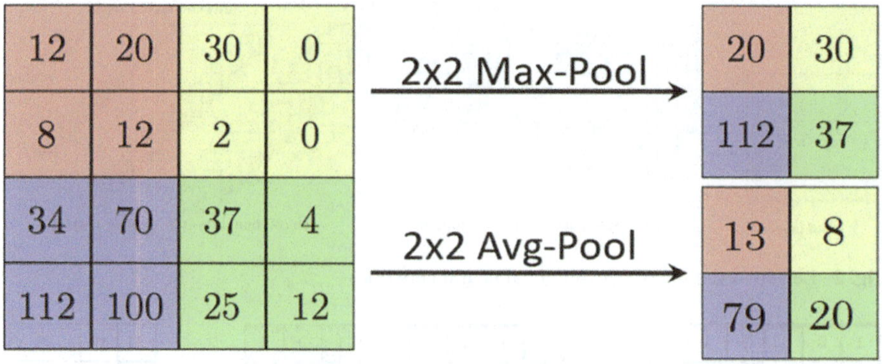

Fig. 4 An example of the max pooling and average pooling layers

Padding is used to preserve information from the edges of the input feature maps, usually zero or one padding is used, i.e. adding rows and columns with values of zero or one.

The pooling layer is used for downsampling the feature maps obtained by a convolutional layer. The number of feature maps will remain the same after the pooling layer, but the size of each feature map will be reduced. Similarly to the convolutional layer, in the pooling layer a filter is convolved with each feature map. The size of the kernel in the pooling layer is smaller than the size of the feature map. Frequently, the size is 2×2 and stride is 2. An example of a pooling layer is shown in Fig. 4.

The operation between a filter and a feature map is specified, it is not a learnable parameter. The most common pooling operators are average and maximum. Average pooling implies finding the average of the elements of the feature map covered by a filter and maximum pooling (also called max pooling) uses the maximal element under the filter as the remaining element in feature map.

Hyperparameters that are used in pooling layers include filter size, stride, padding and the type of pooling layer.

At the end of the CNN, there are one or more fully connected layers also called dense or flattened layers. The first fully connected layer takes the output of the previous layer and flattens it. In other words, it transforms the matrix representation of all feature maps into a single vector. These fully connected layers are the same layers used in artificial, fully connected, neural networks. The last fully connected layer in the CNN is the output layer that gives the final probabilities for each considered class. An example of the fully connected layers in a CNN is presented in Fig. 5.

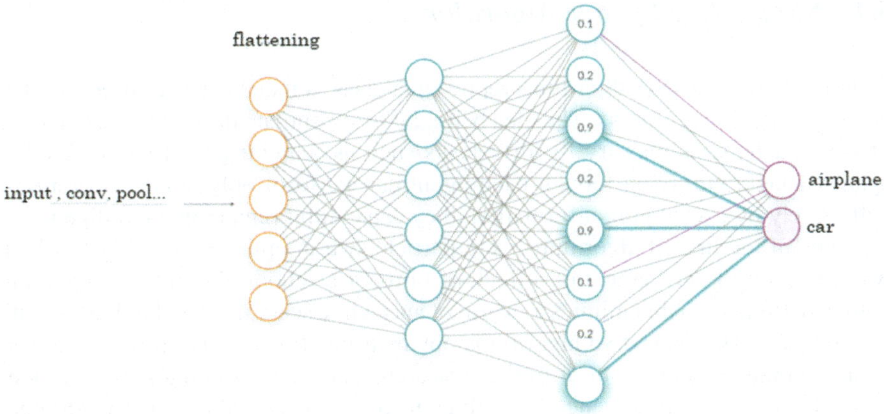

Fig. 5 An example of the fully connected layers in a CNN

3 Hyperparameters Tuning

The aforementioned hyperparameters should be set for each problem individually. Even with little or no hyperparameter tuning, it is possible to achieve high classification accuracy. There are numerous applications of CNNs that were set by guessing the initial values for the hyperparameters and based on previous knowledge to estimate better values. This method is widely used and it was named guestimating. By using a metaheuristic for tuning hyperparameters, the classification accuracy can increase significantly. In machine learning, a frequently used method for tuning hyperparameters is a grid search where for each parameter a set of possible values is chosen and the model is trained by using each possible combination. The problem with the CNN is that the training process requires a lot of time while the number of hyperparameters, as well as the search range for each one of them, is rather large. Current methods do not tune all hyperparameters but rather just a subset of them. In this way, the tuning time is reduced but the quality of the results can still be improved.

Tuning the CNN's hyperparameters is an optimization problem where the fitness function is classification accuracy and the goal is to find the global maximum or the global minimum if the fitness function is classification error. Since the nature of the problem, there is a possibility to apply different optimization metaheuristics. Evolutionary and swarm intelligence algorithms showed promising results in recent years. In this paper we will summarize the state-of-the-art methods in this field.

3.1 Swarm Intelligence Algorithms

Swarm intelligence (SI) algorithms represent a class of metaheuristics used to solve hard optimization problems. The idea behind it is to simulate the collective behavior of simple agents as they try to solve tasks such as searching for food, defending against attack, etc. Even though each agent can perform only simple operations, collectively by exchanging information, they can exhibit remarkable intelligence.

Some of the first SI algorithms were particle swarm optimization (PSO), which was based on the movement of a flock of birds or a school of fish [5, 37] and ant colony optimization that mimics a swarm of ants while searching for the shortest path to food [29, 30]. These two algorithms obtained great results and proved that this idea is promising and worth of further research. The results of almost two decades of work are swarm intelligence algorithms inspired by natural and real-world phenomena such as fireworks [44], brain storming process [36], music harmony [9] and others. Research of the SI algorithms is ongoing but the focus of the research is not on finding new inspiration and algorithms but rather theoretical analysis, the classification of the algorithms and their applications to different real life problems. The list of widely used SI algorithms, also for the optimization of machine learning hyperparameters, includes the bat algorithm [1, 57], the firefly algorithm [12, 48, 56], the fireworks algorithm [44, 51], the brain storm optimization [36, 50], the elephant herding optimization [49, 54], and many more.

In order to use different phenomena for solving optimization problems, some simplifications and assumptions are necessary. Regardless of the inspiration, each swarm intelligence algorithm has the same building units. SI algorithms are iterative. In each iteration, a swarm of solutions is generated and evaluated by the objective function (also called a cost or fitness function). Each algorithm has at least three operators: exploration, an exploitation operator and an operator for generating new solutions based on the previous ones.

Based on the results obtained in one iteration, each algorithm has a rule which solutions will be kept, combined, modified, discarded and/or replaced with a new one. Rules for combining and modifying solutions are made with the goal of exploring the search space in two ways. The first one is to explore the complete search space to find promising areas, i.e. the areas where the best solution could be. The second one is to carefully exploit the promising areas and find the best solution there. The exploration or global search is done at the beginning of the algorithm while the exploitation is increased in the later stages.

In order to apply a SI algorithm to the problem of tuning the CNN's hyperparameters, it is necessary to adjust it first. Since the SI algorithms were, in most cases, originally proposed for real search spaces and real fitness functions. While some of the hyperparameters are integers, it is necessary to adjust the SI algorithm to work with integers. The adjustment of the SI algorithm for the integer search space can be done in different ways and some of them are used for tuning CNNs. The simplest way to adjust the algorithm is to round the obtained real values to the nearest integer and use the algorithm as it is. The problem is that if the range of the search space is

Fig. 6 An example of the images in the MNIST dataset

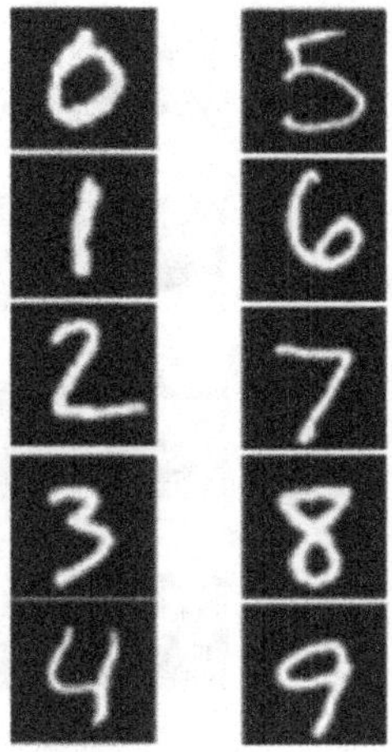

small, the changes of the value of the fitness function can become independent to the changes of the solutions, which will transform a rather powerful search metaheuristic into a basic random search which is not an efficient search for optimization problems with complicated landscape such as CNN hyperparameter tuning. The other solution is to make changes in the fitness function so that reflects the changes of the solution. This adjustment has not yet been studied for the CNN hyperparameter tuning. Another common method of adjusting an SI algorithm for integer search space is to search for the probability of changing integer values rather than the value itself. Current adjustments for the CNN hyperparameters search did not include this method.

Since the tuning of CNN hyperparameters is both an NP-hard optimization problem and a problem with a computationally expensive fitness function, some new adjustments were proposed. Numerous proposed methods combine a grid search and SI algorithms. The set of values for various hyperparameters is determined and the SI algorithm is used as a guided search to determine the best possible combination instead of checking all possible values. This adjustment is specific for tuning a CNN's hyperparameters and there is a high possibility that this method could give results for other similar problems as well.

There are standard benchmark datasets that are used for testing the quality of the CNN. The first dataset named MNIST contains images of handwritten digits [26]. The dataset contains 60000 training images (6000 images of each digit) and 10000 test images (1000 images per digit). All images are grayscale of the 28×28 pixels size. An example of images from the MNIST dataset is given in Fig. 6.

The second standard test dataset is CIFAR-10 that contains 60000 color images of the 32×32 size divided into 10 classes (6000 images per class) [19]. There are 50000 images in the training set and 10000 test images. The CIFAR-10 dataset contains images of airplanes, automobiles, birds, cats, deer, dogs, frogs, horses, ships and trucks. An example of images from the CIFAR-10 dataset are presented in Fig. 7.

Fig. 7 An example of the images in the CIFAR-10 dataset

3.1.1 Convolutional Neural Networks Tuned by SI Algorithms

The particle swarm optimization (PSO) algorithm is one of the earliest and most studied SI algorithms. Most methods that use SI algorithms for the CNN's hyperparameters tuning propose the PSO or its modifications.

In [43], the PSO algorithm is used for optimizing the output layer. Most of the hyperparameters were fixed while the output layer was augmented by δx whose value is determined by the PSO. This is not a classical tuning of hyperparameters as it can be seen in most of the state-of-the-art methods, but rather, it represents a way of optimizing the CNN's classification accuracy using the SI algorithm. The proposed method achieves high classification accuracy, however the main advantage is computational time. The proposed method was tested on the standard MNIST database. The results showed that it is possible to achieve a good classification accuracy (81.08%) even with only 4 epochs.

As mentioned before, many SI-based methods for tuning CNN's hyperparameters are combined with the grid search method. Sinha, Haidar and Verma proposed a PSO based method for finding the optimal input image size, the filter size, a number of feature maps and architecture [40]. Their initial research included a grid search [41] and the goal of the further research was to reduce the training time while keeping the classification accuracy. In [41] the input image had four possible values: $16\times16\times3$, $28\times28\times3$, $32\times32\times3$ and $64\times64\times3$. The original image size is $32\times32\times3$ since the CIFAR-10 dataset was used for testing. The filter or kernel size was chosen between three different values: 3×3, 5×5 or 7×7. The number of feature maps was set to be either 28 or 32. The last element that was included into the grid was the network's

Table 1 The ranges for hyperparameters in [55]

Parameter	Range
Padding	0–7
Filter size conv. layer	2–7
# of feature maps	50–180
Filter size pooling layer	2–7
Pooling algorithm	MAX, AVG

architecture. The choice was between a network with 11 or 13 layers. Details about the order of layers and hyperparameters can be found in [41]. The highest classification accuracy (81% on the test data and 95.8% on the training data) was achieved with the network that had an input image size of $32\times32\times3$, the filter size of 5×5, the number of feature maps was 28 in the network with the 13 layers. It should be mentioned that in both, [40, 41], the filter size and the number of feature maps were used for all convolutional layers. In paper [40], they extend the list of possible values for the filter size and the number of feature maps: the filter size could be $2\times2\times3$, $3\times3\times3$, $4\times4\times5$, $5\times5\times3$, $6\times6\times3$, $7\times7\times3$, $8\times8\times3$, $9\times9\times3$ while the number of feature maps could take the value 4, 8, 16, 28, 32, 36, 64 or 128. Calculating the grid search in this case would be very time consuming which is why the PSO was used to find the best possible combination. There are 512 possible combinations and the proposed PSO had a population size of 4 and iteration number was limited to 10 so 40 CNNs were trained instead of 512.There are not specific details about the solution representation and operator adaptation. The only thing that was said is that one solution contains four numbers that represent the choice for each of the considered hyperparameters. Since swarm intelligence algorithms are not deterministic methods, they have to be started multiple times to determine the true quality of the algorithm for the given problem. In [40], the algorithm was started five times. The best accuracy on a test set was 81.47% while the average accuracy in 5 runs was 81.08%.

Another basic PSO approach was presented by Yamasaki, Honma and Aizawa in [55]. They proposed the PSO for finding optimal values for the kernel sizes, padding, the number of feature maps, and types of pooling. They used fixed architecture based on AlexNet network with 5 convolutional layers followed by three fully connected layers. Since all hyperparameters are integers, they used the simplest PSO adjustment where the obtained real solutions were rounded to the nearest integer. The ranges for each of the parameters are presented in Table 1.

The proposed method was tested on CIFAR-10 and CIFAR-100 which is a dataset similar to the CIFAR-10 but with images classified into 100 classes, with 600 images in each class (500 for training and 100 for testing). The accuracy on CIFAR-10 was 80.15% while for CIFAR-100 it was 53.28%. The accuracy for both datasets increased compared to the original AlexNet network.

A modified PSO version, named linearly decreasing weight PSO (LDW-PSO), was proposed for tuning the hyperparameters of CNN by Serizawa and Fujita in

Table 2 The ranges for hyperparameters in [35]

Parameter	Range
# of feature maps in C1	4–100
# of feature maps in C2	4–100
Filter size in C1	3, 5, 7
Filter size in C2	3, 5, 7
Activation function C1	Sigmoid, relu, tanh
Activation function C2	Sigmoid, relu, tanh
Activation function FC1	Sigmoid, relu, tanh
Activation function FC2	Sigmoid, relu, tanh
Number of neurons in FC1	4–200
Number of neurons in FC2	4–200
Batch size	10–100
Optimizer	Adam, SGD

[35]. The PSO was widely studied and several improved versions were proposed in the years following the original version, among the linearly decreasing weight PSO [8, 38]. The difference is in the formula for generating new solutions based on the old ones. They used the LDW-PSO for finding optimal values for 12 different hyperparameters. The network architecture was based on LeNet network with two convolutional layers. Each one followed by the pooling layer and two flattened layers at the end. The list of hyperparameters and ranges are given in Table 2.

The population size was 10 and the iteration number in LDW-PSO was limited to 10 which means that 100 convolutional neural networks are trained in each run. The proposed method was tested on both, MNIST and CIFAR-10 dataset. The improvement was measured in comparison with the baseline CNN which was the LeNet network with the same architecture and fixed all of the hyperparameters. The details about the baseline network can be found in [35]. After only 5 epochs, the proposed LDW-PSO CNN achieved an accuracy of 98.95% on the test data in the MNIST dataset while the baseline CNN achieved only 94.02%. An even more drastic improvement was achieved by the proposed optimized CNN when trained and tested on CIFAR-10 dataset. After 10 epochs the accuracy of the baseline CNN was only 29.07% while the proposed LDW-PSO CNN had an accuracy of 69.37%. Even though the results obtained on both datasets were not the highest known accuracies, they show the importance of finding optimal values for hyperparameters.

Other similar approaches with the PSO algorithm but with different network architectures and a subset of hyperparameters that were tuned are listed in Table 5.

Beside the PSO, other SI algorithms were also successfully used for tuning convolutional neural network's hyperparameters. The general method of adjusting the SI algorithms for the integer search space and preforming a small number of fitness function evaluations is similar regardless to the chosen algorithm. The only difference is in the mathematical formulation of the exploration and exploitation operators.

Table 3 Accuracies on the test sets

Method	CIFAR-10 (%)	MNIST (%)
PSO (%)	71.55	99.17
FA (%)	71.52	99.16
CS (%)	71.16	99.14
BA (%)	71.43	99.13
BA-OM (%)	71.76	99.19

An enhanced bat algorithm (BA) was proposed for determining the dropout probability in [4]. Since the value of this hyperparameter is a real number between 0 and 1 there was no need for algorithm adjustments. The bat algorithm was combined with an exploration operator from the artificial bee colony algorithm in order to prevent the bat algorithm from becoming stuck in a local optimum. The proposed hybridized BA was compared with the original BA but also with the results presented in [7] where three other SI algorithms were used for the same purpose: particle swarm optimization, the firefly algorithm and the cuckoo search [58]. The proposed metaheuristics were tested on both the MNIST and CIFAR-10 datasets. For the MNIST dataset, LeNet architecture was used, while the CNN architecture for the CIFAR-10 dataset had an additional pair of convolutional and pooling layers and one additional fully connected layer. The obtained results are presented in Table 3.

Two enhanced SI approaches were presented by Bacanin et al. in [3]. One recent SI algorithm, the tree growth algorithm (TGA) with an enhanced exploitation operator and one of the most efficient SI algorithms, the firefly algorithm (FA) was used for setting various CNN hyperparameters. Hyperparameters that were considered were the number of convolutional layers (in range [0, 6]), the number of FC-layer (range [1, 4]), the kernel size (size between 1 and 8), the number of feature maps convolutional layer (between 1 and 128), the FC-layer size [16]. The initial number of fully connected and convolutional layers were 1 or 2. The proposed methods were tested on MNIST dataset and the best result was achieved by the enhanced tree growth algorithm (E-TGA)—99.18% while the enhanced firefly algorithm obtained slightly worse accuracy—99.16%.

A similar approach but a larger subset of hyperparameters was presented in [2] where the monarch butterfly optimization algorithm (MBO) was hybridized by the artificial bee colony (MBO-ABC). The list of hyperparameters is presented in Table 4. The reported error rate when the test error was used as a fitness function instead of training accuracy was 0.34%.

Recent research tries to adapt SI algorithms for the CNN's hyperparameter tuning problem by completely reorganizing the solution coding. Previous methods were used rather simple adjustments that still work, but some of the SI algorithm's efficiencies are lost. In [15] one solution is made of the list of layers, i.e. block units. There are four types of blocks that are implemented: convolutional, max-pooling, avg-pooling and fully connected. For each block there is information about related

Table 4 The list of hyperparameters in [2]

Category	Hyperparameter
Convolutional layer	The number of convolutional layers
	The number of filters
	The filter size
	The activation function
	The pooling layer size
Fully-connected layer	The number of fully-connected layers
	The connectivity pattern
	The number of units
	The activation function
	Weight regularization
	Dropout
General hyperparameters	The batch size
	The learning rule
	The learning rate

hyperparameters: the kernel size for pooling and the convolutional layers, the number of neurons in a fully connected layer and the number of feature maps for a convolutional layer or block. In order to use this type of solution structure, they transformed the exploration/exploitation operator. The calculation of the new solutions based on the previous one resembles the crossover operator in the genetic algorithm. The global best and the particle best solutions were compared on the block bases and a new solution was obtained by keeping the blocks of one or the other solution depending on random probability. If there are fewer blocks in one solution that block is marked with -1. Also there is another special type of blocks, +L block that is used for adding a layer where L represents a type of the new layer. They tested their approach on nine publicly available datasets and MNIST was one of them. The best accuracy of the proposed CNN was 99.68% while the mean accuracy was 99.56%.

Another approach that reorganized the solution space is presented [53] by Wang et al. The representation of the CNN was based on the network IP addresses. For each type of layer there are hyperparameters that mainly affect the final classification accuracy:

- In the convolutional layer kernel size, the number of feature maps and the stride size;
- In the pooling layer kernel and the stride size as well as the type (avg or max);
- In fully connected layer the number of neurons.

Based on the range for each hyperparameter, each layer can be represented as a binary number (e.g. the filter size in convolutional layers can be between 1 and 8 which means 3 bits are needed to write that as a binary number). By a special masking technique that is explained in detail in [53], the binary representation of the CNN can be easily transformed in a more friendly form. They added another type of layer in order to enable disabling some layers. In the proposed method the length of one solution was fixed and the new solutions in each iteration were obtained based on the bit-based transformation operator. Once the transformation between the human friendly and the binary representation of the CNN is coded, this solution representation makes it easy to change the structure of the network while preserving the connection between the changes in the fitness function and solutions. This represents the main contribution of the presented work and the quality of it was proved by achieving the classification accuracy of 98.87%.

There are many other convolutional neural networks optimized by the SI algorithms applied to the real life problems where the results cannot be compared directly with the methods tested on the benchmark datasets. The achieved results were better than conventional methods for the considered problem and/or dataset [6, 10, 17, 21].

Table 5 Classification accuracies from the literature—PSO

Method	Algorithm	Hyperparameters	Accuracy	
			MNIST	CIFAR10
Syulistyo et al. [43]	PSO	Output layer	95.19	NA
Khalifa et al. [16]	PSO	Output layer weights	97.67	NA
Sinha et al. [41]	Grid search	Input image size, filter size, number of feature maps, architecture	NA	81.00
Sinha et al. [40]	PSO, grid search	Input image size, filter size, number of feature maps, architecture	NA	81.08
Yamasaki et al. [55]	PSO	Filter size, padding, number of feature maps, pooling algorithm, kernel size in the pooling layer	NA	80.15
Serizawa and Fujita [35]	LDW-PSO	Filter size in convolutional and pooling layers, number of feature maps in each layer separately, number of neurons in fully connected layers, batch size, optimizer	98.95	69.37
Lorenzo et al. [28]	PSO	Number of convolutional and pooling layers, filter size, number of feature maps, stride in the pooling layers	99.34	59.59
Jiang et al. [14]	MO-PSO	Natural architecture search (number and type of layers, hyperparameters of the layers, etc. [18])	99.60	94.12
Fernandes Junior et al. [15]	Modified PSO	Number of convolutional, pooling and FC layers, kernel size, number of feature maps, number of neurons in FC layer	99.56	NA
Wang et al. [53]	Modified PSO	Filter size, number of feature maps, stride size, type of pooling layer, number of neurons in FC layer	98.87	NA

Table 6 Classification accuracies from the literature

Method	Algorithm	Hyperparameters	Accuracy	
			MNIST	CIFAR10
De Rosa et al. [7]	PSO	Dropout probability	99.17	71.55
	FA		99.16	71.52
	CS		99.14	71.16
Bacanin et al. [4]	Enhanced BA	Dropout probability	99.19	71.76
Bacanin et al. [3]	Enhanced TGA	Number of convolutional layers, number of FC-layer, kernel size, number of feature maps , FC-layer size	99.18	NA
	Enhanced FA		99.16	NA
Strumberger et al. [42]	TGA	Number of convolutional layers, number of FC-layer, kernel size, number of feature maps , FC-layer size	99.77[a]	NA
Bacanin et al. [2]	MBA	Number of convolutional and fully connected layers, number of feature maps, filter size, activation function, pooling size, number of units in FC layer, activation function, dropout, batch size, learning rule, learning rate, connectivity pattern	99.64	NA
	MBA-ABC		99.66	NA
Rosa et al. [33]	HSA	NA	99.23	72.28
	Improved HSA		99.24	71.54
	Global best HSA		99.24	71.86
	Self-adaptive HSA		99.29	72.43
Lee at al. [27]	PSF HSA[b]	The hyperparameters that affect the size of the feature map of the layer in the feature extraction step	99.30	74.8

[a]Fitness function was test error
[b]Parameter-setting-free HSA

Convolutional neural networks are used a lot in computer aided diagnostic systems, e.g. lung disease detection [31, 39], brain tumor segmentation [52], heart disease and breast cancer detection [22], etc.

All results along with the algorithm and the list of hyperparameters that were tuned are summarized in the Tables 5 and 6.

4 Conclusion

Convolutional neural networks are responsible for the recent, significant improvements of artificial intelligence and machine learning methods. The classification accuracy of digital images and other signals increased drastically with this powerful tool. It took a decade before technology enabled the theoretical concept to have real-world applications. The next step is to find a method for setting optimal hyperparameters values. This is an NP-hard optimization problem which combined with a time-consuming fitness function and high computation costs makes the problem considerably more difficult. This paper attempted to summarize recent developments in this field where evolutionary and swarm intelligence algorithms show the best results. Even with simple adjustments of these algorithms, the results can be drastically improved. Based on current state-of-the-art methods, future research will have two branches: The first involves newly-found methods that will be applied to real-life problems. While the second is more theoretical, wherein major transformations of SI algorithms, solution representation and exploration or exploitation operators will be proposed. Tuning CNN's hyperparameters is still an emerging research field with remarkable results thus far, which will motivate researchers to obtain better results.

References

1. Alihodzic, A., Tuba, E., Simian, D., Tuba, V., Tuba, M.: Extreme learning machines for data classification tuning by improved bat algorithm. In: 2018 International Joint Conference on Neural Networks (IJCNN), pp. 1–8. IEEE (2018)
2. Bacanin, N., Bezdan, T., Tuba, E., Strumberger, I., Tuba, M.: Monarch butterfly optimization based convolutional neural network design. Mathematics **8**(6), 936 (2020)
3. Bacanin, N., Bezdan, T., Tuba, E., Strumberger, I., Tuba, M.: Optimizing convolutional neural network hyperparameters by enhanced swarm intelligence metaheuristics. Algorithms **13**(3), 67 (2020)
4. Bacanin, N., Tuba, E., Bezdan, T., Strumberger, I., Jovanovic, R., Tuba, M.: Dropout probability estimation in convolutional neural networks by the enhanced bat algorithm. In: IEEE International Joint Conference on Neural Networks (IJCNN), pp. 1–7 (2020)
5. Eberhart, R., Kennedy, J.: A new optimizer using particle swarm theory. In: Proceedings of the 6th international symposium on micromachine and human science, pp. 39–43. IEEE (1995)
6. Darwish, A., Ezzat, D., Hassanien, A.E.: An optimized model based on convolutional neural networks and orthogonal learning particle swarm optimization algorithm for plant diseases diagnosis. Swarm Evol. Comput. **52** (2020)
7. De Rosa, G.H., Papa, J.P., Yang, X.S.: Handling dropout probability estimation in convolution neural networks using meta-heuristics. Soft. Comput. **22**(18), 6147–6156 (2018)
8. Eberhart, R.C., Shi, Y.: Comparing inertia weights and constriction factors in particle swarm optimization. In: IEEE Congress on Evolutionary Computation (CEC), vol. 1, pp. 84–88 (2000)
9. Geem, Z.W., Kim, J.H., Loganathan, G.V.: A new heuristic optimization algorithm: harmony search. Simulation **76**(2), 60–68 (2001)
10. Haidar, A., Jan, Z.M., Verma, B.: Evolving one-dimensional deep convolutional neural network: a swarm based approach. In: 2019 IEEE Congress on Evolutionary Computation (CEC), pp. 1299–1305 (2019). https://doi.org/10.1109/CEC.2019.8790036

11. Hongtao, L., Qinchuan, Z.: Applications of deep convolutional neural network in computer vision. J. Data Acquisi. Proc. **31**(1), 1–17 (2016)
12. Hrosik, R.C., Tuba, E., Dolicanin, E., Jovanovic, R., Tuba, M.: Brain image segmentation based on firefly algorithm combined with k-means clustering. Stud. Inform. Control **28**, 167–176 (2019)
13. Ishibushi, S., Taniguchi, A., Takano, T., Hagiwara, Y., Taniguchi, T.: Statistical localization exploiting convolutional neural network for an autonomous vehicle. In: IECON 2015-41st Annual Conference of the IEEE Industrial Electronics Society, pp. 001369–001375. IEEE (2015)
14. Jiang, J., Han, F., Ling, Q., Wang, J., Li, T., Han, H.: Efficient network architecture search via multiobjective particle swarm optimization based on decomposition. Neural Netw. **123**, 305–316 (2020)
15. Junior, F.E.F., Yen, G.G.: Particle swarm optimization of deep neural networks architectures for image classification. Swarm Evol. Comput. **49**, 62–74 (2019)
16. Khalifa, M.H., Ammar, M., Ouarda, W., Alimi, A.M.: Particle swarm optimization for deep learning of convolution neural network. In: Sudan Conference on Computer Science and Information Technology (SCCSIT), pp. 1–5 (2017). https://doi.org/10.1109/SCCSIT.2017.8293059
17. Khaw, H.Y., Soon, F.C., Chuah, J.H., Chow, C.O.: High-density impulse noise detection and removal using deep convolutional neural network with particle swarm optimisation. IET Image Proc. **13**(2), 365–374 (2018)
18. Klyuchnikov, N., Trofimov, I., Artemova, E., Salnikov, M., Fedorov, M., Burnaev, E.: Nas-bench-nlp: neural architecture search benchmark for natural language processing. arXiv:2006.07116 (2020)
19. Krizhevsky, A.: Learning multiple layers of features from tiny images. Technical Report (2009)
20. Krizhevsky, A., Sutskever, I., Hinton, G.E.: Imagenet classification with deep convolutional neural networks. Commun. ACM **60**(6), 84–90 (2017)
21. Kumar, P.J., Huan, T.L., Li, X., Yuan, Y.: Panchromatic and multispectral remote sensing image fusion using particle swarm optimization of convolutional neural network for effective comparison of bucolic and farming region. In: Earth Science and Remote Sensing Applications, Series of Remote Sensing/Photogrammetry, vol. 43, pp. 1–30 (2018)
22. Lan, K., Liu, L., Li, T., Chen, Y., Fong, S., Marques, J.A.L., Wong, R.K., Tang, R.: Multi-view convolutional neural network with leader and long-tail particle swarm optimizer for enhancing heart disease and breast cancer detection. Neural Comput. Appl. 1–20 (2020)
23. LeCun, Y., Boser, B., Denker, J.S., Henderson, D., Howard, R.E., Hubbard, W., Jackel, L.D.: Backpropagation applied to handwritten zip code recognition. Neural Comput. **1**(4), 541–551 (1989)
24. LeCun, Y., Boser, B.E., Denker, J.S., Henderson, D., Howard, R.E., Hubbard, W.E., Jackel, L.D.: Handwritten digit recognition with a back-propagation network. In: Advances in Neural Information Processing Systems, pp. 396–404 (1990)
25. LeCun, Y., Bottou, L., Bengio, Y., Haffner, P.: Gradient-based learning applied to document recognition. Proc. IEEE **86**(11), 2278–2324 (1998)
26. LeCun, Y., Cortes, C.: MNIST handwritten digit database (2010), http://yann.lecun.com/exdb/mnist/
27. Lee, W.Y., Park, S.M., Sim, K.B.: Optimal hyperparameter tuning of convolutional neural networks based on the parameter-setting-free harmony search algorithm. Optik **172**, 359–367 (2018)
28. Lorenzo, P.R., Nalepa, J., Kawulok, M., Ramos, L.S., Pastor, J.R.: Particle swarm optimization for hyper-parameter selection in deep neural networks. In: Proceedings of the Genetic and Evolutionary Computation Conference, pp. 481–488 (2017)
29. Dorigo, M., Gambardella, L.M.: Ant colony system: a cooperative learning approach to the traveling salesman problem. IEEE Trans. Evol. Comput. **1**(1), 53–66 (1997)
30. Dorigo, M., Maniezzo, V., Colorni, A.: Ant system: optimization by a colony of cooperating agents. IEEE Trans. Syst. Man Cybern. Part B: Cybern. **26**(1), 29–41 (1996)

31. de Pinho Pinheiro, C.A., Nedjah, N., de Macedo Mourelle, L.: Detection and classification of pulmonary nodules using deep learning and swarm intelligence. Multimedia Tools Appl. **79**(21), 15437–15465 (2020)
32. Qayyum, A., Anwar, S.M., Awais, M., Majid, M.: Medical image retrieval using deep convolutional neural network. Neurocomputing **266**, 8–20 (2017)
33. Rosa, G., Papa, J., Marana, A., Scheirer, W., Cox, D.: Fine-tuning convolutional neural networks using harmony search. In: Iberoamerican Congress on Pattern Recognition, pp. 683–690. Springer (2015)
34. Russakovsky, O., Deng, J., Su, H., Krause, J., Satheesh, S., Ma, S., Huang, Z., Karpathy, A., Khosla, A., Bernstein, M., Berg, A.C., Fei-Fei, L.: ImageNet large scale visual recognition challenge. Int. J. Comput. Vis. (IJCV) **115**(3), 211–252 (2015). https://doi.org/10.1007/s11263-015-0816-y
35. Serizawa, T., Fujita, H.: Optimization of convolutional neural network using the linearly decreasing weight particle swarm optimization. arXiv:2001.05670; Submitted to Neural and Evolutionary Computing (2020)
36. Shi, Y.: Brain storm optimization algorithm. In: International conference in swarm intelligence, pp. 303–309. Springer (2011)
37. Shi, Y., Eberhart, R.: A modified particle swarm optimizer. In: IEEE World Congress on Computational Intelligence (CEC), pp. 69–73. IEEE (1998)
38. Shi, Y., Eberhart, R.C.: Empirical study of particle swarm optimization. IEEE Congress Evol. Comput. (CEC). **3**, 1945–1950 (1999)
39. da Silva, G.L.F., Valente, T.L.A., Silva, A.C., de Paiva, A.C., Gattass, M.: Convolutional neural network-based pso for lung nodule false positive reduction on ct images. Comput. Methods Programs Biomed. **162**, 109–118 (2018)
40. Sinha, T., Haidar, A., Verma, B.: Particle swarm optimization based approach for finding optimal values of convolutional neural network parameters. In: 2018 IEEE Congress on Evolutionary Computation (CEC), pp. 1–6. IEEE (2018)
41. Sinha, T., Verma, B., Haidar, A.: Optimization of convolutional neural network parameters for image classification. In: 2017 IEEE Symposium Series on Computational Intelligence (SSCI), pp. 1–7. IEEE (2017)
42. Strumberger, I., Tuba, E., Bacanin, N., Jovanovic, R., Tuba, M.: Convolutional neural network architecture design by the tree growth algorithm framework. In: International Joint Conference on Neural Networks (IJCNN), pp. 1–8. IEEE (2019)
43. Syulistyo, A.R., Purnomo, D.M.J., Rachmadi, M.F., Wibowo, A.: Particle swarm optimization (PSO) for training optimization on convolutional neural network. Jurnal Ilmu Komputer dan Informasi **9**(1), 52–58 (2016)
44. Tan, Y., Zhu, Y.: Fireworks algorithm for optimization. Advances in Swarm Intelligence, LNCS 6145, pp. 355–364 (2010)
45. Toda, Y., Okura, F., et al.: How convolutional neural networks diagnose plant disease. Plant Phenomics **2019**, 9237136 (2019)
46. Tuba, E., Bacanin, N.: An algorithm for handwritten digit recognition using projection histograms and svm classifier. In: 2015 23rd Telecommunications Forum Telfor (TELFOR), pp. 464–467. IEEE (2015)
47. Tuba, E., Hrosik, R.C., Alihodzic, A., Jovanovic, R., Tuba, M.: Support vector machine optimized by fireworks algorithm for handwritten digit recognition. In: International Conference on Modelling and Development of Intelligent Systems, pp. 187–199. Springer (2019)
48. Tuba, E., Mrkela, L., Tuba, M.: Support vector machine parameter tuning using firefly algorithm. In: 26th International Conference Radioelektronika, pp. 413–418. IEEE (2016)
49. Tuba, E., Stanimirovic, Z.: Elephant herding optimization algorithm for support vector machine parameters tuning. In: Proceedings of the 2017 International Conference on Electronics, Computers and Artificial Intelligence (ECAI), pp. 1–5 (2017)
50. Tuba, E., Strumberger, I., Bezdan, T., Bacanin, N., Tuba, M.: Classification and feature selection method for medical datasets by brain storm optimization algorithm and support vector machine. Proc. Comput. Sci. **162**, 307–315 (2019)

51. Tuba, E., Tuba, M., Dolicanin, E.: Adjusted fireworks algorithm applied to retinal image registration. Studies Inf. Control **26**(1), 33–42 (2017)
52. Vijh, S., Sharma, S., Gaurav, P.: Brain tumor segmentation using otsu embedded adaptive particle swarm optimization method and convolutional neural network. In: Data Visualization and Knowledge Engineering, pp. 171–194. Springer (2020)
53. Wang, B., Sun, Y., Xue, B., Zhang, M.: Evolving deep convolutional neural networks by variable-length particle swarm optimization for image classification. In: 2018 IEEE Congress on Evolutionary Computation (CEC), pp. 1–8. IEEE (2018)
54. Wang, G.G., Deb, S., dos S. Coelho, L.: Elephant herding optimization. In: Proceedings of the 2015 3rd International Symposium on Computational and Business Intelligence (ISCBI), pp. 1–5 (2015)
55. Yamasaki, T., Honma, T., Aizawa, K.: Efficient optimization of convolutional neural networks using particle swarm optimization. In: IEEE Third International Conference on Multimedia Big Data (BigMM), pp. 70–73 (2017)
56. Yang, X.S.: Firefly algorithms for multimodal optimization. Stochastic Algorithms: Foundations and Applications, LNCS 5792, pp. 169–178 (2009)
57. Yang, X.S.: A new metaheuristic bat-inspired algorithm. Studies Comput. Intell. **284**, 65–74 (2010)
58. Yang, X.S., Deb, S.: Cuckoo search via levy flights. In: Proceedings of World Congress on Nature & Biologically Inspired Computing (NaBIC 2009), pp. 210–214 (2009)
59. Yang, Y., Xiang, T., Liu, H., Liao, X.: Convolutional neural network for visual security evaluation. IEEE Trans. Circuits Syst. Video Technol. 1–1 (2020)
60. Yao, X., Wang, X., Wang, S.H., Zhang, Y.D.: A comprehensive survey on convolutional neural network in medical image analysis. Multimedia Tools Appl. 1–45 (2020)

The Case for Quantifying Artificial General Intelligence with Entropy Semifields

Francisco J. Valverde-Albacete and Carmen Peláez-Moreno

Abstract In this chapter we hypotesize that Information Semifields are the underlying calculi that brains operate on and we postulate that strong artificial intelligences should try to imitate them. Information semifields have recently been proposed to describe the calculations that pertain to the manipulation of Renyi entropies. These are semifields that emerge from the pseudo-calculus by the rational choice of generator functions that respect the classical postulates of information theory, and align with the Renyi entropy and divergence. To support our hypothesis we gather evidence in this respect from Neuroscience, Machine Learning, Cognitive Computation, Information Theory and Complex Systems theory. This evidence would also constrain other attempts at describing the workings of an intelligence embodied in a brain.

Keywords Artificial intelligence · Entropy semifield · Pseudo-calculus · Cortex · Stohastic dynamical system · Free energy

1 Introduction: The Need for Bio-inspiration in AI

Perhaps because the improvements of present-day Machine Learning-based approach to Artificial Intelligence (ML/AI) technologies are mostly due to scientific progress made over a decade ago, its core progress seems to have stalled [1].

Most of the DNN distinctive advantages pivot on 40–20 year-old discoveries of perceptual signal processing properties of the neocortex, e.g. for video, audio and text modalities [2]. The advent of DNN came, among other reasons, from advances in training harmoniums [3] that allowed for their use as easily trainable layers in Deep Belief Networks (DBN) [4], one of the first types of DNN. They have since

F. J. Valverde-Albacete (✉) · C. Peláez-Moreno
Department of Signal Theory and Communications, Universidad Carlos III de Madrid Leganés, 28911 Leganés, Spain
e-mail: fva@tsc.uc3m.es

C. Peláez-Moreno
e-mail: carmen@tsc.uc3m.es

E. Pap (ed.), *Artificial Intelligence: Theory and Applications*,
Studies in Computational Intelligence 973,
https://doi.org/10.1007/978-3-030-72711-6_5

been abandoned for production instances of DNN but retain their theoretical and foundational value.

Brain inspiration has been a part of AI since its inception e.g. ANN are point-neuron models only coarsely describing dendritic summation and learning. More modern examples of this inspiration abound (LSTM, attention mechanisms, etc.), the most recent being *neuromorphic computing*,[1] a set of technologies that include hardware and materials developments—e.g. proposing sensors such as *silicon cochleas and retinas* [5]. Of these, SNN—still in their infancy—are powerful algorithms that mimic the spiking mechanisms of neurons to replicate not only their energy efficiency but also their enhanced capabilities using a more nuanced time-encoding of information [6, 7]. Critically, most such developments are still surpassed by ANN in scalability and performance.

In this paper we propose the hypothesis that the next leap in AGI should come from *the explicit modelling of quantifiable, information flows in neural tissue*. For this purpose, in the following sections we will build up the case for this being so, we will end with an explicit statement of the hypothesis and possible avenues for its testing.

2 The Case for Modelling Explicit Information Flows in Neural Tissue

2.1 Intelligence is Embodied, Situated

In this paper we concern ourselves with *situated*—that is, immersed in an environment—*embodied intelligence*—that is of an "animal" or "agent" with a body acting as the seat of a neural systems whose operation brings about such intelligence [8]. This *enactivism* is depicted graphically in Fig. 1 as an embodied neural system senses and acts on its environment and is therefore changed by the latter [9]. Note that for a neural system both the inner medium of the animal—for purposes of homoeostasis—and the outer "world" are part of the environment, and the former is embedded in the latter, that is, the body of the agent also reacts non-sensitively to the accidents of the real world.

For us, the agent realises this intelligence by circulating flows of information to and from the environment by means of effectors and sensors, respectively and "processing" such flows of information, as necessary for its continued operation.

[1]*Neuromorphic* in a wide sense not necessarily hardware systems.

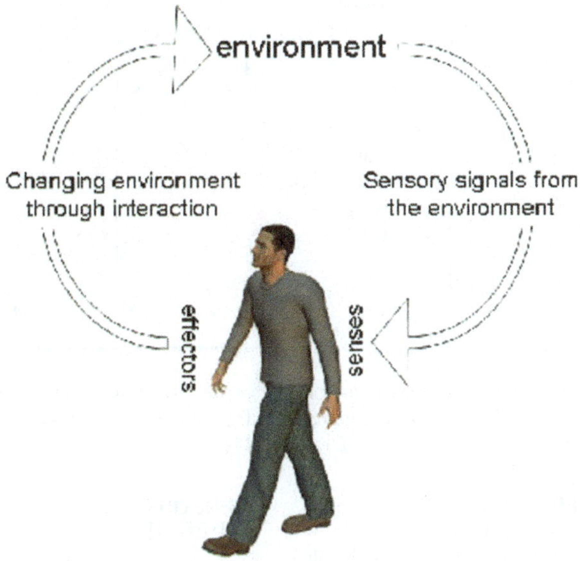

Fig. 1 Situated, embodied neural systems are the backdrop for idiosyncratic intelligence. Different types of information flow in each of the directed edges of this graph, both *to* the agent and *from* it

2.2 Neural Level Hierarchies: The Multi-scale Brain

We take mammalian brain as one of the best examples of neural system generating embodied, situated intelligence: there is mounting anatomical and physiological evidence of *information streams in the human cortex at different spatial and time scales*.

Morphologically, the cortex shows different machinery to support these information streams:

1. *At the micro-scale,* they are supported on the interaction of excitatory and inhibitory populations of the cortical microcircuit [10].
2. *At the meso scale*, they are supported on slowly-operating, lateral connectivity of the feedforward (FF) kind in layer L5 pyramidal cells, and the feedback (FB) kind in L2/3a pyramidal cells.
3. *At the macro scale*, they are supported in the fast, long-range apical connectivity in the FF direction of layer L3b pyramidal cells which supports point-to-point excitation, and the more diffuse, but similar in scope FB connectivity of the L6 pyramidal cells [11].

Recently, two different morphological pathways have been described in the cortex for FF-FB coupled streams, as depicted in Fig. 2. Functionally, the number of levels that can be distinguished is much higher, as schematized in Fig. 3, as well as the fact that these are aggregated in two different, interacting subsytems: the sensory and the motor hierarchies [12], possibly with a decision subsystem mediating between them at higher levels. Note that:

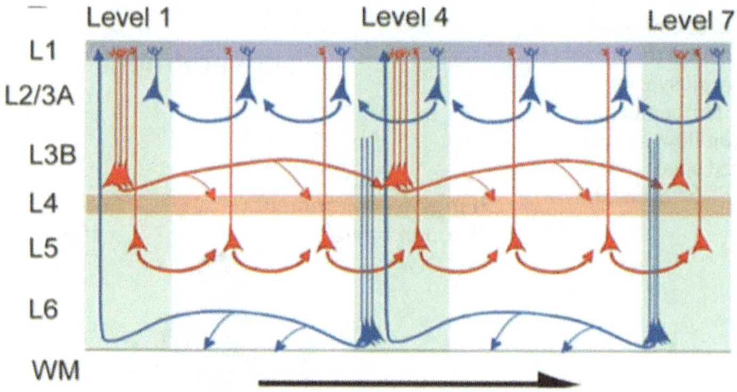

Fig. 2 Information streams in the cortex: local, meso and macro scales. FF (red) and FB (blue) information streams are paired in infra- and supra-granular counter-streams across all levels (only shown levels L_1, L_4 and L_7, from [11])

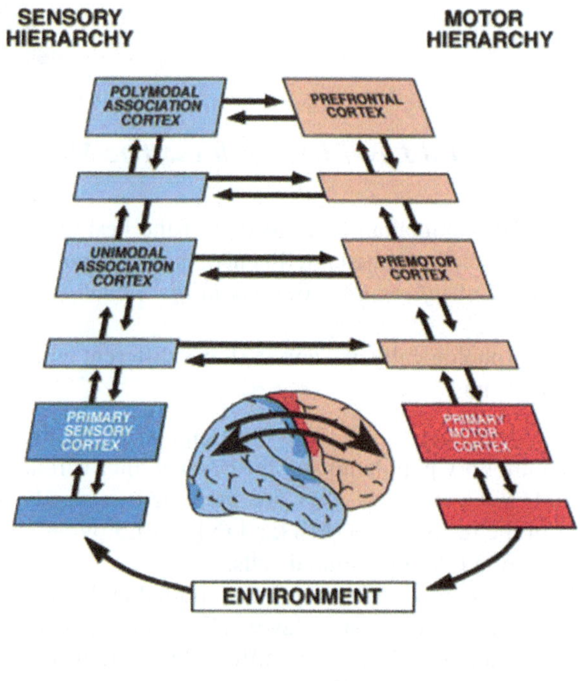

Fig. 3 Functional level systems in cortex architecture with hierarchical (vertical), and heterarchical (horizontal) connections (from [12])

- The putative architecture of the visual subsystem of the sensory hierarchy has provided a number of intuitions and models for the present ML/AI revolution [2].
- The mammalian brain is actually a stack of several "brains" phylogenetically developed for different purposes which have been co-opted and repurposed by later phyletic developments [13].
- Consequently, in this paper, we speak of the *sensory-motor loop* when we refer to low-level behaviour within this hierarchy, while we refer to the *perception-action loop* when the cycle includes higher cognitive cortices [12].

Scale-free, power-law distributed brain dynamics are ubiquitous in neuronal systems of all measurable scales, possibly an emergent property of power law aggregation [14, 15]. For instance, neuronal oscillations are observed in frequency ranges from 0.01 to 200 Hz and are both ubiquitous and pervasive aspects of neuronal processing. Note that supra- and infra-granular FF-FB dual counter-streams have different frequencies of operation [16]. Functionally they regulate the temporal structure of neuronal excitability and thereby control the timing of neuronal communication [17].

Neural oscillations are commonly taken as evidence that brains operate near a critical point of a (second-order) phase transition, i.e. in a regime between synchronization (epilepsy) and randomness (coma), with intermediate synchronization and high levels of power-law structured variability. Other networks working at *criticality* seem endowed with heightened reactiveness to inputs, optimal range operation, dynamical reconfiguration, and emergence of properties [14, 18, 19].

Assumption 1: The cortex implements communications at multiple spatial and temporal scales.
The cortex is in charge of transmitting the computation/activation driven by thalamo-cortical afferents and refined/amplified by L4/L3b neurons, towards cortico-thalamic efferents, using two dual information FF-FB streams operating at criticality.

2.3 The Predictive Mind

One current explanation for the presence of dual information streams in the cortex is *predictive coding theory*, where the L4 populations maintain local states, the FB streams carry predictions throughout the system, and the FF streams carry the error signal between the predictions and the actual states/inputs to each local circuit [20]. This is undertaken by the brain with the goal of maintaining enaction through the *sensory-motor loop* closed by the environment. In this view, different cortical rhythms subserve both the information flow along the dual streams as well as the prediction of the timing of events at different time-scales adapted to environment dynamics [21].

Predictive Coding Theory asserts the existence of a hierarchy of levels L_1 to L_7 of increasing abstraction of sensorial information, from the primary sensory cortex areas towards the polymodal association cortex. Subsequently, the rostral executive stream carries out the fleshing of action commands from the prefrontal cortex towards the primary motor cortex as depicted in Fig. 3. [12].

Assumption 2: The sensory-motor cortex implements *predictive* hierarchical communications.
It carries information to- and from the prefrontal cortex to control its perceptions in a multi-time-scale environment-completed loop.

2.4 Free Energy Models

One possible instantiation of such models of the brain are Stochastic Dynamical Systems (SDS) as described by Infobox 1.

Infobox 1: Stochastic Dynamical Systems
Situated agents modeled by a tuple (X, S, M, A, Ω) with implicit dependences on time t and a set of statistical parameters θ, where we have

(i) *hidden states* $x \in X \subset \mathbb{R}^n$, governed by a nonlinear function $f : X \times A \to R^n$, modeling the outer environment,
(ii) *sensory states* $s \in S \subset \mathbb{R}^p$, governed by another nonlinear function $g : X \times A \to S$, capturing the evolution of perception,
(iii) *(control) actions* $a \in A \subset \mathbb{R}^m$ are designed to drive the dynamics,

$$\dot{x} = f(x, a) + \omega_x \qquad\qquad s = g(x, a) + \omega_s \tag{1}$$

(iv) *internal states* $\mu \in M \subset \mathbb{R}^q$,
(v) Ω is a sample space from which *random fluctuations* $\omega_x : \mathbb{R}^n$ and $\omega_s : \mathbb{R}^p$ are drawn.

Computational Neuroscience abounds with SDS models described by differential equations. In particular, the neuron has well-established models in terms of the differential mechanisms described above, e.g. the original Hodgkin–Huxley model as well as some simplifications and alternatives to it [22, Chap. 2]. But there are also finer models of dendritic trees [23, Chaps. 5–6] or coarser models of aggregates of neurons [24]—e.g. populations—interacting with an environment.

Free-Energy (FE) models are specializations of SDS, Bayesian models developed to reconcile the predictive coding hypothesis with neural architectures in a sensory-motor loop, as graphically depicted in Fig. 4.

A basic FE model [26] is a scale-agnostic model of the behaviour of situated agents for perception and learning that add to the SDS the following components:

- $p(s, x, a \mid m)$ is a *generative density over sensory, hidden & control states* under a generative model m,
- $q(x, a \mid \mu)$ is a *variational density* over hidden states and actions with mean $\eta \in \mathbb{R}$, and
- the *Free Energy* that upper bounds the sensory surprise is

$$\mathcal{F}(s, \mu) = \langle \ln q(\theta \mid \mu) \rangle_q - \langle \ln p(s, |m) \rangle_q \tag{2}$$

so that the solution to (1) that minimizes the FE —the optimal action a^* and internal state μ^*—that describes a predictive coding strategy for a communication channel is

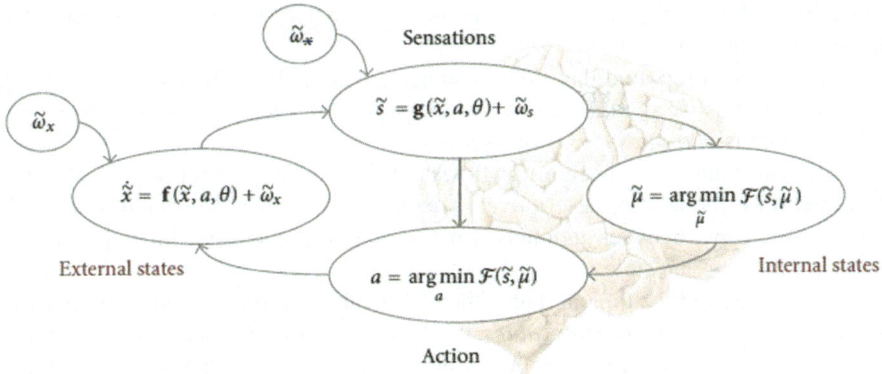

Fig. 4 Basic Situated Complex Dynamical System in the FE framework, with explicit model parameters θ (from [25])

$$\mu^* = \arg\min_{\mu} \mathcal{F}(s, \mu) \qquad a^* = \arg\min_{a} \mathcal{F}(s, \mu). \tag{3}$$

Notice that:

- These models are intended to follow the *Principle of Perceptual Control: agents design their policies to control perceptions, not actions* [27].
- This solution to (3) may be found through a series of technical tricks-of-the-trade—e.g. variational Bayes [28], generalized filtering [29] as borrowed from Statistical Mechanics [30].
- Furthermore, (3) is so important for the framework that in fact it has been raised to a Principle [24, 31].

Therefore it follows that:

Assumption 3: Situated Neural Systems at any scale of analysis can be modelled as SDS that respect the FE framework

2.5 Complex Networks of the Brain

Complex networks (CN) are ubiquitous tools in the study of the brain [32, 33]. Especifically, connection maps are evident at all spatial scales from the cortical micro-circuit to the connectome; *topical (*sometimes *cortical) maps* [34, Chap. 4]—e.g., retinotopic, tonotopic—are phyletic memory networks carrying certain sensory inputs to specific locations in the cortex preserved throughout individuals of a species [35], etc.

The versatility of CN stems from their mathematical underpinnings: networks are weighted graphs through which a "substance" flows, this "substance" being, typically, the "activation state" of the nodes in the network, ultimately describable in terms of neuron activation.

Critically, we propound that this substance are actually averages of information over distributions—e.g. in FE models above—and their temporal flows $H(p, t) = H(p(t))$, that is *entropies and entropy flows*.

On the one hand, note that entropy flows in the cortex are ubiquitous [36, 37]. On the other, if FE models are descriptively adequate, it is easy to see that the FE in (2) is a cross-entropy, between its internal model m and the sensory input s acting as proxy for the external state x.

What is the need for topical maps, then? Arguably, entropy being a *quantitative* property of probability measure *values*, it is "blind" to any particular events or "bins" of the distribution, so brains require topical maps to keep track of the *qualities* quantified by entropy flows from their information sources to their destinations. This is done through the argmax and argmin operations and encoded in the physical and functional connectomes. Hence the next assumption should be fairly uncontroversial.

Assumption 4: Neural processing is the communication of entropy flows though neural networks
Situated neural networks are machines that route entropy—the *quantity of information*—flows between sensors, actuators and the environment and provide interpretation of these—the *quality of information*—in the form of connectivity/topic maps.

2.6 Algebras for Entropy Flows in the Brain

Recently, a new algebraic structure has been introduced for the Rényi entropies whereby they are positive semifields [38]. Recall that a semifield is a semiring whose multiplicative structure $\langle K \setminus \{\epsilon\}, \otimes, e, \cdot^{\circledast}\rangle$ is a group, where $\cdot^{\circledast} : K \to K$ is the function to calculate the inverse such that $\forall u \in K, u \otimes u^{\circledast} = e$.

On the other hand, a *dioid* is a commutative semiring $\mathcal{D}$ where the canonical preorder relation—$a \preccurlyeq b$ if and only if there exists $c \in D$ with $a \oplus c = b$—is actually an order $\langle D, \preccurlyeq\rangle$. In a dioid the canonical order relation is compatible with both $\oplus$ and $\otimes$ [39, Chap. 1, Proposition 6.1.7] and the additive zero ϵ is always the infimum of the dioid or *bottom* hence the notation $\epsilon = \inf D = \bot$. Dioids are all zero sum-free, that is, they have no non-null additive factors of zero: if $a, b \in D, a \oplus b = \epsilon$ then $a = \epsilon$ *and* $b = \epsilon$.

Therefore, a semifield that is also a dioid is called a positive semifield. The non-negative reals $\mathbb{R}_{\geq 0} = \langle [0, \infty[, +, \times, \cdot^{\circledast}, \bot = 0, e = 1\rangle$ or rationals are paragons of these algebras.

To build the *entropy semifields*, let $r \in [-\infty, \infty] \setminus \{0\}$ and $b \in]1, \infty[$ in Rényi's modified averaging function [40] $\varphi(p) = \frac{-1}{r} \log_b p$ and its inverse—these are dual isomorphisms of positive semifields [38]. Then using φ as the g function in Pap's calculus [41] generates the *semifields* $\langle [-\infty, \infty], \oplus_r, \otimes_r, \cdot^{-1}, \bot = \infty, e = 0\rangle$ whose basic operations are:

$$u \oplus_r v = u + v - \log_b \left(b^{ru} + b^{rv}\right)^{\frac{1}{r}} \qquad u \otimes_r v = u + v \qquad u^{\circledast} = -u \tag{4}$$

This can be completed to two dually-ordered positive, entropy semifields:

$$\mathbb{H}_r = \langle [-\infty, \infty], \underset{\cdot}{\oplus}_r, \underset{\cdot}{\otimes}_r, -\cdot, \bot = -\infty, e = 0, \top = \infty \rangle \tag{5}$$

$$-\mathbb{H}_r = \langle [-\infty, \infty], \dot{\oplus}_r, \dot{\otimes}_r, -\cdot, -\bot = \infty, e = 0, -\top = -\infty \rangle \tag{6}$$

whose elements can be considered as emphasized, entropic values [38, Theorem 3]. Note that there is a full-blown calculus available for these [42, 43], whence it is not unreasonable to suppose:

Hypothesis 1: Embodied, Situated Intelligence using Entropy Semifields
There is a procedure that transforms the description of cognitive, situated complex neural networks with entropic-algebraic equations and allows us:

(i) to carry out the analysis and synthesis of information flows through them, and
(ii) to design, analyze and build cognitive agents capable of attaining situated intelligence.

Remark 1 There is shown how Renyi's postulates for entropy lead to Pap's g-calculus where the functions carrying out the domain transformation are Renyi's information function and its inverse. In its turn, Pap's g-calculus under Renyi's information function transforms the set of positive reals into a family of semirings where "standard" product has been transformed into sum and "standard" sum into a power-emphasized sum. Consequently, the transformed product has an inverse whence the structure is actually that of a positive semifield. Instances of this construction lead to idempotent analysis and tropical algebra as well as to less exotic structures.

3 Discussion: The Way Forward

In this position paper, we have addressed the shortcomings of present-day ML/AI by:

- **A deeper inspiration in enaction**—that is, the interaction of neural systems with their environment through the body—, to improve the embodied semantics of AI so it may better relate to human intelligence.
- **A better modeling of the complexity of the brain**, including heterogeneous neural processing spanning multiple spatial and temporal scales with a modular hierarchical structure [17, 37, 44–46], to improve effectiveness.
- **A better understanding and modeling of the nature of neural computations** in terms of the dynamic aggregation and segregation of *entropy flows* [14] using entropy semifields [38, 47].

In the concluding next two sections we sketch steps taken to test our hypothesis.

3.1 Entropic Models Using Tropical Maths

For a proof-of concept, we are aiming at using special entropy semifields. Specifically, the Tropical Algebras are the completed min-plus semifield, or "tropical algebra proper", and the max-plus semifield, or "schedule algebra",

$$\begin{aligned} \overline{\mathbb{R}}_{\min,+} &= \langle [-\infty, \infty], \min, \dot{+}, -\cdot, \infty, 0, -\infty \rangle \\ \overline{\mathbb{R}}_{\max,+} &= \langle [-\infty, \infty], \max, \underset{\cdot}{+}, -\cdot, -\infty, 0, \infty \rangle \end{aligned} \tag{7}$$

which are the entropy algebras of parameter $r \to \pm\infty$.

We believe Tropical Maths are convenient to the modelling of the neural systems because they allow us to carry out non-linear processing in a linear guise [48], what facilitates adoption.

3.2 Test Neural System: The Importance of a Lowly Worm

An anatomically simple yet behaviourally complex, embodied intelligence is that of *Caenorhabditis elegans*. This is a well-studied, transparent, millimetric nematode with two genders: hermaphodite and male. The level of complexity of this animal is optimal for developing low-complexity models: the connectome of the hermaphrodite animal has approximately 300 neurons, yet *C. elegans* offers a wide repertoire of behaviours and taxes [49], and several control levels at different time-scales [50].

It is also rich in neural phenomena what makes it an interesting organism to model: its motor neurons have been found to be endowed with both gradual sinapses and gap junctions [51], but recent results point to the fact that sensory neurons seem to be spiking [52], hence this animals is optimal for trying a number of strategies to cater for several types of neurons and synapses.

The cognitive universe of *C. elegans* is simple, but it is already capable of complex learning experiences [53]. Most importantly it is one of the simplest models of embodied, situated intelligence being studied: the organism is reactive but shows a degree of "free will" e.g. in the exploratory swings of its head while foraging [50].

The importance of *C. elegans* lies in that its simple connectome can be simulated with present-day tools [51] thus lending itself to reproducible science [54]. It should be *almost* straightforward to test our hypotheses on its data, and use it as a stepping stone for more complex neural aggregates. But as the adage goes, *devil is in the details*.

References

1. Hutson, M.: Core progress in AI has stalled in some fields. Science **368**(6494), 927 (2020)
2. Sejnowsky, T.J.: The unreasonable effectiveness of deep learning in artificial intelligence. Proceedings of the National Academy of Sciences (2020)
3. Smolensky, P.: Information processing in dynamical systems: foundations of harmony theory. In: Parallel Distributed Processing, pp. 194–281. MIT Press, Cambridge (1986)
4. Hinton, G.E., Osindero, S., Teh, Y.-W.: A fast learning algorithm for deep belief nets. Neural Comput. **18**(7), 1527–1554 (2006)
5. Rajendran, B., Sebastian, A., Schmuker, M., Srinivasa, N., Eleftheriou, E.: Low-power neuromorphic hardware for signal processing applications: a review of architectural and system-level design approaches. IEEE Signal Process. Mag. **36**(6), 97–110 (2019)
6. Sevuktekin, N.C., Varshney, L.R., Hanumolu, P.K., Singer, A.C.: Signal processing foundations for time-based signal representations: neurobiological parallels to engineered systems designed for energy efficiency or hardware simplicity. IEEE Signal Process. Mag. **36**(6), 38–50 (2019)
7. Li, G., Deng, L., Chua, Y., Li, P., Neftci, E.O., Li, H.: Editorial: spiking neural network learning, benchmarking, programming and executing. Front. Neurosci. **14**, 276 (2020)
8. Russell, S.J., Norvig, P.: Artificial Intelligence - A Modern Approach (3. internat. ed.) (2010)
9. Varela, F.J., Thompson, E., Rosch, E.: The Embodied Mind. Cognitive Science and Human Experience, revised edn. MIT Press, Cambridge (2017)
10. Capone, F., Paolucci, M., Assenza, F., Brunelli, N., Ricci, L., Florio, L., Di Lazzaro, V.: Canonical cortical circuits: current evidence and theoretical implications. Neurosci. Neuroeconomics 1–8, (2016)
11. Markov, N.T., Vezoli, J., Chameau, P., Falchier, A., Quilodran, R., Huissoud, C., Lamy, C., Misery, P., Giroud, P., Ullman, S., Barone, P., Dehay, C., Knoblauch, K., Kennedy, H.: Anatomy of hierarchy: feedforward and feedback pathways in macaque visual cortex. J. Comp. Neurol. **522**(1), 225–259 (2013)
12. Fuster, J.M.: Upper processing stages of the perception-action cycle. Trends Cogn. Sci. **8**(4), 143–145 (2004)
13. Marcus, G.: Kluge: the haphazard evolution of the human mind. Houghton MIfflin Company, Boston (2008)
14. Werner, G.: Fractals in the nervous system: conceptual implications for theoretical neuroscience. Front Physiol. **1**, 1–28 (2010)
15. Anderson, R.B.: The power law as an emergent property. Mem. Cogn. **29**(7), 1061–1068 (2001)
16. Bastos, A.M., Vezoli, J., Bosman, C.A., Schoffelen, J.-M., Oostenveld, R., Dowdall, J.R., De Weerd, P., Kennedy, H., Fries, P.: Visual areas exert feedforward and feedback influences through distinct frequency channels. Neuron **85**(2), 390–401 (2015)
17. Palva, S., Palva, J.M.: Roles of brain criticality and multiscale oscillations in temporal predictions for sensorimotor processing. Trends Neurosci. **41**(10), 729–743 (2018)
18. Joglekar, M.R., Mejias, J.F., Yang, G.R., Wang, X.-J.: Inter-areal balanced amplification enhances signal propagation in a large-scale circuit model of the primate cortex. Neuron **98**(1), 222–234 (2018)
19. Kinouchi, O., Copelli, M.: Optimal dynamical range of excitable networks at criticality. Nat. Phys. **2**(5), 348–352 (2006)
20. Rao, R.P.N., Ballard, D.H.: Predictive coding in the visual cortex: a functional interpretation of some extra-classical receptive-field effects. Nat. Neurosci. (1999)
21. Arnal, L.H., Giraud, A.-L.: Cortical oscillations and sensory predictions. Trends Cognit. Sci. **16**(7), 390–398 (2012)
22. Izhikevich, E.M.: Dynamical Systems in Neuroscience. MIT Press, Cambridge (2007)
23. Dayan, P., Abbot, L.F.: Theoretical Neuroscience. Computational and Mathematical Modeling of Neural Systems. MIT Press, Cambridge (2005)
24. Razi, A., Friston, K.J.: The connected brain: causality, models, and intrinsic dynamics. IEEE Signal Proc. Mag. **33**(3), 14–35 (2016)

25. Friston, K., Ao, P.: Free energy, value, and attractors. Comput. Math. Methods Med. **5**, 1–27 (2012)
26. Bogacz, R.: A tutorial on the free-energy framework for modelling perception and learning. J. Math. Psychol. **76**, 198–211 (2017)
27. Wiiliam, T.: Powers. Living Control Systems. Selected Papers, CSG (1989)
28. MacKay, D.J.C.: Information Theory. Inference and Learning Algorithms. Cambridge University Press, Cambridge (2003)
29. Friston, K., Stephan, K., Li, B., Daunizeau, J.: Generalised filtering. Math. Probl. Eng. **2010**(3), 1–34 (2010)
30. MacKay, D.J.C.: Free energy minimisation algorithm for decoding and cryptanalysis. Electron. Lett. **31**, 446–447 (1995)
31. Friston, K.: The free-energy principle: a unified brain theory? Nat. Rev. Neurosci. **11**(2), 127–138 (2010)
32. Gu, S., Pasqualetti, F., Cieslak, M., Telesford, Q.K., Yu, A.B., Kahn, A.E., Medaglia, J.D., Vettel, J.M., Miller, M.B., Grafton, S.T., Bassett, D.S.: Controllability of structural brain networks. Nat. Commun. **6**, 8414 (2015)
33. Schmidt, M., Bakker, R., Hilgetag, C.C., Diesmann, M., van Albada, S.J.: Multi-scale account of the network structure of macaque visual cortex. Brain Struct. Funct. **223**(3), 1409–1435 (2018)
34. Ballard, D.H.: Brain Computation as Hierarchical Abstraction. MIT Press, Cambridge (2015)
35. Fox, K.: Barrel Cortex. Cambridge University Press, Cambridge (2008)
36. Sayood, K.: Information theory and cognition: a review. Entropy **20**(9), 1–19 (2018)
37. Deco, G., Kringelbach, M.L.: Hierarchy of information processing in the brain: a novel 'Intrinsic Ignition' framework. Neuron **94**(5), 961–968 (2017)
38. Valverde-Albacete, F.J., Peláez-Moreno, C.: The Rényi entropies operate in positive semifields. Entropy **21**(8) (2019)
39. Gondran, M., Minoux, M.: Graphs, Dioids and Semirings. New Models and Algorithms. Operations Research Computer Science Interfaces series. Springer, Berlin (2008)
40. Renyi, A.: Probability Theory. Courier Dover Publications, Mineola (1970)
41. Pap, E.: g-calculus. Zbornik Radova Prirodno-Matematichkog Fakulteta. Serija za Matematiku. Review of Research. Faculty of Science. Mathematics Series **23**(1), 145–156 (1993)
42. Grossman, M., Katz, R.: Non-Newtonian Calculus. Lee Press, Pigeon Cove (1972)
43. Mesiar, R., Pap, E.: Idempotent integral as limit of g-integrals. Fuzzy Sets Syst. **102**(3), 385–392 (1999)
44. Palva, J.M., Palva, S.: Roles of multiscale brain activity fluctuations in shaping the variability and dynamics of psychophysical performance. Prog. Brain Res. **193**, 335–350 (2011)
45. Siegel, M., Donner, T.H., Engel, A.K.: Spectral fingerprints of large-scale neuronal interactions. Nat. Rev. Neurosci **13**(2), 121–134 (2012)
46. Zhigalov, A., Arnulfo, G., Nobili, L., Palva, S., Palva, J.M.: Modular co-organization of functional connectivity and scale-free dynamics in the human brain. Netw. Neurosci. **1**(2), 143–165 (2017)
47. Valverde-Albacete, F.J., Peláez-Moreno, C.: The case for shifting the Renyi entropy. Entropy **21**(1) (2019)
48. Butkovič, P.: Max-linear Systems. Theory and Algorithms. Monographs in Mathematics. Springer, Berlin (2010)
49. Kato, S., Kaplan, H.S., Schrödel, T., Skora, S., Lindsay, T.H., Yemini, E., Lockery, S., Zimmer, M.: Global brain dynamics embed the motor command sequence of caenorhabditis elegans. Cell **163**(3), 656–669 (2015)
50. Kaplan, H.S., Thula, O.S., Khoss, N., Zimmer, M.: Nested neuronal dynamics orchestrate a behavioral hierarchy Across timescales. Neuron **105**(3), 562–576.e9 (2020)
51. Gleeson, P., Lung, D., Grosu, R., Hasani, R., Larson, S.D.: c302: a multiscale framework for modelling the nervous system of Caenorhabditis elegans. Philos. Trans. R. Soc. Lond. Ser. B Biol. Sci. **373**(1758), 20170379 (2018)

52. Liu, Q., Kidd, P.B., Dobosiewicz, M., Bargmann, C.I.: C. elegans AWA olfactory neurons fire calcium-mediated all-or-none action potentials. Cell **175**(1), 57–70.e17 (2018)
53. Pradhan, S., Quilez, S., Homer, K., Hendricks, M.: Environmental programming of adult foraging behavior in C. elegans. Curr. Biol. **29**(17), 2867–2879.e4 (2019)
54. Estévez-Albuja, I., Valverde-Albacete, F.J., Peláez-Moreno, C.: Replication of two computational models of locomotion of *Caenorhabditis elegans* in Brian 2. In: Neuromatch Conference (2020)

Fuzzy Metrics and Its Applications in Image Processing

Nebojša Ralević

Abstract This paper is a review of recent research, in which fuzzy metrics and their applications in image processing are studied. The notions of the fuzzy T-metric and the fuzzy S-metric are presented, after which, examples of known fuzzy metrics are provided, along with theorems that enable algorithms to develop new metrics. Two applications of fuzzy metrics in image processing are illustrated: Image filtering and Copy-move forgery detection. Image filtering reduces the amount of noise, while maintaining satisfactory image quality. The aim was to improve the sharpness and quality of the image, measured by the image quality indices UIQI and CPBD. It is illustrated that the image filtered with this modified algorithm has better quality and greater sharpness than images filtered with the median filter. The fuzzy metric parameters that produce images with the best quality and sharpness are determined experimentally. The digital image falsification obtained by copying and pasting part of the original image into another part of the same image is also considered. Copy-move forgery detection (CMFD) is one of the methods used to detect such forgeries in images. A cluster algorithm that successfully solves this problem is presented.

Keywords Aggregation function · Triangular norm · Triangular conorm · Fuzzy sets · Fuzzy metric · Fuzzy T-metric · Fuzzy S-metric · Image processing

1 Introduction

The notion of a distance over a non-empty set X comprises a wide mapping class d from X^2 to $[0, \infty[$. By imposing or relaxing some constraints on the distance d, various new notions are obtained, such as similarity, resemblance, metrics, semi metrics, …[6]. The notion of distance can be generalized so that the distance between

---

N. Ralević (✉)
Department of Fundamental Sciences, Faculty of Technical Sciences, University of Novi Sad, Trg Dositeja Obradovića 6, 21000 Novi Sad, Serbia
e-mail: nralevic@uns.ac.rs

E. Pap (ed.), *Artificial Intelligence: Theory and Applications*,
Studies in Computational Intelligence 973,
https://doi.org/10.1007/978-3-030-72711-6_6

two elements of a set does not have to be a non-negative number, for example, it can be a function or a fuzzy set [3].

There is evidence that the notion of a probabilistic metric space [31] plays a crucial role in regulating the generalization of a metric space in which the distance between x and y is replaced by the distribution function $F_{x,y}$, and where the value of that distribution function at some point a is interpreted as a probability, in which case the distance between x and y is smaller than a. The introduction of the concept of fuzzy sets, and many of their key aspects, was proposed by L. Zadeh in 1965.

The characterization of fuzzy sets is important for our deeper understanding of fuzzy logic and its implications, and in this manner the theory of fuzzy metric spaces has evolved, where the distance between the elements x and y is a fuzzy set which certain properties. There are several important areas where this theory makes an original contribution to finding wider applications, particularly in specific practical problems, one of which is image processing.

The paper is organized as follows. The second section presents the basic notions of a triangular norm and a conorm, a fuzzy complement and an aggregation operation (see [8, 14, 15, 31]).

In the third section, the fuzzy S-metric and the fuzzy T-metric are defined and relevant examples of their use are illustrated. Their properties were examined, with special emphasis on the construction of new fuzzy metrics.

The reason for this is that the fuzzy metric has a number of advantages over the classical one, which are reflected in a wider range of possibilities of application in problems from practice. Namely, by its definition, a fuzzy metric also includes a parameter that enables better adaptability, which is very useful in improving performance.

In the Sect. 4, the values of the described parameters are experimentally determined. Ultimately, the main aim of this study is to investigate the differences between the quality of image filtering, utilizing our algorithm and filtering using VMF, both of which were calculated by means of the UIQI and CPBD indices, which have been defined in [24, 33].

In practice, examples of digital image falsification can often be found as obtained by copying and pasting a part of the original image into another part of the same image. The purpose of such changes is to duplicate the content of the image or hide certain image content. Copy-move forgery detection (CMFD) is one of the methods to detect such forgeries in images. In the fifth section, a new clustering algorithm for edited images which was proposed in [18, 30] is covered.

Section 6 consists of conclusions about the application of fuzzy metrics in image processing. In addition, some additional remarks and ideas for future research are stated.

For the purpose of illustrating the applications of fuzzy metrics in image processing, the image from the article shown in in Fig. 1 serves as an example, which was obtained from a public database https://www2.eecs.berkeley.edu/Research/Projects/CS/vision/bsds/, *Berkeley University of California, Department of Electrical Engineering and Computer Sciences, Computer Vision Group, segmentation dataset—training images* [17].

2 Preliminaries

To make the paper self-contained, we give the definitions of some well-known operations on $[0, 1]^n$ (see e.g. monographs [8, 14, 15]).

Definition 1 The *triangular norm*, abbreviation t-norm (*triangular conorm*, abbreviation t-conorm) is a binary operation $\mathsf{T} : [0, 1]^2 \to [0, 1]$ ($\mathsf{S} : [0, 1]^2 \to [0, 1]$) which satisfies the following conditions: monotonicity, commutativity, associativity and has a neutral element which is 1 (0). The t-(co)norm is said to be an *Archimedean* t-(co)norm, if T (S) is a continuous function and $\mathsf{T}(a, a) < a$ ($\mathsf{S}(a, a) > a$) is satisfied, for arbitrary $a \in]0, 1[$.

If a property holds for both t-norms and t-conorms, then we will put the word norm in the property instead of these expressions and use the notation N.

If, in the definition of the norm, instead of the axiom of monotonicity, a strict monotonicity is valid, i.e.

$$a_1 < a_2 \wedge b_1 < b_2 \Rightarrow \mathsf{N}(a_1, b_1) < \mathsf{N}(a_2, b_2),$$

for all $a_1, a_2, b_1, b_2 \in [0, 1]$, then the norm is *strict*.

The power of the norm N is given by formulas ($n \geqslant 2$):

$$\mathsf{N}^1(a_1, a_2) = \mathsf{N}(a_1, a_2) \quad \mathsf{N}^n(a_1, \ldots, a_n, a_{n+1}) = \mathsf{N}(\mathsf{N}^{n-1}(a_1, \ldots, a_n), a_{n+1}).$$

Definition 2 A function $\mathsf{c} : [0, 1] \to [0, 1]$ is a *fuzzy complement*, if the following conditions are satisfied:

(c_1) $\mathsf{c}(0) = 1$ and $\mathsf{c}(1) = 0$, (boundary conditions)
(c_2) $(\forall a, b \in [0, 1])\ a \leqslant b \Rightarrow \mathsf{c}(a) \geqslant \mathsf{c}(b)$ (monotonicity).

If $\mathsf{c}(\mathsf{c}(a)) = a$ holds for all $a \in [0, 1]$, then a function c is *involutive*. If c is a continuous function, then we say that c *is a continuous fuzzy complement*.

The triangular norm T and the triangular conorm S are *dual* with respect to the fuzzy complement c if

$$\mathsf{c}(\mathsf{T}(a, b)) = \mathsf{S}(\mathsf{c}(a), \mathsf{c}(b)) \text{ and } \mathsf{c}(\mathsf{S}(a, b)) = \mathsf{T}(\mathsf{c}(a), \mathsf{c}(b)).$$

$(\mathsf{T}, \mathsf{S}, \mathsf{c})$ is called a *dual triple*.

Definition 3 An n-ary *aggregation function* is a function $\mathsf{A}_{[n]} : I^n \to I$ such that

(i) $\mathsf{A}_{[n]}(x_1, \ldots, x_n) \leqslant \mathsf{A}_{[n]}(y_1, \ldots, y_n)$ whenever $x_i \leqslant y_i$ for all $i \in \{1, \ldots, n\}$ (A is a monotonically increasing function in all its arguments).
(ii) $\inf_{x \in I^n} \mathsf{A}_{[n]}(x) = \inf I$ and $\sup_{x \in I^n} \mathsf{A}_{[n]}(x) = \sup I$, $x = (x_1, \ldots, x_n)$ (boundary condition).

An *aggregation function* is a function $\mathsf{A} : \bigcup_{n\in\mathbb{N}} I^n \to I$ such that for $n = 1$ holds $\mathsf{A}(x) = x$, for all $x \in I$, whose restriction is $\mathsf{A}|_{I^n} = \mathsf{A}_{[n]}$, for any $n \in \mathbb{N}$.

Remark 1 It is customary to take $I = [0, 1]$. In this case, the boundary condition reduces to $\mathsf{A}(0, \ldots, 0) = 0$ and $\mathsf{A}(1, \ldots, 1) = 1$.

3 Fuzzy Metrics

This section presents the findings of the research, which focuses on the fuzzy S-metric and the fuzzy T-metric. Of the many features of fuzzy metrics, those used in image processing are listed, and they are also proven in the papers: [10, 11, 26–28, 30].

Definition 4 ([26]) Let $X \neq \emptyset$, S be a continuous t-conorm, T a continuous t-norm, and $\boldsymbol{d}$ a fuzzy set defined on $X \times X \times]0, \infty[$. Some of the possible properties of d are:

1. (i) $\boldsymbol{d}(x, y, \alpha) \in [0, 1[$,
   (ii) $\boldsymbol{d}(x, y, \alpha) \in]0, 1]$;
2. (i) $\boldsymbol{d}(x, y, \alpha) = 0 \Leftrightarrow x = y$,
   (ii) $\boldsymbol{d}(x, y, \alpha) = 1 \Leftrightarrow x = y$;
3. (i), (ii) $\boldsymbol{d}(x, y, \alpha) = \boldsymbol{d}(y, x, \alpha)$;
4. (i) $\mathsf{S}(\boldsymbol{d}(x, y, \alpha), \boldsymbol{d}(y, z, \beta)) \geqslant \boldsymbol{d}(x, z, \alpha + \beta)$,
   (ii) $\mathsf{T}(\boldsymbol{d}(x, y, \alpha), \boldsymbol{d}(y, z, \beta)) \leqslant \boldsymbol{d}(x, z, \alpha + \beta)$;
5. (i), (ii) $\boldsymbol{d}(x, y, \_) :]0, \infty[\to [0, 1]$ is a continuous function;
   for all $x, y, z \in X, \alpha, \beta > 0$.

The fuzzy set $\boldsymbol{d}$ is called

(i) a *fuzzy S-metric* and a triple $(X, \boldsymbol{d}, \mathsf{S})$ is a *fuzzy S-metric space* if $\boldsymbol{d}$ satisfies axioms (1–5)(i);
(ii) a *fuzzy T-metric* and a triple $(X, \boldsymbol{d}, \mathsf{T})$ is a *fuzzy T-metric space* if $\boldsymbol{d}$ satisfies axioms (1–5)(ii).

If instead of 1., it holds that $\boldsymbol{d}(x, y, \alpha) \in [0, 1]$, the fuzzy set $\boldsymbol{d}$ is a *fuzzy S-metric (fuzzy T-metric) in the broader sense*, and $(X, \boldsymbol{d}, \mathsf{S})$ $((X, \boldsymbol{d}, \mathsf{T}))$ is a *fuzzy S-metric (fuzzy T-metric) space in the broader sense.*

We will denote the fuzzy S-metric with $\boldsymbol{s}$ and the fuzzy T-metric with $\boldsymbol{t}$, and we will write the symbol $\boldsymbol{d}$ if a statement is valid in both cases and use the term fuzzy metric.

Fuzzy metric $\boldsymbol{d}$ is *stationary* on X if $\boldsymbol{d}$ does not depend on α, i.e., if for all fixed $x, y \in X$, the function $\boldsymbol{d}_{x,y}(\alpha) = \boldsymbol{d}(x, y, \alpha)$ is a constant.

Example 1 ([10, 26]) The mapping $\mathbf{t}_K : \mathbb{R}^+ \times \mathbb{R}^+ \to \mathbb{R}$ defined by

$$\mathbf{t}_K(x, y) = \frac{\min\{x, y\} + K}{\max\{x, y\} + K},$$

where $K > 0$, is a fuzzy T-metric with respect to multiplication, and

$$\mathbf{s}_K(x, y) = \frac{|x - y|}{\max(x, y) + K}$$

is a fuzzy S-metric with respect to the algebraic sum, $\mathsf{S}(x, y) = 1 - (1 - x)(1 - y) = x + y - xy$, dual to T with respect to the standard fuzzy complement.

Example 2 Let $f : X \to \mathbb{R}^+$ be a one-to-one function and let $g : \mathbb{R}^+ \to [0, \infty]$ be an increasing continuous function. The mapping

$$\mathbf{t} : X \times X \times \mathbb{R}^+ \to [0, 1]$$

defined by

$$\mathbf{t}(x, y, \alpha) = \left(\frac{\frac{(f(x))^p + (f(y))^p}{2} + g(\alpha)}{(\max\{f(x), f(y)\})^p + g(\alpha)} \right)^q, \tag{1}$$

where $p, q > 0$ fixed, is a fuzzy T-metric with respect to multiplication.

The next example illustrates a special case of Example 2.

Example 3 ([26]) The mapping $\mathbf{t}_K : \mathbb{R}^+ \times \mathbb{R}^+ \to \mathbb{R}$ defined by

$$\mathbf{t}_K(x, y) = \frac{\frac{x+y}{2} + K}{\max\{x, y\} + K},$$

where $K > 0$, is a fuzzy T-metric with respect to multiplication, and

$$\mathbf{s}_K(x, y) = \frac{|x - y|}{2(\max(x, y) + K)}$$

is the fuzzy S-metric with respect to the algebraic sum, dual to T with respect to the standard fuzzy complement.

Example 4 ([26]) The mapping $\mathbf{t}_p : \mathbb{R}^+ \times \mathbb{R}^+ \to \mathbb{R}$, $p > 0$, defined by

$$\mathbf{t}_p(x, y) = \frac{\sqrt[p]{\frac{x^p + y^p}{2}}}{\max\{x, y\}}, \tag{2}$$

is a fuzzy T-metric with respect to multiplication.

Example 5 ([30]) Mapping $\mathbf{t} : \mathbb{R}^+ \times \mathbb{R}^+ \to \mathbb{R}$, $p \geqslant 1$, defined by

$$\mathbf{t}(x, y, \mathsf{K}) = \mathbf{t}_{\mathsf{K}}(x, y) = \frac{\sqrt[p]{\frac{x^p + y^p}{2}} + \mathsf{K}}{\max\{x, y\} + \mathsf{K}}, \tag{3}$$

where $\mathsf{K} > 0$, is a fuzzy T-metric with respect to multiplication.

Example 6 ([10]) Let $g : \mathbb{R}^+ \to [0, \infty]$ be an increasing continuous function. If (X, d) is a metric space then the mapping $\mathbf{t} : X \times X \times \mathbb{R}^+ \to \mathbb{R}$ defined by

$$\mathbf{t}(x, y, t) = \frac{g(t)}{g(t) + d(x, y)}, \tag{4}$$

is a fuzzy T-metric with respect to the multiplication. Especially, for $f(\alpha) = \alpha$:

$$\mathbf{t}(x, y, t) = \frac{t}{t + d(x, y)}, \tag{5}$$

and its dual (with respect to the standard fuzzy complement)

$$\mathbf{s}(x, y, t) = 1 - \mathbf{t}(x, y, t) = \frac{d(x, y)}{t + d(x, y)}$$

is a fuzzy S-metric with respect to the algebraic sum.

Remark 2 Fuzzy S-metric $\boldsymbol{s}(x, y, \_) :]0, \infty[\to [0, 1]$ is non-decreasing function, and fuzzy T-metric $\boldsymbol{t}(x, y, \_) :]0, \infty[\to [0, 1]$ is non-increasing function.

Theorem 1 *If $\boldsymbol{d} : X \times X \times]0, \infty[\to \mathbb{R}$ is a fuzzy T-metric with respect to the norm T, then polygonal inequality, for $n \geqslant 2$, hold:*

$$\mathsf{T}^{n-1}(\boldsymbol{d}(x_1, x_2, \alpha_1), \boldsymbol{d}(x_2, x_3, \alpha_2), \ldots, \boldsymbol{d}(x_n, x_{n+1}, \alpha_n))$$
$$\leqslant \boldsymbol{d}(x_1, x_{n+1}, \alpha_1 + \alpha_2 + \cdots + \alpha_n). \tag{6}$$

For the case S-metric with respect to the conorm S, for $n \geqslant 2$, hold:

$$\mathsf{S}^{n-1}(\boldsymbol{d}(x_1, x_2, \alpha_1), \boldsymbol{d}(x_2, x_3, \alpha_2), \ldots, \boldsymbol{d}(x_n, x_{n+1}, \alpha_n))$$
$$\geqslant \boldsymbol{d}(x_1, x_{n+1}, \alpha_1 + \alpha_2 + \cdots + \alpha_n). \tag{7}$$

Theorem 2 ([26]) *If (X, s, S) is a fuzzy S-metric space and the T is a t-norm dual to the t-conorm S with respect to the continuous involutive fuzzy complement c, then $(X, \mathsf{c} \circ s, \mathsf{T})$ is a fuzzy T-metric space.*

If $(X, \boldsymbol{t}, \mathsf{T})$ is a fuzzy T-metric space and S is a t-conorm dual to the norm T with respect to a continuous involutive fuzzy complement c, then $(X, \mathsf{c} \circ \boldsymbol{t}, \mathsf{S})$ is a fuzzy S-metric space.

The theorem is also valid for fuzzy metric spaces in the broader sense.

Theorem 3 ([27]) *Let $f : X \to]0, 1]$ be a one-to-one function and let $g : \mathbb{R}^+ \to (0, 1]$ be an increasing continuous function. The mapping $\boldsymbol{d} : X \times X \times]0, \infty[\to \mathbb{R}$ defined by*

$$\boldsymbol{d}(x, y, \alpha) = \begin{cases} 1, & x = y \\ \mathsf{T}(\mathsf{T}(f(x), f(y)), g(\alpha)), & x \neq y \end{cases} = \begin{cases} 1, & x = y \\ \mathsf{T}^2(f(x), f(y), g(\alpha)), & x \neq y \end{cases},$$

is a fuzzy T-metric with respect to the continuous t-norm T.

The following properties hold (see [26]) which enable the construction of new fuzzy metrics.

Theorem 4 *Let $\boldsymbol{d}$ be a stationary fuzzy metric (in the broader sense) with respect to the norm N. If N is an Archimedean norm and g its corresponding generator, then $d = g \circ \boldsymbol{d}$ is a standard metric.*

Theorem 5 *If $\boldsymbol{d}_1 : X \times X \times]0, \infty[\to \mathbb{R}$ and $\boldsymbol{d}_2 : X \times X \times]0, \infty[\to \mathbb{R}$ are fuzzy metrics, with respect to the strict norm N, then the mapping $\sigma(\boldsymbol{d}_1, \boldsymbol{d}_2) : X \times X \times]0, \infty[\to \mathbb{R}$ defined by*

$$\sigma(\boldsymbol{d}_1, \boldsymbol{d}_2)(x, y, \alpha) = \mathsf{N}(\boldsymbol{d}_1(x, y, \alpha), \boldsymbol{d}_2(x, y, \alpha))$$

is also a fuzzy metric with respect to the norm N. If N is not a strict norm, then $\sigma(\boldsymbol{d}_1, \boldsymbol{d}_2)$ is a fuzzy metric in a broader sense.

Theorem 6 *If $\boldsymbol{d}_i : X_i \times X_i \to \mathbb{R}$, $i = 1, \ldots, n$, $n \in \mathbb{N}$, are fuzzy metrics with respect to the strict norm N, then $\boldsymbol{d} : X^2 \to \mathbb{R}$, $X = X_1 \times \cdots \times X_n$, defined with*

$$\boldsymbol{d}(x, y) = \mathsf{N}^{n-1}(\boldsymbol{d}_1(x_1, y_1), \boldsymbol{d}_2(x_2, y_2), \ldots, \boldsymbol{d}_n(x_n, y_n)),$$

$$x = (x_1, \ldots, x_n), \quad y = (y_1, \ldots, y_n),$$

is a fuzzy metric with respect to the norm N. If N is not a strict norm, then $\boldsymbol{d}$ is a fuzzy metric in a broader sense.

The next example illustrates a special case of Theorem 6.

Example 7 If $\boldsymbol{t}_i : X_i \times X_i \to]0, 1]$, $i = 1, \ldots, n$, $n \in \mathbb{N}$, are fuzzy T-metrics with respect to the product, then $\boldsymbol{t} : X^2 \to]0, 1]$, $X = X_1 \times \cdots \times X_n$ defined with

$$\boldsymbol{t}(x, y) = \prod_{i=1}^{n} \boldsymbol{t}_i(x_i, y_i), \quad x = (x_1, \ldots, x_n), \quad y = (y_1, \ldots, y_n),$$

is a fuzzy T-metric with respect to the product.

Theorem 7 ([28]) *If $\boldsymbol{d}_i : X_i \times X_i \to [0, 1[$, $i \in I = \{1, \ldots, n\}$, $n \in \mathbb{N}$, are fuzzy S-metrics with respect to the conorm* max, *and $\kappa_i \in]0, 1]$, $i \in I$, then $\boldsymbol{d}$ defined with*

$$\boldsymbol{d}(x, y, \alpha) = \max\{\min\{\kappa_1, \boldsymbol{d}_1(x_1, y_1, \alpha), \ldots, \min\{\kappa_n, \boldsymbol{d}_n(x_n, y_n, \alpha)\}\}, \quad (8)$$

$$x = (x_1, \ldots, x_n),\ y = (y_1, \ldots, y_n) \in X = X_1 \times \cdots \times X_n,$$

is the fuzzy S-metric with respect to the conorm max.

The features of the fuzzy metric highlighted in this article are those that cover the construction of new metrics, with particular emphasis on the applicable ones.

There is a large number of works devoted to studying fuzzy metric spaces and their properties e.g. completeness, fixed point theorems, etc. For example conditions for the completeness of space $(X, \boldsymbol{d})$, from the previous Theorem 7, are shown in [28].

4 Image Filtering

In this section the application of fuzzy metrics in filtering color images is given. The fuzzy metrics embody a multitude of concepts which are influenced by image pixels. Each image pixel (i, J_i) ("position", "color") can be characterized by spatial coordinates of the pixel i_1, i_2 (points $i = (i_1, i_2) \in I \times I$, $I = \{0, 1, \ldots, n-1\}$ from the screen), and by the vector $J_i = (J_i^1, J_i^2, J_i^3)$. The first coordinate represents, the quantity of the color red, the second the quantity of the color green and the third the quantity of the color blue. Collectively they are referred to as RGB.

The task of image filtering is to replace the pixel which represents noise by the pixel without noise, which can be achieved by replacing the central pixel $(\bar{i}, \bar{J}_i)$ in the window $W = \{(i, J_i) | i \in I_1 \times I_2\}$, $(\bar{i} = (\bar{i}_1, \bar{i}_2) \in I_1 \times I_2$, and the size from the window is an odd number, with the pixel that represents the other pixels from W in the best possible way, i.e. by a pixel which is most similar in color and spatial distance to all the other pixels in W.

Selection bias is another potential concern because it is of significant importance to choose a good criterion for selecting such a pixel without noise, which will replace the pixel with noise in a given window W, because the choice of pixels affects the image quality, i.e. affects the degree of noise reduction.

A key issue is the safe disposal of criterion selection. Specifically, the choice of a criterion will be conditioned by a good selection of fuzzy T-metric $\boldsymbol{c}$. All pixels in the same window W will be ordered according to the order relation induced by the metric $\boldsymbol{c}$. This order relation will be used to compare the pixels (the 'position", the "color") in the image and to choose a pixel that differs the least from all the other pixels in the window, i.e. which is the most similar to all other the pixels in W (in terms of color and distance). The central pixel the given window W will be replaced by the pixel found using the algorithm which is applied on each sliding window.

In the algorithm for filtering the image, the mapping $\boldsymbol{c} : W \times W \to \mathbb{R}$ will be used, defined on the window W, with

$$\boldsymbol{c}((i, J_i), (j, J_j)) = \mathsf{T}(\boldsymbol{\partial}(J_i, J_j), \boldsymbol{d}(i, j)) \tag{9}$$

$$\left(c((i, J_i), (j, J_j)) = \mathsf{S}(\partial(J_i, J_j), d(i, j)\right), \tag{10}$$

where ∂ and d are fuzzy T-metrics (fuzzy S-metrics) with respect to a triangular norm T (a triangular conorm S). It follows from Theorem 6 that c is a fuzzy T-metric (a fuzzy S-metric).

The fuzzy metric used to measure similarity in colors between pixels is denoted by ∂. It is defined in the following way:

$$\partial(J_i, J_j) = \mathsf{T}^2(\partial_1(J_i^1, J_j^1), \partial_2(J_i^2, J_j^2), \partial_3(J_i^3, J_j^3)) \tag{11}$$

$$\left(\partial(J_i, J_j) = \mathsf{S}^2(\partial_1(J_i^1, J_j^1), \partial_2(J_i^2, J_j^2), \partial_3(J_i^3, J_j^3))\right). \tag{12}$$

The mappings ∂_1, ∂_2, and ∂_3 are fuzzy metrics, then from Theorem 6 it follows that ∂ is also a fuzzy metric with respect to the same norm.

The process of changing the pixels with noise would proceed as follows. For each pixel in the W filter window, the accumulated measure against the others in W must be calculated:

$$A_k = \sum_{(j, J_j) \in W,\ j \neq k} c((k, J_k), (j, J_j)). \tag{13}$$

The values A_k are arranged in descending order (in which the value is placed at that r-position marked with $A_{(r)}$):

$$A_{(0)} \geqslant A_{(1)} \geqslant \cdot \geqslant A_{(n^2-1)}.$$

If c is defined with (9), then the (central) pixel with noise $(\bar{i}, \overline{J}_i)$ changes with the pixel $((0), J_{(0)})$. If c is defined with (10), then the (central) pixel with noise $(\bar{i}, \overline{J}_i)$ changes with the pixel $((n^2 - 1), J_{(n^2-1)})$. In the above procedure, using parameters in fuzzy metrics can adjust the importance of the spatial criterion or color criteria, which can be seen in the test example.

4.1 *Measures of Image Quality*

4.1.1 UIQI Indices

The metric used for comparing the quality of the images was UIQI, it is defined in [33]. Using that idea we will define the generalized UIQI.

Let $\mathsf{A}_i : I \times I \to I, i = 1, 2$ be two binary aggregation functions defined on the interval $I \subseteq [0, \infty[$. Let $x = \{x_i | i = 1, 2, \cdots, n\}$ and $x^* = \left\{x_i^* | i = 1, 2, \cdots, n\right\}$ be the original and the test nonconstant image signals, whose values are from the interval $I \subseteq [0, \infty[$. With standard labels

$$\overline{x} = \frac{1}{n}\sum_{i=1}^{n} x_i, \quad \overline{x^*} = \frac{1}{n}\sum_{i=1}^{n} x_i^*,$$

$$\sigma_x^2 = \frac{1}{n-1}\sum_{i=1}^{n} (x_i - \overline{x})^2, \quad \sigma_{x^*}^2 = \frac{1}{n-1}\sum_{i=1}^{n} \left(x_i^* - \overline{x^*}\right)^2,$$

$$\sigma_{xx^*} = \frac{1}{n-1}\sum_{i=1}^{n} (x_i - \overline{x})\left(x_i^* - \overline{x^*}\right),$$

we define the *generalized UIQI index* by the formula

$$AQ = \frac{\sigma_{xx^*}}{\sigma_x \sigma_{x^*}} \cdot \mathsf{A}_1(\overline{x}, \overline{x^*}) \cdot \mathsf{A}_2(\sigma_x, \sigma_{x^*}), \tag{14}$$

Of particular concern is the first component, which is the correlation coefficient between x and x^*. Then, the second component shows how close the pixel brightness values of x and x^* are. Finally, the third component measures how similar the contrasts between x and x^* are. The particular metric of quality generalized UIQI is based on the fact that every image distortion is viewed as a combination of three factors: the loss of correlation, luminance distortion and contrast distortion. The generalized UIQI index can take values from the interval $[-1, 1]$. The closer it is to 1, the better the image quality, and that is achieved when dependence between the signals is linear.

If we assume that $\mathsf{A}_1, \mathsf{A}_2$ are harmonic means, we get the image quality index UIQI from [33], which is defined by the following formula:

$$Q = \frac{4\sigma_{xx^*}\overline{x}\,\overline{x^*}}{\left(\sigma_x^2 + \sigma_{x^*}^2\right)\left[(\overline{x})^2 + (\overline{x^*})^2\right]}. \tag{15}$$

4.1.2 CPBD Metric

One of the ways to check image quality is the image sharpness [24]. It is based on the cumulative probability of image blur detection (CPBD). We present a discretized version of the CPBD metric. The computation of the CPBD sharpness metric is performed through the following steps: The image is first divided into 64×64 blocks and then each block is characterized as an edge block or a non-edge block. The non-edge blocks are not processed further, whereas, for each edge block, the width of each edge in the block is determined. The probability of blur detection at each edge is estimated using:

$$P_{BLUR} = \mathbb{P}(e_i) = 1 - \exp\left(-\left|\frac{w(e_i)}{w_{JNB}(e_i)}\right|\right), \tag{16}$$

where $w_{JNB}(e_i)$ is the JNB edge width which depends on the local contrast C (of the edge block to which the edge belongs) and $w(e_i)$ is the measured width of the edge e_i. It should be noted that when $w(e_i) = w_{JNB}(e_i)$ then $P_{BLUR} = 63\% = P_{JNB}$. It follows that the blur is not detected at an edge if $P_{BLUR} \leq P_{JNB}$. Finally, the *cumulative probability of blur detection* (CPBD) is calculated as:

$$\text{CPBD} = \mathbb{P}(P_{BLUR} \leqslant P_{JNB}) = \sum_{P_{BLUR}=0}^{P_{JNB}} \mathbb{P}(P_{BLUR}) \tag{17}$$

where $\mathbb{P}(p_{BLUR})$ denotes the value of the probability distribution function at a given p_{BLUR}.

It has been shown [24] that the proposed metrics are able to predict the quality of tested images well. Compared to existing sharpness metrics, the proposed CPBD-based metrics are found to correlate significantly better with subjective scores.

4.2 Experiments

The process of removing noise from an image using fuzzy metrics, was tested on the image "Coral" given in the jpg format. "Its size is 256 × 171 pixels. The filtered image given below is contaminated with 10% salt and pepper noise. The values of UIQI are calculated for each of the three colors with 10% noise. In the following example, in which the image quality measure UIQI is calculated, the window of size 5 is chosen.

In the procedure, we will use the fuzzy T-metric with respect to the t-norm $\cdot$.

The metrics ∂_k, $k = 1, 2, 3$ in (12) are fuzzy T-metric (3) with parameter $p = 1$. Accordingly, the metric ∂ from (9) is:

$$\partial(J_i, J_j) = \partial_1(J_i^1, J_j^1) \cdot \partial_2(J_i^2, J_j^2) \cdot \partial_3(J_i^3, J_j^3) = \prod_{l=1}^{3} \frac{\frac{J_i^l + J_j^l}{2} + K}{\max\{J_i^l, J_j^l\} + K}.$$

Fuzzy T-metric that considers spatial distance between pixels, which is marked with $\boldsymbol{d}$, is a metric (Figs. 2 and 3):

$$\boldsymbol{d}(i, j) = \frac{t}{t + |i_1 - j_1| + |i_2 - j_2|}. \tag{18}$$

The values of UIQI for each color of the image filtered by our filter (with the parameters of used fuzzy metric: $K = 1125$, $t = 0.1$) are equal to:

$$\text{UIQI}: \quad (0.549591726350782, 0.530464341924132, 0.515141897245744).$$

Fig. 1 The original image Coral, dimension 256 × 171, format jpg

Fig. 2 The image Coral with noise, 10% salt and pepper

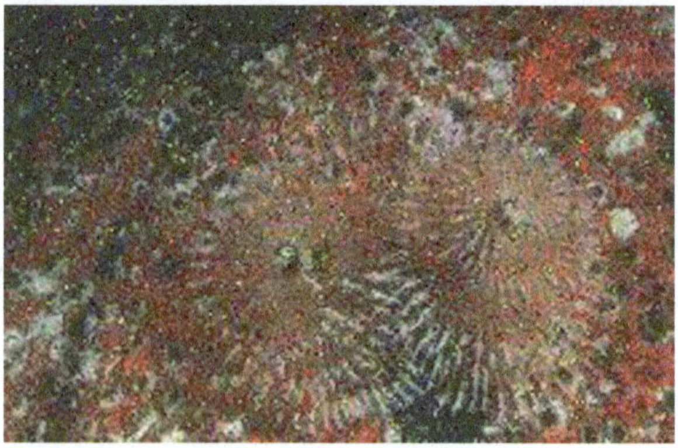

Fig. 3 The filtered image Coral, fuzzy metrics with parameters $K = 1125$, $t = 0.1$

Fig. 4 The filtered image Coral, fuzzy metrics with parameters $K = 1250$, $t = 0.7$

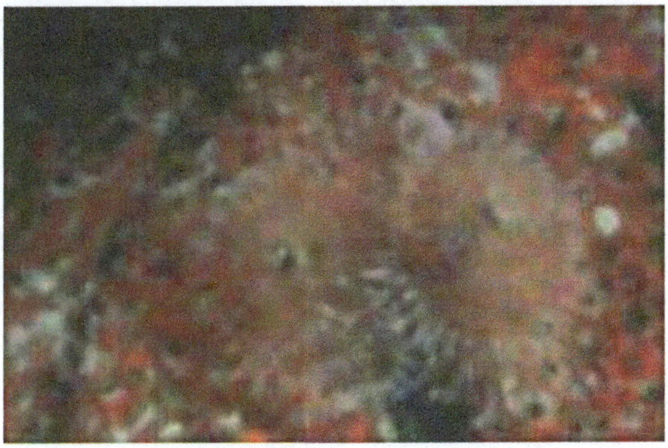

Fig. 5 The filtered image Coral by median filter

The sharpness for the image filtered by our filter is 0.9843 (Fig. 4).

The values of UIQI for each color of the image filtered by our filter (with the parameters of used fuzzy metric: $K = 1125$, $t = 0.1$) are equal to: $K = 1250$, $t = 0.7$) are equal to:

$$\text{UIQI}: \quad (0.548409569519015, 0.529808666671712, 0.514841313470943).$$

The sharpness for the image filtered by our filter is 0.9876 (Fig. 5).

The values of UIQI for each color of the image filtered by a median filter (VMF) with:

$$\text{UIQI}: \quad (0.49820127717114, 0.459097282441571, 0.437852787277136).$$

The sharpness for the image filtered by VMF is 0.5406.

By comparison the values of UIQI for the corresponding colors (red, green and blue, respectively), it can be concluded that all indices of the images filtered by the method tested in this paper are greater than the corresponding indices of the images filtered by the median filter [2]. As those indices are closer to one, it can be concluded that the image quality is better. Also, images filtered using this method have higher sharpness.

For additional literature on fuzzy filtering, author recommend the following list: [9, 21–23, 32].

5 Copy-Move Forgery Detection in Images

In practice, examples of digital image falsification are often found as a result of parts of an original image being copied and then pasted into another segment of the same image. The purpose of such changes is to duplicate the content of the image or hide some image content. Copy-move forgery detection (CMFD) is one of the methods used to detect such forgeries in images. This section contains a clustering algorithm for the edited images proposed in [18, 30].

A picture is divided into non-overlapping blocks. Fuzzy metrics are used for calculating the distance between blocks. Metaheuristic search methods such as the Variable Neighborhood Search (VNS) and Bee Colony Optimization (BCO) are used to classify blocks. The goal of classification is to divide the picture into modified and original blocks. It is suggested that the algorithm is implemented in the Python programming language. An illustration of the algorithm was made on the modified image Coral.jpg.

5.1 Metaheuristics

The metaheuristics Variable neighborhood search (VNS) [19, 20] is based on the following facts:

- the local minimum in relation to one environment, and does not have to be a local minimum with respect to another environment; the global minimum is the local minimum with respect to all environments;
- for most problems, local minima in different environments are close to each other.

In this chapter, due to the problem of detecting changes in the image, we applied: BVNS (The basic method of Variable neighborhood search) and RVNS (The reduced Variable neighborhood search method).

```
The pseudocode BVNS\\
 Generate initial solution x;\\
  do {
  k=1;
  while(k <= k_max) {
    x'=Shaking (x, k);
    x''=LocalSearch (x');
    if(f(x'') < f(x)) {
      x=x'';
      k=1;
    }
    else
      k=k+1;
  }
} while(!STOP);
```

```
The pseudocode RVNS\\
Generate initial solution x;
 x_opt = x; f_opt = f(x);
 do{
  k=1;
  do{
    x'=Shaking (x, k);
    if(f(x')<f(x_optimal)){
      x_opt=x';
      f(x_opt)=f(x');
      k=1;
    }
    else
      k=k+1;
  }while(k!=kmax);
} while(!STOP);
```

The basic variable neighbourhood search (BVNS) is the most widespread variant of the environment variable method because it provides more preconditions for obtaining better solutions. For the basic VNS method, the basic steps are contained in a loop in which we change the index of the environment k, then determine the random solution from that environment (function Shaking()), perform the local search function (function LocalSearch()) and check the quality of the local minimum obtained. We repeat these steps until one of the stopping criteria is satisfied. Each time we select an environment, k the initial solutions are randomly generated to ensure that different regions are searched the next time they are deployed. We distinguish the environments by the number of transformations (distances) or by the type of transformations (metrics).

The reduced variable neighbourhood search method does not contain a local search phase. It is a method based solely on staging, and involves systematically changing the environment and selecting one random solution in each environment. The decision-making steps are based on that one random solution. Its advantage is the speed of the execution as it avoids a detailed search, often of extremely large environments. This method is useful the for large-scale examples or for obtaining initial high-quality solutions for another variant of the VHS method. The usual stop criterion (STOP) is the maximum number of iterations between the two enhancements [5].

RVNS and BVNS were applied to the problem of clustering image blocks as follows. First, the distances between the blocks are calculated by applying formula (3). In the pre-processing phase, the types of distance matrices (as well as the corresponding block indices) were sorted in non-descending order. This data was used for a more efficient implementation of the stirring operator. Namely, as each solution of the character is a set of centroids (the centroid represents a block with its index, where the index represents the block label), the shuffling operator involves replacing the appropriate number of centroids.

More precisely, the shifting in the environment k implies that the centroids are replaced by randomly selected objects that were not centroids, and they are the furthest away from the centroids when they change. In each step, the replacement of all centroids is considered, and in the event of replacement, the object closest to that centroid will not be chosen by chance (it created that centroid). The local search consists of the systematic replacement of a single centroid of an object other than a centroid. It starts from the solution of the obtained stirring and is performed according to the principle of best improvement as long as there are improvements.

5.1.1 BCO

The Bee Colony Optimization (BCO) is a metaheuristic developed to solve complex combinatorial optimization problems. Lučić and Teodorović were among the first to have used the basic principles of collective bee intelligence in dealing with combinatorial optimization problems (see [16]). The basic idea behind BCO is to build a multi agent system (a colony of B artificial bees) that will search for good solutions combinatorial optimization problems, exploring the principles used by honey bees during the nectar collection process. We implemented an improved version of the BCO algorithm, denoted by the BCO in [4]. This means that each bee is assigned a complete feasible solution of the clustering problem and performs some transformations that should improve its quality. In order to improve the efficiency of the proposed BCO, we introduced a preprocessing phase that starts with calculating the distances between the objects using the metrics. In this way, a distance matrix is obtained along with the values of all entries. The next step of the pre-processing phase is to sort the distance matrix (as well as the corresponding objects' indices) into non-decreasing order. This means that in the rows of the sorted matrix, the position of objects correspond to their distance from the object marked by the row index. Each iteration of BCO starts with the initialization of the population consisting of B bees by some randomly selected solutions. This means that K random objects are selected to be initial centroids. In order to obtain the corresponding value of the objective function, the K-means heuristic (see [12]) is applied to each of the initial solutions. During the forward pass, a transformation of the current solution held by each bee is performed. This transformation consists of replacing the current centroids with some other objects (blocks). Namely, for each of the centroids a new object is selected randomly with a probability that decreases as the distance between the centroid and the new object increases. It is possible to select 0 as an index of the new centroid meaning that this centroid will not be replaced in the transformed solution. The procedure to select a new object consists of the following steps. First, a random number $\text{rand}(1, m)$ between 1 and the number of objects m is selected. Next, we select $k_1 = \text{rand}(0, k)$. If $k_1 = 0$, the corresponding centroid will not be changed. The probability to select an object k', $k' = 1, \ldots, k_1$ is calculated as $p(k') = 1 - k'/k_1$. This means that the object closest to the current centroid has the highest probability of being selected, while the selection probability for an object at distance $k_1 - 1$ is the lowest. The object to replace the current centroid is selected using a roulette wheel. This procedure is repeated for all K centroids and then the transformation of the current solution is completed. The value of the objective function is again determined by the application of K-means heuristic. After transforming all the B solutions, the BCO performs a backward pass that is implemented in a standard way (see [4]).

5.2 The Experiment

We used metaheuristics, namely the VNS (BVNS and RVNS) method, which is based on the fuzzy metric proposed in the section above. The size of the blocks, as far as the algorithm is concerned, can be resized, and we used at 8×8 and 16×16 pixels. We clustered them into two groups and analyzed the validity and the quality of the clustering using false positive FP (False Positive) and false negative FN (False Negative) detected blocks. When an image was loaded in RGB format, it was divided into non-overlapping blocks, squared shapes, fixed dimensions, $n \times n$, where n is 8 or 16. The block division is usually done to reduce the computing time and the complexity required for the pixel matching process. that is, blocks. For testing purposes, all of the images were of 256×256 resolution, which does not affect the generality, but allows for greater transparency of the method. The method can be applied to the images with arbitrary resolutions, say $2n \times 2n$. Given the resolution of 256×256, each image is divided into 128 dimensions 16×16 or 256 dimensions 8×8. The fuzzy distance from (3) is then calculated and for $p = 1$ and $p = 2$. The distances are sorted and VNS is applied. The RVNS and BVNS applied to the image block clustering problem are implemented as follows [5]. First, the distances between the blocks were calculated using the formula in (3). In the preprocessing stage, the types of the distance matrices were sorted into descending order. Those data were used to increase the efficiency of the deployment operator. Specifically, as each solution is characterized by a set of centroids, the deployment operator consists of replacing the corresponding number of centroids. More specifically, a k-environment in k means that the centroids (block) are replaced by randomly selected non-centroid objects. They are mostly k places a change. Each step considers the replacement of all centroids, with no replacement occurring if the randomly selected object is closest to that centroid (in fact, it is the centroid itself). The local search consists of systematically replacing one centroid with a non-centroid object. It starts from the solution obtained by rolling and is executed on the principle of best improvement as long as there are improvements [7].

The performance of the proposed methods is most often measured in terms of precision and revocation (recall). The precision indicates the probability that the blocks which have been changed, have really been detected. The revocation indicates the probability (possibility) of detecting the altered blocks in an image. The true positive (TP) is the number of blocks that have been modified, which have been classified as modified. The false positive (FP) represents the number of original (authentic) blocks that have been classified as modified, while the false negative (FN) represents the number of blocks that have been modified but classified as original (authentic):

Fig. 6 Image tested

$$\text{Precision} = \text{TP}/(\text{TP} + \text{FP}) \tag{19}$$

$$\text{Revocation} = \text{TP}/(\text{TP} + \text{FN}) \tag{20}$$

Specifically, the precision is calculated as the quotient of the number of modified blocks that are classified as modified, and sums the number of modified blocks that are classified as the modified and the number of original blocks that are classified as modified. On the other hand, the recall is calculated as a quotient of the number of modified blocks that are classified as modified, and the sum of the number of modified blocks that are classified as modified and the number of modified blocks that are classified as original.

For the fuzzy metric in our method for CMFD, we will take (3) with parameter $K = 1$.

Let us apply the method to the given Fig. 6 with division into 8×8 blocks. For the metric parameter $p = 1$ and VNS metaheuristics we get Precision = 100% and Revocation = 98.99% while for BCO metaheuristics, Precision = 100% and Revocation = 99.05%. For the metric parameter $p = 2$ and VNS metaheuristics we get Precision = 100% and Revocation = 98.99% while for BCO metaheuristics, Precision = 100% and Revocation = 99.00%.

Let us apply the method to the given Fig. 6 with division into 16×16 blocks. For the metric parameter $p = 1$ and VNS metaheuristics we get Precision = 100% and Revocation = 99.60% while for BCO metaheuristics, Precision = 100% and Revocation = 100%. For the metric parameter $p = 2$ and VNS metaheuristics we get Precision = 100% and Revocation = 100% while for BCO metaheuristics, Precision = 100% and Revocation = 100%.

These results show a remarkable characteristic of this algorithm. As the considered image has not been tested by other authors (see for example [1]) we cannot make a comparison. The superiority of our method, when it comes to precision and recall, as seen in paper [18] is shown in almost all pictures. The dependency of precision and recall on p, as well on dividing the image into blocks of different dimensions, can be seen in [18].

6 Conclusion

The fuzzy metric defined in this paper provides a wide range of possibilities for further application in image processing. Depending on the type of the problem, i.e., on the expected performance, the appropriate t-norm or t-conorm can be chosen to contribute to achieve desired performance. In this paper, a type of fuzzy metric was applied when constructing an image filtering algorithm. We concluded that for parameters in a certain range, using the UIQI image quality metric, the quality of the image filtered by the operation from this paper is better than the image quality obtained by filtering with the median filter.

Further work on testing the quality of the filtered image consists, among other things, of using the generalized UIQI index introduced here and comparing it with other filters. Different types of noise will be added to the considered images. Testing will be performed on a large number of medical images using machine learning methods.

In the future, we plan to use the T-metric to filter the image, replace it with the appropriate fuzzy S-metric, and select the appropriate parameter domains to obtain the adequate filtering results. It is also possible to construct new fuzzy metrics by applying aggregation operators on existing fuzzy metric and to apply them in image filtering, copy-move forgery detection image in images or image copy-move forgery detection ...) and image segmentation as in the article [25, 29].

Acknowledgements The author has been supported by the Ministry of Education, Science and Technological Development through the project no. 451-03-68/2020-14/200156: "Innovative scientific and artistic research from the FTS (activity) domain".

References

1. Alkawaz, M.H., Sulong, G., Saba, T., Rehman, A.: Detection of copymove image forgery based on discrete cosine transform. Neural Comput. Appl. 1–10 (2016). https://doi.org/10.1007/s00521-016-2663-3
2. Astola, J., Haavisto, P., Neuvo, Y.: Vector median filters. Proc. IEEE **78**(4), 678–689 (1990). https://doi.org/10.1109/554807
3. Bloch, I.: On fuzzy distances and their use in image processing under imprecision. Pattern Recognit. Lett. **32**, 1873–1895 (1999)
4. Davidović, T., Teodorović, D., Šelmić, M.: Bee colony optimization part i: the algorithm overview. Yugoslav J. Oper. Res. **25**(1), 33–56 (2015). https://doi.org/10.2298/YJOR131011017D
5. Davidović, T., Glišović, N., Rašković, M.: Bee colony optimization for clustering incomplete data. In: The 7th International Conference on Optimization Problems and Their Applications OPTA-2018, pp. 8–14 (2018)

6. Deza, M.M., Deza, E.: Encyclopedia of Distances. Springer, Berlin (2009)
7. Glišović, N., Davidović, T., Rašković, M.: Clustering when data are missing using the environment variable method. In: XLIV Symposium on Operations Research SYMOPIS, pp. 158–165, Zlatibor, September 25–28, ISBN 978-86-7488- 135-4 (2017)
8. Grabisch, M., Marichal, J.L., Mesiar, R., Pap, E.: Aggregation functions. Cambridge University Press, Cambridge (2009)
9. Gregori, V., Morillas, S., Roig, B., Sapena, A.: Fuzzy averaging filter for impulse noise reduction in colour images with a correction step. J. Vis. Commun. Image Represent. **55**, 518–528 (2018). https://doi.org/10.1016/j.jvcir.2018.06.025
10. Gregori, V., Morillas, S., Sapena, A.: Examples of fuzzy metrics and applications. Fuzzy Sets Syst. **170**(1), 95–111 (2011). https://doi.org/10.1018/f.fss.2010.10.019
11. Gregori, V., Romaguera, S.: Some properties of fuzzy metric spaces. Fuzzy Sets Syst. **115**(3), 485–489 (2000). https://doi.org/10.1016/S0165-0114(98)00281-4
12. Jain, A.K.: Data clustering: 50 years beyond $K-$means. Pattern Recognit. Lett. **31**(8), 651–666 (2010). https://doi.org/10.1016/j.patrec.2009.09.011
13. Karaklić, D., Gajić, Lj., Ralević, N.M.: Some fixed point results in a strong probabilistic metric spaces. Filomat **33**(8), 2201–2209 (2019). https://doi.org/10.2298/FIL1908201K
14. Klement, E.P., Mesiar, R., Pap, E.: Triangular Norms. Kluwer Academic Publishers, Dordrecht (2000)
15. Klir, G.J., Yuan, B.: Fuzzy Sets and Fuzzy Logic: Theory and Applications. Prentice Hall, New Jersey (1995)
16. Lučić, P., Teodorović, D.: Bee system: modeling combinatorial optimization transportation engineering problems by swarm intelligence. In: Preprints 340 of the TRISTAN IV Triennial Symposium on Transportation Analysis, Sao Miguel, Azores Islands, Portugal, pp. 441–445 (2001)
17. Martin, D., Fowlkes, C., Tal, D., Malik, J.: A database of human segmented natural images and its application to evaluating segmentation algorithms and measuring ecological statistics. In: Proceedings eighth IEEE international conference on computer vision. ICCV 2001, pp. 416–423 (2001)
18. Milosavljević, N.S., Ralević, N. M.: Fuzzy Metaheuristics Algorithm for Copy-Move Forgery Detection in Images. 20th International Workshop on Combinatorial Image Analysis, Novi Sad, Serbia, 16–18 Jul, 2020, Springer Nature Switzerland AG 2020 T. Lukić et al. (Eds.): IWCIA 2020, LNCS 12148, pp. 273–281 (2020). https://doi.org/10.1007/978-3-030-51002-2_20
19. Mladenović, N., Hansen, P.: Variable neighborhood searchs. Comput. Oper. Res. **24**(11), 1097–1100 (1997)
20. Mladenović, N., Brimberg, J., Hansen, P., Moreno-Perez, J.: The $p-$median problem: a survey of metaheuristic approaches. Eur. J. Oper. Res. **179**(3), 927–939 (2007). https://doi.org/10.1016/j.ejor.2005.05.034
21. Morillas, S., Gregori, V., Peris-Fajarnes, G., Latorre, P.: A fast impulsive noise color image filter using fuzzy metrics. Real-Time Imaging **11**(5–6), 417–428 (2005). https://doi.org/10.1016/j.rti.2005.06.007
22. Morillas, S., Gregori, V., Peris-Fajarnes, G., Latorre, P.: A new vector median filter based on fuzzy metrics. In: Kamel, M., Campilho, A. (Eds.). Image Analysis and Recognition - ICIAR2005, Lecture Notes in Computer Science, vol. 3656, pp. 81–90. Springer, Berlin (2005). https://doi.org/10.1117/1.2767335
23. Morillas, S., Gregori, V., Peris-Fajarnes, G., Sapena, A.: New adaptive vector filter using fuzzy metrics. J. Electron. Imaging **16**(3), 033,007:1–15 (2007). https://doi.org/10.1116/1.2767335
24. Narvekar, N.D., Karam, L.J.: An improved no-reference sharpness metric based on the probability of blur detection. In: Conference Proceedings 2009 International Workshop on Video Processing and Quality Metrics or Consumer Electronics (VPQM) (2010)
25. Nedovic, L., Ralevic, N., Pavkov, I.: Aggregated distance functions and their application in image processing. Soft Comput. **22**(14), 4723–4739 (2018). https://doi.org/10.1007/s00500-017-2657-9

26. Ralević, N.M., Karaklić, D., Pištinjat, N.: Fuzzy metric and its applications in removing the image noise. Soft Comput. **23**(22), 12049–12061 (2019). https://doi.org/10.1007/s00500-019-03762-5
27. Ralević, N.M., Paunović, M., Iričanin, B.: Fuzzy metric spaces and applications in imahe processing. Math. Montisnigri **48**, 103–117 (2020). https://doi.org/10.20948/mathmontis-2020-48-9
28. Ralević, N.M., Gajić, Lj.: Max-min combination fuzzy $S-$metrics. In: The Proceedings of the Fifth Conference on Mathematics in Engineering: Theory and Applications. META: Faculty of Technical Sciences, May 9–10th, 2020. Novi Sad, Serbia (2020)
29. Ralević, N., Paunović, M.: Applications of the Fuzzy Metrics in Image Denoising and Segmentation. Technical Gazette **28**(3) (2021). https://doi.org/10.17559/TV-20200305075136
30. Ralević, N. M., Milosavljević, N.S.: Fuzzy metric and its applications in Copy-move forgery detection in image. Iranian journal of fuzzy systems (submitted)
31. Schweizer, B., Sklar, A.: Probabilistic Metric Spaces. Elsevier North-Holland, New York (1983)
32. Smolka, B., Szczepanski, M., Plataniotis, K.N., Venetsanopoulos, A.N.: On the fast modified vector median filter. Can. Conf. Electr. Comput. Eng. **2**(2), 1315–1320 (2001). https://doi.org/10.1109/CCECE.2001.93.23636
33. Wang, Z., Bovik, A.C.: A universal image quality index. IEEE Signal Proc. Lett. **9**(3), 81–84 (2002). https://doi.org/10.1109/97.995823

Aggregation Operators and Distributivity Equations

Dragan Jočić and Ivana Štajner-Papuga

Abstract Aggregation operators are an important mathematical tool in a number of areas and disciplines of both pure and applied mathematics. For both theoretical and practical reasons, aggregation operators with an annihilator and aggregation operators with a neutral element are of special interest for researchers. The issue of distributivity of aggregation operators is crucial for many different areas such as decision making theory and integration theory. This chapter covers the characterization of all pairs (F, G) of aggregation operators that satisfy distributivity law, on both whole and restricted domains, where F is a T-uninorm in $\mathcal{U}_{\max}$ or a nullnorm with the annihilator $a \in]0, 1[$, and G is a t-conorm or a uninorm from the classes $\mathcal{U}_{\min}$ or $\mathcal{U}_{\max}$.

Keywords Aggregation operator · Triangular norm · Triangular conorm · Uninorm · Nullnorm · Distributivity equation · Conditional distributivity

1 Introduction

The applicability of aggregation operators is recognized in many research areas, both theoretical and practical, from natural sciences and mathematics to economics and social sciences. Of special interest is the study of additional properties of aggregation operators, that are usually derived from the solutions of functional equations involving these kind of functions. Some of those properties include idempotency, associativity, modularity, migrativity, distributivity, etc. The investigation of dis-

D. Jočić
Mathematical Institute of the Serbian Academy of Sciences and Arts, Kneza Mihaila 36, 11000 Belgrade, Serbia
e-mail: jocicndragan@gmail.com

I. Štajner-Papuga (✉)
Department of Mathematics and Informatics, Faculty of Sciences, University of Novi Sad, Trg D. Obradovića 4, 21000 Novi Sad, Serbia
e-mail: ivana.stajner-papuga@dmi.uns.ac.rs

E. Pap (ed.), *Artificial Intelligence: Theory and Applications*,
Studies in Computational Intelligence 973,
https://doi.org/10.1007/978-3-030-72711-6_7

tributivity comes from [1] and, lately, the focus is on t-norms and t-conorms [16], as well as on their generalizations such as uninorms, nullnorms, semi-nullnorms, semi-uninorms, semi-t-operators, uni-nullnorms, 2-uninorms, etc. (see [2, 7, 9, 13, 19, 20, 23, 25, 26, 29]). Additionally, a number of research is being conducted on conditional or restricted distributivity. This type of distributivity is observed under additional conditions imposed on the domain of aggregation operators and it ensures a wider variety of solutions due to the elimination of rigid points for the domain ([6, 8, 11, 12, 15–18, 24, 27, 28]). The significance of this topic is due not only to theoretical considerations, but also due to its applicability in integration theory [16, 24] and in utility theory [3, 5, 10].

This chapter gives an overview of the results for both research directions. The first part is based on [12, 19], and it is focused on distributivity law on the whole domain. The second part is based on [8, 12], where the distributivity equation on a restricted domain was investigated. Therefore, the main concern of this chapter is to present solutions of functional equations

$$F(x, G(y, z)) = G(F(x, y), F(x, z)), \quad x, y, z \in [0, 1],$$

and

$$F(x, G(y, z)) = G(F(x, y), F(x, z)), \quad x, y, z \in [0, 1], \quad G(y, z) < 1,$$

where F is a T-uninorm in $U_{\max}$ or a nullnorm with an annihilator $a \in]0, 1[$, and G is a t-conorm or a uninorm from the class $\mathsf{U}_{\min} \cup U_{\max}$. The motivation for this line of investigation lies in the possibility of obtaining new pairs of aggregation operators that can be further applied in utility theory for modelling some specific problems, e.g. problems that deals with the existence of a veto.

Some preliminary notions concerning aggregation operators, uninorms, aggregation operators with an annihilator, and both types of distributivity equations are given in Sect. 2. The results on distributivity laws on the whole domain are presented in the third section. Distributivity on a restricted domain, i.e., conditional distributivity, is the topic of the fourth section. Some concluding remarks are given in the fifth section.

2 Preliminaries

An overview of the notions relevant for the research that is presented in this chapter is given in this section (see [4, 5, 16, 21, 22]).

Definition 1 ([5]) An aggregation operator in $[0, 1]^n$ is a function $\mathsf{A}^{(n)} : [0, 1]^n \to [0, 1]$ that is nondecreasing in each variable and that fulfills the boundary conditions

$$\mathsf{A}^{(n)}(0, \dots, 0) = 0 \quad \text{and} \quad \mathsf{A}^{(n)}(1, \dots, 1) = 1.$$

The previous definition can be extended to an arbitrary real interval, not necessarily closed. As expected, the integer n represents the number of input values of the aggregation in question. Since the research presented in this chapter focuses on binary aggregation operators, hereinafter the simple notation A will be used instead of $\mathsf{A}^{(2)}$.

2.1 Uninorms

Definition 2 ([30]) A uninorm $\mathsf{U} : [0, 1]^2 \to [0, 1]$ is a binary aggregation operator that is commutative, associative, and for which there exists a neutral element $e \in [0, 1]$, i.e., $\mathsf{U}(x, e) = x$ for all $x \in [0, 1]$.

Recall that for $e = 1$ uninorm U becomes a t-norm denoted by T, for $e = 0$, U is a t-conorm denoted by S, and the following holds for $e \in]0, 1[$.

Theorem 1 ([4]) *Let* U *be a uninorm with a neutral element* $e \in]0, 1[$. *Then there exists a t-norm* T_U, *a t-conorm* S_U, *and an increasing commutative operator* $C : [0, e[\times]e, 1]\cup]e, 1] \times [0, e[\to [0, 1]$ *that fulfils* $\min \leqslant C \leqslant \max$, *such that* U *is given by*

$$\mathsf{U}(x, y) = \begin{cases} e\mathsf{T}_\mathsf{U}\left(\frac{x}{e}, \frac{y}{e}\right) & \text{if } (x, y) \in [0, e]^2, \\ e + (1 - e)\mathsf{S}_\mathsf{U}\left(\frac{x-e}{1-e}, \frac{y-e}{1-e}\right) & \text{if } (x, y) \in [e, 1]^2, \\ C(x, y) & \text{otherwise.} \end{cases} \quad (1)$$

The t-norm T_U and t-conorm S_U from the previous theorem are called the underlying t-norm and the underlying t-conorm of U, respectively.

For all uninorms it holds $\mathsf{U}(0, 1) \in \{0, 1\}$. If $\mathsf{U}(0, 1) = 0$, the uninorm U is conjunctive and, if $\mathsf{U}(0, 1) = 1$ the uninorm U is called disjunctive. Unfortunately, the previous result provides only the necessary condition for the representation of uninorms. For the given t-norm T_U, t-conorm S_U and the neutral element $e \in]0, 1[$ determining the corresponding uninorm U is equivalent to finding an operator C of such form that (1) holds. In general, it is still an open problem. Under additional assumptions this problem is solved, completely or partially, and thus some well known classes of uninorms are obtained (see [4, 21]). Some of those classes are:

- Uninorms in $\mathsf{U}_{\min}$ ($\mathsf{U}_{\max}$), i.e., uninorms given by minimum (maximum) in $I(e) = [0, e[\times]e, 1]\cup]e, 1] \times [0, e[$.
- Representable uninorms, i.e., uninorms that have an additive generator.
- Uninorms continuous in the open unit square $]0, 1[^2$.
- Uninorms with continuous underlying t-norms and t-conorms.
- Idempotent uninorms, i.e., uninorms that satisfy $\mathsf{U}(x, x) = x$ for all $x \in [0, 1]$.
- Locally internal uninorms, i.e., uninorms such that $\mathsf{U}(x, y) \in \{x, y\}$ for all $(x, y) \in I(e)$.

Example 1 The first uninorms considered by Yager and Rybalov [30] are idempotent uninorms U_e^{min} and U_e^{max} from the classes $\mathsf{U}_{\min}$, and $\mathsf{U}_{\max}$, respectively, of the following form

$$\mathsf{U}_e^{\min} = \begin{cases} \max \text{ on } [e, 1]^2, \\ \min \text{ otherwise}, \end{cases} \tag{2}$$

and

$$\mathsf{U}_e^{max} = \begin{cases} \min \text{ on } [0, e]^2, \\ \max \text{ otherwise}. \end{cases} \tag{3}$$

The only idempotent t-norm and t-conorm are the operators minimum and maximum, respectively.

2.2 Associative a-CAOA

Additional properties, such as the existence of an annihilator, are often required from aggregation operators in order to successfully model particular real life problems. An element $a \in [0, 1]$ is an annihilator of aggregation operator A if $\mathsf{A}(a, x) = \mathsf{A}(x, a) = a$ for all $x \in [0, 1]$. The research presented here is focused on commutative aggregation operators with an annihilator a, denoted with a-CAOA, that are studied in [22]. Further on, for any binary operator $\mathsf{A} : [0, 1]^2 \to [0, 1]$, and any element $c \in [0, 1]$, the section $\mathsf{A}_c : [0, 1] \to [0, 1]$ of the form $\mathsf{A}_c(x) = \mathsf{A}(c, x)$ will be denoted by A_c. The continuity (discontinuity) of sections A_0 and A_1 has a key role in the classification of the associative a-CAOA operators.

Definition 3 ([22]) Let $\mathsf{A} : [0, 1]^2 \to [0, 1]$ be an associative a-CAOA.

- A is called S-uninorm if:
  - A_0 is continuous, and A_1 is not;
  - there is an IFC element, i.e., there is an idempotent element $e \in]0, 1[$ for A, such that A_e is continuous and $\mathsf{A}_e(1) = 1$.
- A is called T-uninorm if:
  - A_1 is continuous, and A_0 is not;
  - there is an IFC element, i.e., there is an idempotent element $e \in]0, 1[$ for A, A_e is continuous and $\mathsf{A}_e(0) = 0$.
- A is called bi-uninorm if:
  - A_0 and A_1 are not continuous;
  - there are IFC elements, i.e., there are idempotent elements $e_0, e_1 \in]0, 1[$ for A, such that A_{e_0} and A_{e_1} are continuous, $\mathsf{A}_{e_0}(0) = 0$, and $\mathsf{A}_{e_1}(1) = 1$.
- A is called nullnorm (t-operator) if:

– A_0 and A_1 are continuous.

As expected, T-uninorms are dual to S-uninorms, i.e., A is a T-uninorm if and only if A^d is an S-uninorm, where A^d is given by

$$\mathsf{A}^d(x, y) = 1 - \mathsf{A}(1 - x, 1 - y).$$

On the other hand, the class of bi-uninorms, as well as the classes of uninorms and nullnorms, are closed under duality.

The form of a T-uninorm is based on a t-norm T and on a uninorm, while an S-uninorm consists of a t-conorm S and a uninorm, hence the names of the operators in question. Since the focus of this chapter is on T-uninorms and nullnorms, an overview of the results from [22] for those two classes of associative a-CAOA follows. Further on let $I_a = [0, a] \times [a, 1] \cup [a, 1] \times [0, a]$.

- A binary aggregation operator $\mathsf{A} : [0, 1]^2 \to [0, 1]$ is a T-uninorm if and only if there exists $a \in]0, 1]$, a t-norm T, and a disjunctive uninorm U' with the neutral element $e' \in]0, 1[$ such that

$$\mathsf{A}(x, y) = \begin{cases} a\mathsf{U}'\left(\frac{x}{a}, \frac{y}{a}\right) & \text{if } (x, y) \in [0, a]^2, \\ a + (1 - a)\mathsf{T}\left(\frac{x-a}{1-a}, \frac{y-a}{1-a}\right) & \text{if } (x, y) \in [a, 1]^2, \\ a & \text{if } (x, y) \in I_a. \end{cases} \tag{4}$$

  If e is the IFC element of a T-uninorm, then $e' = \frac{e}{a}$. The IFC element e is the neutral element for a T-uninorm on $[0, a]^2$.
- A binary aggregation operator $\mathsf{A} : [0, 1]^2 \to [0, 1]$ is a nullnorm with an annihilator $a \in]0, 1[$ if and only if there exists a t-norm T and a t-conorm S such that A is given by

$$\mathsf{A}(x, y) = \begin{cases} a\mathsf{S}\left(\frac{x}{a}, \frac{y}{a}\right) & \text{if } (x, y) \in [0, a]^2, \\ a + (1 - a)T\left(\frac{x-a}{1-a}, \frac{y-a}{1-a}\right) & \text{if } (x, y) \in [a, 1]^2, \\ a & \text{if } (x, y) \in I_a. \end{cases} \tag{5}$$

  For $a = 0$, the nullnorm is a t-norm, while for $a = 1$ it is a t-conorm.

Remark 1 Let $\mathsf{A} : [0, 1]^2 \to [0, 1]$ be a T-uninorm given by (4).

- For $a = 1$ it becomes a disjunctive uninorm, i.e., $\mathsf{A} = \mathsf{U}'$.
- $a \neq 0$ in order to ensure the discontinuity of A_0, and $e < a$ since $\mathsf{A}_e(0) = 0$.
- If the uninorm U' is from the class $\mathsf{U}_{\max}$, then a T-uninorm A is called a T-uninorm in $\mathsf{U}_{\max}$, and this is the class of T-uninorms that is covered in this chapter. Since the uninorm in (4) has to be disjunctive, a T-uninorm can not be constructed with a uninorm from $\mathsf{U}_{\min}$.

2.3 Distributivity Equations

Functional equations that are called the left and right distributivity laws ([1], p. 318) are given by the following definition.

Definition 4 Let $F, G : [0, 1]^2 \to [0, 1]$ be two operators. F is distributive over G, if the following two laws hold:

(LD) - F is left distributive over G, i.e.,

$$F(x, G(y, z)) = G(F(x, y), F(x, z)), \quad \text{for all } x, y, z \in [0, 1]$$

(RD) - F is right distributive over G, i.e.,

$$F(G(y, z), x) = G(F(y, x), F(z, x)), \quad \text{for all } x, y, z \in [0, 1]$$

For commutative operators F, laws (LD) and (RD) coincide.

Since the problem of distributivity of a t-norm over a t-conorm gives us only a trivial solution, i.e., t-conorm has to be $\mathsf{S}_M = \max$, a relaxed version of this concept was introduced in the following manner (see [16]).

Definition 5 ([16]) A t-norm T is conditionally distributive (CD) over a t-conorm S if for all $x, y, z \in [0, 1]$ the following holds

$$\text{(CD)} \qquad \mathsf{T}(x, \mathsf{S}(y, z)) = \mathsf{S}\left(\mathsf{T}(x, y), \mathsf{T}(x, z)\right), \text{ whenever } \mathsf{S}(y, z) < 1.$$

The previous definition can be extended to more general aggregation operators.

Definition 6 ([8, 12]) Let F be a T-uninorm or a nullnorm with annihilator $a \in]0, 1[$, and G be a t-conorm or $G \in \mathsf{U}_{min} \cup \mathsf{U}_{max}$. F is conditionally distributive (CD) over G if for all $x, y, z \in [0, 1]$ the following holds

$$\text{(CD)} \qquad F\left(x, G(y, z)\right) = G\left(F(x, y), F(x, z)\right), \text{ whenever } G(y, z) < 1.$$

The previous equation is also known as restricted distributivity (see [5]) and, although the domain is only weakly restricted, the class of pairs of operators that fulfill (CD) is much wider. This can be seen from the following.

Theorem 2 ([16]) *A continuous t-norm* T *and a continuous t-conorm* S *satisfy (CD), if and only if exactly one of the following two cases is fulfilled:*

1. $\mathsf{S} = \mathsf{S}_M$;
2. *There is a strict t-norm* T^* *and a nilpotent t-conorm* S^* *such that additive generator* s *of* S^* *satisfying* $s(1) = 1$ *is also multiplicative generator of* T^**, and there is a* $c \in [0, 1[$ *such that for some continuous t-norm* T^{**} *we have*

$$\mathsf{S}(x,y)=\begin{cases} c+(1-c)\mathsf{S}^*\left(\frac{x-c}{1-c},\frac{y-c}{1-c}\right) & \textit{if } (x,y)\in[c,1]^2,\\ \max(x,y) & \textit{otherwise},\end{cases} \tag{6}$$

and

$$\mathsf{T}(x,y)=\begin{cases} c\mathsf{T}^{**}\left(\frac{x}{c},\frac{y}{c}\right) & \textit{if } (x,y)\in[0,c]^2,\\ c+(1-c)\mathsf{T}^*\left(\frac{x-c}{1-c},\frac{y-c}{1-c}\right) & \textit{if } (x,y)\in[c,1]^2,\\ \min(x,y) & \textit{otherwise}.\end{cases} \tag{7}$$

Remark 2 Due to isomorphisms between strict t-norms and product t-norm $\mathsf{T}_P(x,y)=x\cdot y$ and nilpotent t-conorms and the Lukasiewicz t-conorm $\mathsf{S}_L(x,y)=\min\{x+y,1\}$ (see [16]), the previous result is often used in the form where $\mathsf{S}_*=\mathsf{S}_L$ and $\mathsf{T}_*=\mathsf{T}_P$.

Generalizations of Theorem 2 can be found in [17] where T is a uninorm, in [24] where T is a pseudo-multiplication, in [8] where T is a nullnorm, in [11] where T is a semi-t-operator, and in [12] where T is a T-uninorm.

3 Distributivity on the Complete Domain

This section contains the results for the distributivity of aggregation operators with an annihilator with respect to aggregation operators with a neutral element. Although the domain is untouched, the number of solutions is smaller with respect to the next section. Also, the form is simpler. The observed aggregation operators with an annihilator are T-uninorms in $\mathsf{U}_{\max}$ and nullnorms, while aggregation operators with a neutral element are t-norms, t-conorms and uninorms from the class $\mathsf{U}_{\min}\cup\mathsf{U}_{\max}$.

3.1 Distributivity for T-Uninorms in $\mathsf{U}_{\max}$

Let F be a T-uninorm in $\mathsf{U}_{\max}$ with an annihilator a, and the IFC element e. Let G be a t-conorm, or a t-norm, or a uninorm from the class $\mathsf{U}_{\min}\cup\mathsf{U}_{\max}$. Since the distributivity of F over G for $a=1$, i.e., for F being a uninorm from $\mathsf{U}_{\max}$, is considered in [19, 20], it is further assumed that $a\in]0,1[$. The first result is for t-conorms and it provides only an idempotent solution.

Theorem 3 ([12]) *Let F be a T-uninorm in $\mathsf{U}_{\max}$ and S be a t-conorm. F is distributive over S if and only if $\mathsf{S}=\max$.*

The result on distributivity of T-uninorms over t-norms is given by the following theorem. As can be seen, the sufficient and necessary condition requires some additional starting assumptions.

Theorem 4 ([12]) *Let F be a T-uninorm in $\mathsf{U}_{\max}$ and T be a t-norm.*

1. *If F is distributive over* T *then* $\mathsf{T}(x, x) = x$ *for all* $x > e$.
2. *Let the function* $t(x) = \mathsf{T}(x, x)$ *be right-continuous at the point* $x = e$. *Then* F *is distributive over* T *if and only if* $\mathsf{T} = \min$.

This research is also extended to uninorms from the class $\mathsf{U}_{\max}$.

Theorem 5 ([12]) *Let F be a T-uninorm in $\mathsf{U}_{\max}$ and U be a uninorm from the class $\mathsf{U}_{\max}$ with a neutral element $e_1 \in]0, 1[$ and underlying t-norm T such that $\mathsf{T}(x, x)$ is right-continuous at the point $x = e$. F is distributive over U if and only if $e_1 < a$ and exactly one of the following cases is fulfilled:*

1. *$e_1 = e$, and U is an idempotent uninorm, i.e., $\mathsf{U} = \mathsf{U}_{e_1}^{\max}$,*
2. *$e < e_1$, $\mathsf{U} = \mathsf{U}_{e_1}^{\max}$, and F is given by*

$$F(x, y) = \begin{cases} e\mathsf{T}_1\left(\frac{x}{e}, \frac{y}{e}\right) & \text{if } (x, y) \in [0, e]^2, \\ e + (e_1 - e)\mathsf{S}_1'\left(\frac{x-e}{e_1-e}, \frac{y-e}{e_1-e}\right) & \text{if } (x, y) \in [e, e_1]^2, \\ e_1 + (a - e_1)\mathsf{S}_1''\left(\frac{x-e_1}{a-e_1}, \frac{y-e_1}{a-e_1}\right) & \text{if } (x, y) \in [e_1, a]^2, \\ a + (1 - a)\mathsf{T}'\left(\frac{x-a}{1-a}, \frac{y-a}{1-a}\right) & \text{if } (x, y) \in [a, 1]^2, \\ a & \text{if } (x, y) \in I_a, \\ \max(x, y) & \text{otherwise,} \end{cases} \tag{8}$$

where T_1 and T' are t-norms, and S_1' and S_1'' are t-conorms.

Remark 3 The first case from the previous theorem also holds without the assumption of right-continuity for the function $\mathsf{T}(x, x)$ at $x = e$. The second case, according to Theorem 4, has the following form: *if F is distributive over U then $\mathsf{U}(x, x) =$ for $x > e$ and F is given by (8).*

Remark 4 The previous result for $a = 1$, i.e., for T-uninorm being a uninorm from $\mathsf{U}_{\max}$, has been shown in [19, 20]. The first case generalizes Proposition 6.2 from [19, 20], and the second one Proposition 6.3 from [19].

The next result is on T-uninorms in $\mathsf{U}_{\max}$ and uninorms from $\mathsf{U}_{\min}$.

Theorem 6 ([12]) *Let F be a T-uninorm in $\mathsf{U}_{\max}$ and U be a uninorm from the class $\mathsf{U}_{\min}$ with a neutral element $e_1 \in]0, 1[$ and underlying t-norm T such that $\mathsf{T}(x, x)$ is right-continuous at the point $x = e$. F is distributive over U if and only if $a < e_1$, $\mathsf{U} = \mathsf{U}_{e_1}^{\min}$ and F is given by*

$$F(x, y) = \begin{cases} a\mathsf{U}'\left(\frac{x}{a}, \frac{y}{a}\right) & \text{if } (x, y) \in [0, a]^2, \\ a + (e_1 - a)\mathsf{T}_1'\left(\frac{x-a}{e_1-a}, \frac{y-a}{e_1-a}\right) & \text{if } (x, y) \in [a, e_1]^2, \\ e_1 + (1 - e_1)\mathsf{T}_2'\left(\frac{x-e_1}{1-e_1}, \frac{y-e_1}{1-e_1}\right) & \text{if } (x, y) \in [e_1, 1]^2, \\ a & \text{if } (x, y) \in I_a, \\ \min(x, y) & \text{otherwise,} \end{cases} \tag{9}$$

where T_1', T_2' are t-norms, U' is a uninorm from the class $\mathsf{U}_{\max}$.

Remark 5 If F is a T-uninorm in $\mathsf{U}_{\max}$ with annihilator $a = 1$, that is if F is a uninorm from the class $\mathsf{U}_{\max}$, there is no uninorm U from the class $\mathsf{U}_{\min}$ such that F is distributive over U (see [19], Lemma 6.1). The previous theorem shows that for annihilator $a \in]0, 1[$ a solution exists.

From the previous results, using duality between T-uninorms and S-uninorms, t-norms and t-conorms, uninorms from the class $\mathsf{U}_{\max}$ and uninorms from $\mathsf{U}_{\min}$, the corresponding results for distributivity of S-uninorms can be obtained (see [14]).

3.2 Distributivity for Nullnorms

Now, let F be a nullnorm. The first result in this part is for distributivity over t-norms and t-conorms. Again, only idempotent solutions are obtained.

Theorem 7 ([19]) *Let F be a nullnorm with an annihilator $a \in]0, 1[$, and G be a t-norm or a t-conorm. F is distributive over G if and only if $G = \min$ or $G = \max$.*

The previous result also holds for $a \in \{0, 1\}$, i.e., when the nullnorm F degenerates into a t-norm or a t-conorm (see [19]).

Theorem 8 ([19]) *Let F be a nullnorm with an annihilator $a \in]0, 1[$, and U be a uninorm from the class $\mathsf{U}_{\max}$ with a neutral element $e_1 \in]0, 1[$. F is distributive over U if and only if $e_1 < a$, $\mathsf{U} = \mathsf{U}_{e_1}^{\max}$, and F is given by*

$$F(x, y) = \begin{cases} e_1 \mathsf{S}_1'\left(\frac{x}{e_1}, \frac{y}{e_1}\right) & \textit{if } (x, y) \in [0, e_1]^2, \\ e_1 + (a - e_1)\mathsf{S}_2'\left(\frac{x-e_1}{a-e_1}, \frac{y-e_1}{a-e_1}\right) & \textit{if } (x, y) \in [e_1, a]^2, \\ a + (1 - a)\mathsf{T}'\left(\frac{x-a}{1-a}, \frac{y-a}{1-a}\right) & \textit{if } (x, y) \in [a, 1]^2, \\ a & \textit{if } (x, y) \in I_a, \\ \max(x, y) & \textit{otherwise}, \end{cases} \tag{10}$$

where $\mathsf{S}_1', \mathsf{S}_2'$ are t-conorms and T' is a t-norm.

If $a = 1$, F is a t-conorm and the previous result holds. The problem for $a = 0$ was solved in [23].

Theorem 9 ([19]) *Let F be a nullnorm with an annihilator $a \in]0, 1[$, and U be a uninorm from the class $\mathsf{U}_{\min}$ with a neutral element $e_1 \in]0, 1[$. F is distributive over U if and only if $e_1 > a$, $\mathsf{U} = \mathsf{U}_{e_1}^{\min}$, and F is given by*

$$F(x, y) = \begin{cases} a\mathsf{S}'\left(\frac{x}{a}, \frac{y}{a}\right) & \textit{if } (x, y) \in [0, a]^2, \\ a + (e_1 - a)\mathsf{T}_1'\left(\frac{x-a}{e_1-a}, \frac{y-a}{e_1-a}\right) & \textit{if } (x, y) \in [a, e_1]^2, \\ e_1 + (1 - e_1)\mathsf{T}_2'\left(\frac{x-e_1}{1-e_1}, \frac{y-e_1}{1-e_1}\right) & \textit{if } (x, y) \in [e_1, 1]^2, \\ a & \textit{if } (x, y) \in I_a, \\ \min(x, y) & \textit{otherwise}, \end{cases} \tag{11}$$

where T_1', T_2' *are t-norms, and* S' *is a t-conorm.*

The previous result also holds for $a = 0$, while the case $a = 1$ was solved in [23].

Remark 6 Theorems 8, 9 also hold for a uninorm U from the class of idempotent uninorms (see [23]). More on distributivity of nullnorms over uninorms can be found in [19, 23].

4 Distributivity on the Restricted Domain

As seen in the previous section, the distributivity law is a rather strong condition that simplifies the structure of the inner operator considerably, i.e., the inner operator is reduced to an idempotent operator. Thus, the relaxed version of distributivity law, as given in Definition 6, is investigated in this section.

4.1 Conditional Distributivity for T-Uninorms in $\mathsf{U}_{\max}$

This section contains the counterparts of Theorems 3, 5, 6 for the restricted domain. In order to obtain a characterization of all pairs (F, G) that satisfy (CD) condition, according to Theorem 2, a form of continuity for F and G is needed.

Let $I_a = [0, a] \times [a, 1] \cup [a, 1] \times [0, a]$, $I_{a,c} = [a, c] \times [c, 1] \cup [c, 1] \times [a, c]$ and $I_{e_1} = [0, e_1[\times[0, 1] \cup [0, 1] \times [0, e_1[$. Also, T_P is the product t-norm and S_L is the Lukasiewicz t-conorm.

Theorem 10 ([12]) *Let* F *be a* T*-uninorm in* $U_{\max}$ *with continuous underlying t-norm* T'*, and* S *a continuous t-conorm.* F *is conditionally distributive over* S *if and only if exactly one of the following cases is fulfilled:*

1. $\mathsf{S} = \mathsf{S}_M$*;*
2. *there is a* $c \in [a, 1[$ *such that* S *is of the form (6) and* F *is given by*

$$F(x, y) = \begin{cases} a\mathsf{U}'\left(\frac{x}{a}, \frac{y}{a}\right) & \text{if } (x, y) \in [0, a]^2, \\ a + (c - a)\mathsf{T}_1\left(\frac{x-a}{c-a}, \frac{y-a}{c-a}\right) & \text{if } (x, y) \in [a, c]^2, \\ c + (1 - c)\mathsf{T}_P\left(\frac{x-c}{1-c}, \frac{y-c}{1-c}\right) & \text{if } (x, y) \in [c, 1]^2, \\ a & \text{if } (x, y) \in I_a, \\ \min(x, y) & \text{otherwise,} \end{cases} \tag{12}$$

where T_1 *is a continuous t-norm, and* U' *is a uninorm from the class* $\mathsf{U}_{\max}$.

Theorem 11 ([12]) *Let* F *be a* T*-uninorm in* $\mathsf{U}_{\max}$ *with a continuous underlying t-norm* T'*, and* U *be a uninorm from the class* $\mathsf{U}_{\max}$ *with a neutral element* $e_1 \in]0, 1[$ *and continuous underlying t-norm and t-conorm.* F *is conditionally distributive over* U *if and only if* $e_1 < a$ *and exactly one of the following cases is fulfilled:*

1. $e_1 = e$, *and* U *is an idempotent uninorm, i.e.,* $\mathsf{U} = \mathsf{U}^{\max}_{e_1}$,
2. $e_1 = e$, *and there is a* $c \in [a, 1[$ *such that* F *and* U *are given by*

$$\mathsf{U}(x, y) = \begin{cases} \min(x, y) & \textit{if } (x, y) \in [0, e_1]^2, \\ c + (1-c)\mathsf{S}_L\left(\frac{x-c}{1-c}, \frac{y-c}{1-c}\right) & \textit{if } (x, y) \in [c, 1]^2, \\ \max(x, y) & \textit{otherwise} \end{cases} \tag{13}$$

and

$$F(x, y) = \begin{cases} a\mathsf{U}'\left(\frac{x}{a}, \frac{y}{a}\right) & \textit{if } (x, y) \in [0, a]^2, \\ a + (c-a)\mathsf{T}_2\left(\frac{x-a}{c-a}, \frac{y-a}{c-a}\right) & \textit{if } (x, y) \in [a, c]^2, \\ c + (1-c)\mathsf{T}_P\left(\frac{x-c}{1-c}, \frac{y-c}{1-c}\right) & \textit{if } (x, y) \in [c, 1]^2, \\ a & \textit{if } (x, y) \in I_a, \\ \min(x, y) & \textit{otherwise,} \end{cases} \tag{14}$$

3. $e < e_1$, $\mathsf{U} = \mathsf{U}^{\max}_{e_1}$, *and* F *is given by (8)*
4. $e < e_1$, *and there is a* $c \in [a, 1[$ *such that* U *is given by (13), and* F *is given by*

$$F(x, y) = \begin{cases} e\mathsf{T}_1\left(\frac{x}{e}, \frac{y}{e}\right) & \textit{if } (x, y) \in [0, e]^2, \\ e + (e_1 - e)\mathsf{S}'_1\left(\frac{x-e}{e_1-e}, \frac{y-e}{e_1-e}\right) & \textit{if } (x, y) \in [e, e_1]^2, \\ e_1 + (a - e_1)\mathsf{S}''_1\left(\frac{x-e_1}{a-e_1}, \frac{y-e_1}{a-e_1}\right) & \textit{if } (x, y) \in [e_1, a]^2, \\ a + (c-a)\mathsf{T}_3\left(\frac{x-a}{c-a}, \frac{y-a}{c-a}\right) & \textit{if } (x, y) \in [a, c]^2, \\ c + (1-c)\mathsf{T}_P\left(\frac{x-c}{1-c}, \frac{y-c}{1-c}\right) & \textit{if } (x, y) \in [c, 1]^2, \\ \min(x, y) & \textit{if } (x, y) \in I_{a,c}, \\ a & \textit{if } (x, y) \in I_a, \\ \max(x, y) & \textit{otherwise,} \end{cases} \tag{15}$$

where U' *is a uninorm from the class* $\mathsf{U}_{\max}$, $\mathsf{S}'_1, \mathsf{S}''_1$ *are t-conorms,* T_1 *is a t-norm, and* $\mathsf{T}_2, \mathsf{T}_3$ *are continuous t-norms.*

Theorem 12 ([12]) *Let* F *be a* T*-uninorm in* $\mathsf{U}_{\max}$ *with a continuous underlying t-norm* T', *and* U *be a uninorm from the class* $\mathsf{U}_{\min}$ *with a neutral element* $e_1 \in]0, 1[$ *and continuous underlying t-norm and t-conorm.* F *is conditionally distributive over* U *if and only if* $a < e_1$ *and exactly one of the following cases is fulfilled:*

1. $\mathsf{U} = \mathsf{U}^{\min}_{e_1}$, *and* F *is given by (9);*
2. *there is a* $c \in [e_1, [)$ *such that* F *and* U *are given by*

$$\mathsf{U}(x, y) = \begin{cases} \min(x, y) & \textit{if } (x, y) \in I_{e_1}, \\ c + (1-c)\mathsf{S}_L\left(\frac{x-c}{1-c}, \frac{y-c}{1-c}\right) & \textit{if } (x, y) \in [c, 1]^2, \\ \max(x, y) & \textit{otherwise,} \end{cases} \tag{16}$$

and

$$F(x,y)=\begin{cases} a\mathsf{U}'\left(\frac{x}{a},\frac{y}{a}\right) & \text{if } (x,y)\in[0,a]^2,\\ a+(e_1-a)\mathsf{T}_1'\left(\frac{x-a}{e_1-a},\frac{y-a}{e_1-a}\right) & \text{if } (x,y)\in[a,e_1]^2,\\ e_1+(c-e_1)\mathsf{T}_1''\left(\frac{x-e_1}{c-e_1},\frac{y-e_1}{c-e_1}\right) & \text{if } (x,y)\in[e_1,c]^2,\\ c+(1-c)\mathsf{T}_P\left(\frac{x-c}{1-c},\frac{y-c}{1-c}\right) & \text{if } (x,y)\in[c,1]^2,\\ a & \text{if } (x,y)\in I_a,\\ \min(x,y) & \text{otherwise,}\end{cases} \tag{17}$$

where U' *is a uninorm from* $\mathsf{U}_{\max}$, *and* T_1', T_1'' *are continuous t-norms.*

4.2 Conditional Distributivity for Nullnorms

Finally, the counterparts of the results from Sect. 3.2 for the restricted domain, are given in this section. As in Sect. 4.1 the assumption of some form of continuity for F and G is required.

Theorem 13 ([8]) *Let* F *be a continuous nullnorm with an annihilator* $a\in]0,1[$ *and* S *be a continuous t-conorm.* F *is conditionally distributive over* S *if and only if exactly one of the following cases is fulfilled:*

1. $\mathsf{S}=\mathsf{S}_M$;
2. *there is* $c\in[a,1[$ *such that* S *is of the form (6), and* F *is given by*

$$F(x,y)=\begin{cases} a\mathsf{S}_1\left(\frac{x}{a},\frac{y}{a}\right) & \text{if } (x,y)\in[0,a]^2,\\ a+(c-a)\mathsf{T}_1\left(\frac{x-a}{c-a},\frac{y-a}{c-a}\right) & \text{if } (x,y)\in[a,c]^2,\\ c+(1-c)\mathsf{T}_P\left(\frac{x-c}{1-c},\frac{y-c}{1-c}\right) & \text{if } (x,y)\in[c,1]^2,\\ a & \text{if } (x,y)\in I_a,\\ \min(x,y) & \text{otherwise,}\end{cases} \tag{18}$$

where S_1 *is a continuous t-conorm and* T_1 *is a continuous t-norm.*

Theorem 14 ([8]) *Let* F *be a continuous nullnorm with an annihilator* $a\in]0,1[$ *and* U *be a uninorm from the class* $\mathsf{U}_{\max}$ *with a neutral element* $e_1\in]0,1[$, *and continuous underlying t-norm and t-conorm.* F *is conditionally distributive over* U *if and only if* $e_1<a$ *and exactly one of the following cases is fulfilled:*

1. F *and* U *are as in Theorem 8;*
2. *there is* $c\in[a,1[$ *such that* U *is given by (13), and* F *is given by*

$$F(x,y)=\begin{cases} e_1 \mathsf{S}'_1\left(\frac{x}{e_1},\frac{y}{e_1}\right) & \textit{if } (x,y)\in[0,e_1]^2,\\ e_1+(a-e_1)\mathsf{S}'_2\left(\frac{x-e_1}{a-e_1},\frac{y-e_1}{a-e_1}\right) & \textit{if } (x,y)\in[e_1,a]^2,\\ a+(c-a)\mathsf{T}'_1\left(\frac{x-a}{c-a},\frac{y-a}{c-a}\right) & \textit{if } (x,y)\in[a,c]^2,\\ c+(1-c)\mathsf{T}_P\left(\frac{x-c}{1-c},\frac{y-c}{1-c}\right) & \textit{if } (x,y)\in[c,1]^2,\\ a & \textit{if } (x,y)\in I_a,\\ \min(x,y) & \textit{if } (x,y)\in I_{a,c},\\ \max(x,y) & \textit{otherwise,} \end{cases} \tag{19}$$

where $\mathsf{S}'_1, \mathsf{S}'_2$ *are continuous t-conorms, and* T'_1 *is a continuous t-norm*

Theorem 15 ([8]) *Let* F *be a continuous nullnorm with an annihilator* $a \in]0,1[$ *and* U *be a uninorm from the class* $\mathsf{U}_{\min}$ *with a neutral element* $e_1 \in]0,1[$, *and continuous underlying t-norm and t-conorm.* F *is conditionally distributive over* U *if and only if* $e_1 > a$ *and exactly one of the following cases is fulfilled:*

1. F *and* U *are the same as in Theorem 9;*
2. *there is* $c \in [e_1, 1[$ *such that* U *is given by (16), and* F *is given by*

$$F(x,y)=\begin{cases} a\mathsf{S}'_1\left(\frac{x}{a},\frac{y}{a}\right) & \textit{if } (x,y)\in[0,a]^2,\\ a+(e_1-a)\mathsf{T}'_1\left(\frac{x-a}{e_1-a},\frac{y-a}{e_1-a}\right) & \textit{if } (x,y)\in[a,e_1]^2,\\ e_1+(c-e_1)\mathsf{T}'_2\left(\frac{x-e_1}{c-e_1},\frac{y-e_1}{c-e_1}\right) & \textit{if } (x,y)\in[e_1,c]^2,\\ c+(1-c)\mathsf{T}_P\left(\frac{x-c}{1-c},\frac{y-c}{1-c}\right) & \textit{if } (x,y)\in[c,1]^2,\\ a & \textit{if } (x,y)\in I_a,\\ \min(x,y) & \textit{otherwise,} \end{cases} \tag{20}$$

where S'_1 *is a continuous t-conorm, and* $\mathsf{T}'_1, \mathsf{T}'_2$ *are continuous t-norms.*

Remark 7 The importance of conditionally distributive pairs of aggregation operators for Utility Theory can be illustrated by the results from papers [3] and [10].

Paper [3] contains an example of an optimistic hybrid utility function that is constructed from a continuous t-norm T and a continuous t-conorm S that have to fulfill conditional distributivity law.

This approach has been extended in [10] with an example of a utility function obtained from a continuous nullnorm F with an annihilator $k \in]0,1[$, and a continuous t-conorm S that also satisfies the conditional distributivity law. The annihilator for F allows a Decision Maker to predetermine a threshold for a certain decision process. Additionally, the behavioral characteristics of the obtained utility function are investigated in [10].

5 Conclusions

This chapter presents an overview of distributivity equations, on both complete and restricted domains, involving special classes of aggregation operators with an annihilator and special classes of uninorms. The distributivity and conditional distributivity of T-uninorms in U_{max} and nullnorms with an annihilator $a \in]0, 1[$, over t-conorms and uninorms from $\mathsf{U}_{min} \cup \mathsf{U}_{max}$ are the focus of this chapter. It can be concluded that the classical distributivity law is a strong condition, since it simplifies the structure of the inner operator considerably, i.e., it is being reduced to an idempotent operator. This can be seen from Sect. 3. Consequently, Sect. 4 presents a characterization of all pairs (F, G) satisfying the weaker form of the distributivity law. Now, this conditional distributivity produces a larger variety of solutions. In the forthcoming work, an analogous study will be done for the other classes of associative a-CAOA and uninorms. Also, further research will investigate the applicability of the obtained structures in utility theory since this restricted setting is useful for the modelling behavior of decision makers.

Acknowledgements The authors acknowledge financial support of the Ministry of Education, Science and Technological Development of the Republic of Serbia.

References

1. Aczél, J.: Lectures on Functional Equations and their Applications. Academic, New York (1966)
2. Drewniak, J., Drygaś, P., Rak, E.: Distributivity between uninorms and nullnorms. Fuzzy Set. Syst. **159**, 1646–1657 (2008)
3. Dubois,D., Pap, E., Prade, H.: Hybrid probabilictic-possibilistic mixtures and utility functions. In: Preferences and Decisions under Incomplete Knowledge, Studies in Fuzziness and Soft Computing, vol. 51, pp. 51–73, Springer, Berlin (2000)
4. Fodor, J.C., Yager, R.R., Rybalov, A.: Structure of uninorms. Int. J. Uncertain. Fuzz. **5**, 411–427 (1997)
5. Grabisch, M., Marichal, J., Mesiar, R., Pap, E.: Aggregations Functions. Cambridge University Press, Cambridge (2009)
6. Hou, W., Li, G.: The conditional distributivity of nullnorms over uninorms. Aequ. Math. **94**, 355–368 (2020)
7. Jočić, D., Drygaś, P., Štajner-Papuga, I.: Left and right distributivity equations in the class of semi-t-operators. Fuzzy Set. Syst. **372**, 62–80 (2019)
8. Jočić, D., Štajner-Papuga, I.: Restricted distributivity for aggregation operators with absorbing element. Fuzzy Set. Syst. **224**, 23–35 (2013)
9. Jočić, D., Štajner-Papuga, I.: Distributivity equations and Mayor's aggregation operators. Knowl. Based Syst. **52**, 194–200 (2013)
10. Jočić, D., Štajner-Papuga, I.: Some implications of the restricted distributivity of aggregation operators with absorbing elements for utility theory. Fuzzy Set. Syst. **291**, 54–65 (2016)
11. Jočić, D., Štajner-Papuga, I.: On the conditional distributivity of continuous semi-t-operators over uninorms. Fuzzy Set. Syst. **334**, 110–125 (2018)
12. Jočić, D., Štajner-Papuga, I.: Distributivity and conditional distributivity for T-uninorms. Inf. Sci. **424**, 91–103 (2018)

13. Jočić, D., Štajner-Papuga, I.: On distributivity between aggregation operators with annihilator and Mayor's aggregation operators. Filomat **32**(4), 1475–1489 (2018)
14. Jočić, D., Štajner-Papuga, I.: On distributivity and conditional distributivity of S-uninorms over uninorms. Acta Polytech. Hungarica **16**(1), 209–226 (2019)
15. Jočić, D., Štajner-Papuga, I.: Conditional distributivity of continuous semi-t-operators over disjunctive uninorms with continuous underlying t-norms and t-conorms. Fuzzy Set. Syst. (2020). https://doi.org/10.1016/j.fss.2020.10.012
16. Klement, E.P., Mesiar, R., Pap, E.: Triangular Norms. Kluwer Academic Publishers, Dordrecht (2000)
17. Li, G., Liu, H.-W.: Distributivity and conditional distributivity of uninorm with continuous underlying operators over a continuous t-conorm. Fuzzy Set. Syst. **287**, 154–171 (2016)
18. Li, W.-H., Qin, F.: Conditional distributivity equation for uninorms with continuous underlying operators. IEEE Trans. Fuzzy Syst. **28**(8), 1664–1678 (2020)
19. Mas, M., Mayor, G., Torrens, J.: The distributivity condition for uninorms and t-operators. Fuzzy Set. Syst. **128**, 209–225 (2002)
20. Mas, M., Mayor, G., Torrens, J.: Corrigendum to "The distributivity condition for uninorms and t-operators" [Fuzzy Sets and Systems 128 (2002), 209–225]. Fuzzy Set. Syst. **153**, 297–299 (2005)
21. Mas, M., Massanet, S., Ruiz-Aguilera, D., Torrens, J.: A survey on the existing classes of uninorms. J. Intell. Fuzzy Syst. **29**, 1021–1037 (2015)
22. Mas, M., Mesiar, R., Monserat, M., Torrens, J.: Aggregation operations with annihilator. Int. J. Gen. Syst. **34**, 1–22 (2005)
23. Qin, F., Zhao, B.: The distributive equations for idempotent uninorms and nullnorms. Fuzzy Set. Syst. **155**, 446–458 (2005)
24. Sander, W., Siedekum, J.: Multiplication, distributivity and fuzzy-integral I, II, III. Kybernetika **41** 397–422; 469–496; 497–518 (2005)
25. Su, Y., Liu, H.-W., Ruiz-Aguilera, D., Vicente Riera, J., Torrens, J.: On the distributivity property for uninorms. Fuzzy Set. Syst. **287**, 184–202 (2016)
26. Sun, X.R., Liu, H.-W.: The left (right) distributivity of semi-t-operators over 2-uninorms. Iran. J. Fuzzy Syst. **17**(3), 103–116 (2020)
27. Wang, G., Qin, F., Li, W.-H.: Distributivity and conditional distributivity for uni-nullnorms. Fuzzy Set. Syst. **372**, 34–49 (2019)
28. Wang, G., Qin, F., Li, W.-H.: Conditional distributivity for uni-nullnorms with continuous and Archimedean underlying t-norms and t-conorms. J. Intell. Fuzzy Syst. **36**, 2773–2785 (2019)
29. Wang, Y.M., Liu, H.-W.: On the distributivity equation for uni-nullnorms. Kybernetika **55**(1), 24–43 (2019)
30. Yager, R.R., Rybalov, A.: Uninorm aggregation operators. Fuzzy Set. Syst. **80**, 111–120 (1996)

The Use of Fuzzy Logic in Various Combinatorial Optimization Problems

Darko Drakulić, Aleksandar Takači, and Miroslav Marić

Abstract Conventional models of many combinatorial optimization problems rarely encompass real-life problems, because real-life problems usually contain a high degree of uncertainty. These uncertainties can be modeled using various methods, including fuzzy sets. Apart from the precise description from the problems' nature, fuzzy variables can describe the problem better, improve the solution and reduce costs for decision makers. In this chapter we show how fuzzy logic can be used for modeling uncertainties in combinatorial problems and improve their quality. The focus will be on the Location Set Covering Problem (LSCP), the Maximal Covering Location Problem (MCLP) and the Minimal Covering Location Problem (MinCLP) as a special modification of the MCLP, but the same method could be applied to other problems. These problems are applicable in searching for optimal places for desired and undesired facilities under the given conditions. Each problem will be formally described with its own mathematical model and some of their instances will be solved. Firstly, small-size instances of the problems will be solved with an exact algorithm using the CPLEX optimizer tool, and when a dimension becomes too big for exact solving, the instances will be then solved with a Particle Swarm Optimization (PSO) meta-heuristic.

Keywords Artificial intelligence · Aggregation function · Triangular norm · Ordered weighted sum · Fuzzy logic · Fuzzy sets · Combinatorial optimization · Covering location problem · Swarm optimization

---

D. Drakulić
NAVER LABS Europe, Meylan, France
e-mail: darko.drakulic@naverlabs.com

A. Takači (✉)
Faculty of Technology, University of Novi Sad, Novi Sad, Serbia
e-mail: atakaci@uns.ac.rs

M. Marić
Faculty of Mathematics, University of Belgrade, Belgrade, Serbia
e-mail: maricm@matf.bg.ac.rs

E. Pap (ed.), *Artificial Intelligence: Theory and Applications*,
Studies in Computational Intelligence 973,
https://doi.org/10.1007/978-3-030-72711-6_8

1 Introduction

Covering location problems represent an important class of combinatorial optimization problems with significant application for solving real-life problems. Simply speaking, the goal of location problems is to find optimal positions for the facilities on the given set of locations. This type of problem defines the objective function and its optimality (maximum or minimum), as well as the number of facilities (fixed or not). The set of locations is usually represented by a complete weighted graph in which the nodes are locations and the edges are connection between them. This graph is directed and edge weights are usually computed by the Euclidean distance between locations, travel time or transportation cost. The choice of weight does not reduce the generality of the problem or affect its modeling. On the other hand, travel time between locations is most often used in modeling real life problems because the optimization graph is built from a road network.

In the covering location problems, the number of nodes is the problem size and each problem has a predefined coverage radius which determines the objective function. A location is covered by a facility if the travel time (distance) between them is less than the given covering radius. Depending on the type of problem, the task may be to minimize or maximize the objective function.

Location problems are widely used in solving practical problems [2, 7, 24, 29, 30]—finding optimal positions for facilities like shops, public institutions (schools, hospitals, etc.), emergency services, various transmitters (like mobile telephony, internet, radio, etc.), but also in finding optimal position for unwanted facilities, such as landfills, power plants, noise sources, etc. It is clear that in the first class of problems the goal is to maximize the objective function (in other words, facilities should cover as many locations as possible), while in the second class the goal is to minimize the objective function (facilities should affect as few locations as possible). In this chapter, it is assumed that all locations have the same importance; however the requirement that facilities have their own coefficient of importance does not diminish the generality of the problem and all of the models that will be presented here can be very easily extended to that case. It is just necessary to multiply each facility with its own coefficient of the importance.

Three typical representatives of covering location problems are the Location Set Covering Problem (LSCP), the Maximal Covering Location Problem (MCLP), and the Minimal Covering Location Problem (MinCLP).

In the standard models of these problems, the covering radius and travel times between the locations are modeled with real numbers. The real-life problems contain a high degree of uncertainty but classical models of these problems cannot model any on them. Therefore, the authors developed different extensions of standard models for describing and modeling these uncertainties.

Like most location problems, LSCP, MCLP and MinCLP belong to the class of NP-hard problems [16]. The usual approach with NP-hard combinatorial problems is solving small-sized instances using an exact method, however, when the dimension becomes too high for exact algorithms, metaheuristic methods are employed to solve them and prove that metaheuristics gives the same results as exact methods on solvable instances.

There are several different approaches to modeling uncertainties in location covering problems, but the most common involve using probabilities [6, 22, 23, 27, 28] and fuzzy sets [3, 4, 9, 10, 17, 25]. A review of the recent results in using fuzzy sets in covering location problems can be found in [1].

The content of this chapter is mainly based on [11–13], where the authors presented new models for covering location problems with a fuzzy covering radius and fuzzy travel times. The authors solved all models using the IBM CPLEX exact solver for solvable instances and a discrete version of the Particle Swarm Optimization (PSO) meta-heuristic for solving large-size instances.

1.1 Particle Swarm Optimization Metaheuristic

Particle swarm optimization (PSO) is a nature-based metaheuristic method, developed by Kennedy and Eberhart in [18, 20]. As the name says, this method is inspired by the social behavior of particles in swarms, like birds in flocks. The original PSO was later adapted for solving discrete problems [19] and this version is known as Discrete Particle Swarm Optimization (DPSO). This discrete version is appropriate for solving the problems presented in this chapter.

Initially, the algorithm randomly creates a swarm of size S containing n instances (particles) and then iteratively moves them through the solution space with the given intelligence: each particle remembers its best position so far and looks on the best position of particles from its neighbourhood. Using this information, each particle decides how to make a step through the solution space.

Formally, each particle is represented with positional and velocity vectors $x_i = (x_{i1}, x_{i2}, \ldots, x_{id}) \in [\mathbf{0}, \mathbf{1}]^d$ and $v_i = (v_{i1}, v_{i2}, \ldots, v_{id})$, $x, v \in \mathbb{R}^d$. Vector x represents the position of the particle in the solution space (i.e. the current solution to the problem), while the velocity vector determines the position of the particle in the next iteration.

The velocity vector for each particle must be recalculated at the end of each iteration, and it is done by particle's best position so far (b_i) and the best position (c_i) from its neighbourhood $N(i) \subseteq S$: [20]

$$v_i^k = c_1\xi v_i^{k-1} + c_2\xi(b_i - x_i^{k-1}) + c_3\xi(g_i - x_i^{k-1}). \tag{1}$$

Hyperparameters c_1, c_2 and c_3 represent a degree of confidence for each addend (current position, best position so far and best position in the neighbourhood) and they are usually tuned empirically. The parameter ξ is an independent random variable generated in each iteration with uniform distributions in the interval [0, 1]. Velocity values v_i are usually limited to the given interval, to avoid extreme values:

$$|v_{ij}| < V_{max}. \tag{2}$$

In practice, PSO achieves the best performance for $V_{max} \approx 6.0$.

The obtained velocity vector v_{ij} represents a probability that the j-th binary value of the particle i obtains the value 0 or 1, and that probability is computed using a sigmoid function:

$$x_{ij} = \frac{1}{1 + \exp(-v_{ij})}. \tag{3}$$

2 Covering Location Problems

In this section, three conventional covering location problems will be presented, with a brief descriptions and mathematical models.

For all three models, the following notation will be used:

$$I \text{ - the set of locations (indexed by } i), \tag{4}$$

$$J \subset I \text{ - set of eligible facility sites (indexed by } j), \tag{5}$$

$$x_j = \begin{cases} 1, \text{ if facility is located at location } j, \\ 0, \text{ otherwise,} \end{cases} \tag{6}$$

$$t_{ij} \text{ - travel time from location i to location } j, \tag{7}$$

$$S \text{ - the radius of coverage,} \tag{8}$$

$$P \text{ - the number of facilities.} \tag{9}$$

2.1 *Location Set Covering Problem*

The first problem from the class of covering location problems is the Location Set Covering Problem (LSCP). LSCP finds the minimal number of facilities such that all locations will be covered with at least one facility. Be covered means that distance between location and some facility is lower than a given radius. As explained previously, radius is not necessarily a geometric representation, it can also be determined by travel time.

The typical application of the LSCP involves finding the minimal number and the locations for emergency services, such that every patient is covered by the service. In this problem, residential buildings represent locations and emergency services represent facilities. Together, the buildings and locations for the emergency services represent graph vertices and the paths between them are edges. Some applications of the LSCP are listed in [26].

The mathematical model for LSCP [29], with notations (4)–(9) is:

$$P = \min \sum_{j \in J} x_j, \tag{10}$$

subject to: (11)

$$\sum {}_{j \in N_i} x_j \geqslant 1, \forall i \in I, \tag{12}$$

$$x_j \in \{0, 1\}, \forall j \in J, \tag{13}$$

where

$N_i = \{j | t_{ij} \leqslant S\}$ - set of all facilities j which cover the location i.

2.2 Maximal Covering Location Problem

The Maximal Covering Location Problem (MCLP) was defined in 1974 by Church and ReVelle [5]. The aim of MCLP is to find optimal positions for a given number of facilities such that they cover as many locations as possible. The MCLP can be used for modeling and solving many real-life problems, like finding the best positions for a given number of shops, emergency services, heating plants etc.

With notations (4)–(9), the mathematical model of the MCLP is [5]

$$\max \sum_{i \in I} y_i,$$

subject to:

$$\sum {}_{j \in J} x_j = P,$$

$$\sum {}_{j \in N_i} x_j \geqslant y_i, \forall i \in I,$$

$$x_j \in \{0, 1\}, \forall j \in J,$$

$$y_i \in \{0, 1\}, \forall i \in I,$$

where

$N_i = \{j | t_{ij} \leqslant S\}$ - the set of all facilities j which cover location i.

2.3 Minimal Covering Location Problem

The Minimal Covering Location Problem (MinCLP) is the opposite problem of MCLP. The aim of the model is to place a given number of facilities in such a way that their impact on the locations is minimized. The location has impacted by the facility if the distance between them is less than a predetermined impact radius.

This problem is utilized for finding the optimal locations for any type of unwanted facilities, like power plants, chemical industry sites, incinirators etc.

This problem requires an additional demand—a minimum distance between any two facilities, it is clear that without this requirement, an optimal solution will place all facilities in the same position.

Using notations (4)–(9), together with the additional variable

$$D - \text{the minimal distance between facilities.} \tag{14}$$

A mathematical model of the MinCLP is as follows:

$$\begin{aligned} & \min \sum y_i \\ \text{subject to:} & \\ & \sum_{j \in J} x_j = P, \\ & \max_{j \in J} x_j \cdot c_{ij} \geqslant y_i, \forall i \in I, \\ & d(x_i, x_j) \geqslant D, \forall i \neq j, \\ & x_i \in \{0, 1\}, \\ & y_i \in [0, 1]. \end{aligned}$$

3 Fuzzy Covering Location Problems

All of the models from previous section have a deterministic (given by real number) covering radiuses and travel times. Although this approach is legitimate in theory, it does not model real life problems. Firstly, travel time always contains some degree of uncertainty due to traffic conditions, weather and traffic lights. On the other hand, the real number gives a very rough boundary for the covering radius. That implies that a location just a few centimeters outside the radius has the same importance to the result as the locations kilometers outside the radius. In both cases, these bounds should be defined in a less strict way. For instance "travel time is between 10 and 15 min" or "the location is fully covered if the distance between it and a facility is less than 1 km and partially covered up to 2 km". Although this change may seem small, it can totally change the optimal facility positions and the quality of the solution.

Let us illustrate using the small-size instance of finding optimal locations for hospitals (Figs. 1 and 2). For the sake of simplicity, travel times will be deterministic. In the standard model (Fig. 1), the covering radius determines the maximum survival time of a patient. On the other case, if the covering radius has two borders—the first on, where the probability of survival is 100% and another, external radius which determines a border at which there is a probability of survival somewhere between 0% and 100%. It is clear that this probability decreases as the distance increases. Figure 2

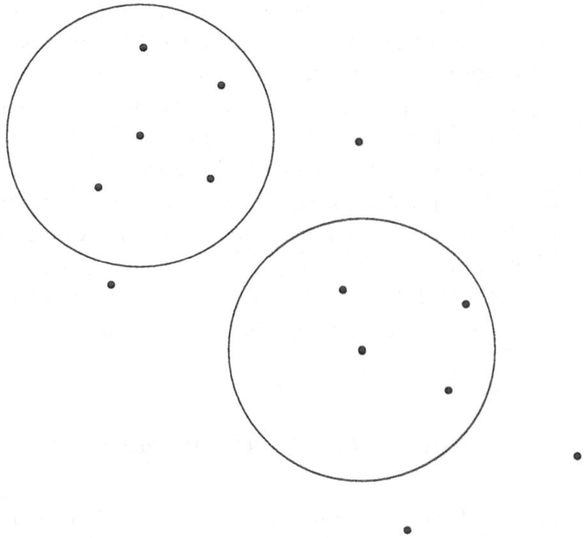

Fig. 1 Standard model of MCLP

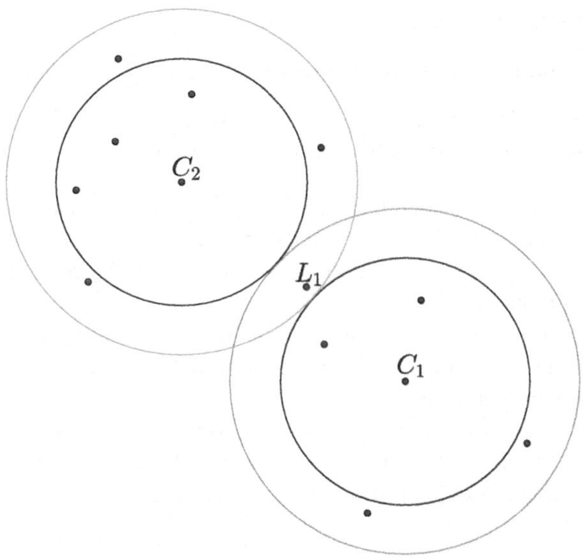

Fig. 2 MCLP with extensible radius

illustrates the optimal solution for a model defined in this way. In the classical model, the optimal solution covers 9 locations, while 4 are uncovered with 3 being very far from the covering bound. On the other hand, the modified model fully covers 7 locations but all the other locations are partially covered, very close to the coverage bounds, with a high probability of getting help in time.

Another question that remains is how to treat locations partially covered by more than one facility (location L_1 on Fig. 2). This depends directly on the type of the problem. For example, if the problem is to find optimal positions for medical services, quality of the solution will not be better if location is partially covered by more than one facility, if a patient will be treated by just one medical service whereas other close services do not have any importance. On the other hand, in some problems multiple partial coverage can improve the quality of the solution, e.g. when the problem is to locate fire stations, if a location is partially covered by multiple fire stations, it is safer. In the end, computing the quality of the solution in some models can be more complex and there is a possibility to use another metric.

The following subsections will show how to model the described uncertainties related to the radius and travel time by using fuzzy sets and how to use t-conorms for computing the overall quality of the solution.

3.1 Definitions and Preliminaries

Fuzzy Set Theory was introduced by L. Zadeh in [31]. There are three basic types of fuzzy sets: triangular, trapezoidal and a left(right) shoulder which are usually called fuzzy numbers.

Definition 1 A fuzzy number on the set $\mathbb{R}$ is a mapping $u : \mathbb{R} \to [0, 1]$ such that:

- u is monotonically non-decreasing on $]-\infty, b[$,
- u is monotonically non-increasing on $]c, \infty[, \ c \geqslant b$.

If $b = -\infty$ a left shoulder is obtained, if $c = \infty$ a right shoulder is obtained. If the function is linear and strictly increasing on $]a, b[$ and linear and strictly decreasing on $]c, d[$ the fuzzy number is trapezoidal. At the end, a triangular fuzzy number is obtained for $b = c$.

It is not hard to conclude that the triangular fuzzy number can model travel times with uncertainty, e.g. 5 ± 2, while the right shoulder fuzzy numbers can model the covering radius.

The main question is how to use both fuzzy measures simultaneously, e.g. how to compute the interaction between them. A simple, but very effective solution is presented in [11], where the authors have introduced the equation:

$$c_{ij} = \begin{cases} 1, t_{ij} \leqslant R, \\ 0, t_{ij} > R + r, \\ \frac{1}{2}(\mu(v_1) + \mu(v_2)), \text{otherwise.} \end{cases} \tag{15}$$

where $\mu(v_1)$ and $\mu(v_2)$ represent the intersection values of fuzzy numbers for the coverage radius and travel time (Fig. 3).

Definition 2 A is an aggregation operator on the unit interval $\mathsf{A} : \bigcup_{n \in \mathbb{N}} [0, 1]^n \to [0, 1]$ if the following conditions hold:

(A1) $\mathsf{A}(0, \ldots, 0) = 0$,

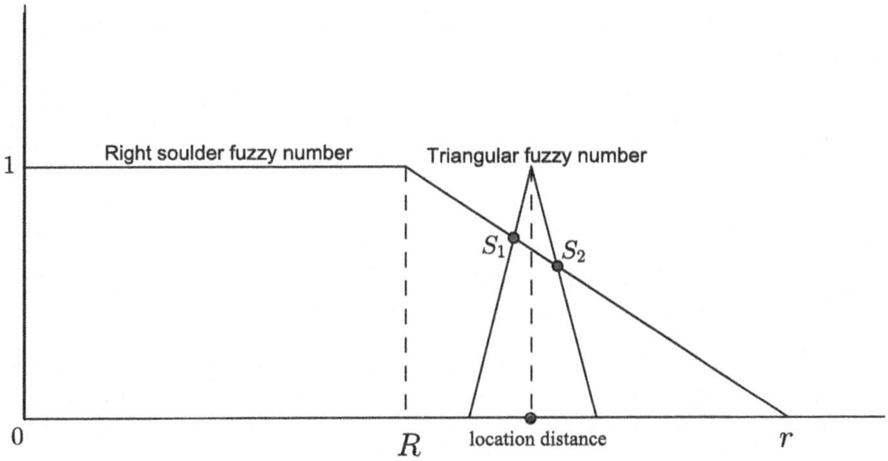

Fig. 3 Intersection of fuzzy numbers for modeling radius and distance between locations

(A2) $\mathsf{A}(1, \ldots, 1) = 1$,
(A3) if $(\forall i = 1, 2, \ldots, n)(x_i \leqslant y_i)$ then $\mathsf{A}(x_1, x_2, \ldots, x_n) \leqslant \mathsf{A}(y_1, y_2, \ldots, y_n)$.

Conditions **(A1)** and **(A2)** are called boundary conditions, and **(A3)** resembles the monotonicity property of the operator A. Some of of the most common aggregation operators are t-norms and t-conorms.

Definition 3 ([21]) A mapping $T : [0, 1]^2 \to [0, 1]$ is called a t-norm if the following conditions are satisfied for all $x, y, z \in [0, 1]$:
(T1) $\mathsf{T}(x, y) = \mathsf{T}(y, x)$,
(T2) $\mathsf{T}(x, \mathsf{T}(y, z)) = \mathsf{T}(\mathsf{T}(x, y), z)$,
(T3) if $y \leqslant z$ then $\mathsf{T}(x, y) \leqslant \mathsf{T}(x, z)$,
(T4) $\mathsf{T}(x, 1) = x$.

One of the most common t-norms is the $\mathsf{T}_\mathbf{M}(x, y) = \min(x, y)$. Besides $\mathsf{T}_\mathbf{M}$, two t-norms that are often used are the product t-norm $\mathsf{T}_\mathbf{P}(x, y) = x \cdot y$ and the Lukasiewicz t-norm $\mathsf{T}_\mathbf{L}(x, y) = \max(x + y - 1, 0)$.

Definition 4 A mapping $\mathsf{S} : [0, 1]^2 \to [0, 1]$ is called an s-norm or a t-conorm if the following conditions **(T1)**, **(T2)**, **(T3)** from definition 3 and condition
(S4) $\mathsf{S}(x, 0) = x$
are satisfied for all $x, y, z \in [0, 1]$.

All these operators are dual i.e. for each t-norm T there exists a t-conorm such that

$$\mathsf{T}(x, y) = 1 - \mathsf{S}(1 - x, 1 - y). \tag{16}$$

Thus, the dual t-conorm for $\mathsf{T_M}$ is $\mathsf{S}_M(x, y) = max(x, y)$, for $\mathsf{T_P}$ is the probabilistic sum $\mathsf{S}_P(x, y) = x + y - xy$ and finally, the dual for $\mathsf{T_L}$ is the limited sum $\mathsf{S}_L(x, y) = \min(x + y, 1)$.

Since t-norms and t-conorms are associative, they can be extended to n-ary operators. For each t-norm $\mathsf{T} : [0, 1]^2 \rightarrow [0, 1]$ its n-ary extension is obtained for each $n \in \mathbb{N}$:

$$\mathsf{T}_n(x_1, \ldots, x_n) = \begin{cases} x_1, \text{if } n = 1, \\ \mathsf{T}(x_1, x_2), \text{if } n = 2, \\ \mathsf{T}(\mathsf{T}_{n-1}(x_1, \ldots, x_{n-1}), x_n). \end{cases} \tag{17}$$

In the models described in the following subsections, the quality of the models' solutions will be computed by using n-ary maximum t-conorm S_M, the limited sum t-conorm S_L and an ordered weighted sum (OWS) from the next definition.

Definition 5 ([12]) An ordered weighted sum (OWS) n-dimensional operator is a mapping $F : \mathbb{R}^n \rightarrow \mathbb{R}$ that has an associated collection of weights $W = [w_1, \ldots, w_n]$ lying in the unit interval and with $F(a_1, \ldots, a_n) = \sum_{j=1}^{n} w_j b_j$ where b_j is the j-th largest of the a_i. In case $w_1 = \cdots = w_n = \frac{1}{n}$ the arithmetic mean is obtained.

Before presenting fuzzy covering location problems, it is necessary to redefine the notations used in standard models. The following notation will be used in the fuzzy models:

$$I \text{ - the set of locations (indexed by } i), \tag{18}$$

$$n = card\{I\} \text{ - the number of locations }, \tag{19}$$

$$J \subset I \text{ - the set of eligible facility sites (indexed by } j), \tag{20}$$

$$x_j = \begin{cases} 1, \text{if the facility is located at the location } j, \\ 0, \text{otherwise}, \end{cases} \tag{21}$$

$$t_{ij} \pm t \text{ - fuzzy travel time from location } i \text{ to location } j, \tag{22}$$

$$R + f_r \text{ - the fuzzy coverage radius }, \tag{23}$$

$$c_{ij} = \begin{cases} 1, t_{ij} \leqslant R, \\ 0, t_{ij} > R + f_r, \\ \frac{1}{2}(\mu(v_1) + \mu(v_2)), \text{otherwise}, \end{cases} \tag{24}$$

as introduced in equation (15).

3.2 *Fuzzy Location Set Covering Problem*

As previously described, the goal of the Fuzzy Location Set Covering Problem (FLSCP) is to incorporate a triangular fuzzy number for the travel times (22) and the right shoulder fuzzy number for the coverage radius (23). In the LSCP each location must be fully covered with at least one facility, a limited sum conorm S_L should be used. The limited sum conorm ensures that the degree of coverage for each location is 1, and full coverage could be accomplished by a single facility or by partial coverages of more facilities.

Considering the aforementioned, the mathematical model of FLSCP is:

$$\min \quad \sum_{j \in J} x_j,$$

subject to:

$$\begin{aligned} \mathsf{S}_L(x_j, c_{ij}) &= 1, \forall i \in I, \forall j \in J, \\ x_j \quad &\in \{0, 1\}, \forall j \in J, \end{aligned}$$

3.3 Fuzzy Maximal Covering Location Problem

Unlike the FLSCP, the Fuzzy Maximal Covering Location Problem (FMCLP) allows us to use all three conorms, S_M, S_L, and OWS. The reason for this is described in the introductory part of this section.

The following are mathematical models of all three versions of the FMCLP.

The FMCLP model with the maximum conorm:

$$\max \quad \sum_{i \in I} y_i,$$

subject to:

$$\begin{aligned} \textstyle\sum_{j \in J} x_j \quad &= P, \\ \mathsf{S}_M(x_j, c_{ij}) &\geqslant y_i, \forall i \in I, \forall j \in J, \\ x_i \quad &\in \{0, 1\}, \\ y_i \quad &\in [0, 1]. \end{aligned}$$

The MCLP model with the limited sum conorm:

$$\max \sum y_i$$

subject to:

$$\begin{aligned} &\sum {}_{j\in J} x_j = P, \\ &\mathsf{S}_L \ (x_j, c_{ij}) = 1, \forall j \in J, \forall i \in I, \\ &x_i \ \in \{0, 1\}, \\ &y_i \ \in [0, 1]. \end{aligned}$$

The FMCLP model with the OWS operator:

$$\max \quad \sum y_i$$

subject to:

$$\begin{aligned} \sum\nolimits_{j\in J} x_j &= P, \\ F(x_1, x_2, \ldots, x_n) &\geqslant y_i, \forall i \in I, \\ x_i &\in \{0, 1\}, \\ y_i &\in [0, 1]. \end{aligned}$$

where F in operator defined in Definition 5 with weight coefficients $W = [c_{1j}, c_{2j}, \ldots, c_{nj}], \forall j$.

All of the models have two differences from the original MCLP model. The main difference is that here variable y_i is no longer binary. Now it is continuous on the unit interval and allows partial coverage. One more difference is the second constraint that ensures the coverage of location i by a facility. In the standard model, it is simply an inequality, here it is one of n-ary operators, S_M, S_L and OWS.

3.4 *Fuzzy Minimal Covering Location Problem*

In the case of the Fuzzy Minimal Covering Location Problem (FMinCLP), all three aggregation operators can be used. For example, the total impact caused by air pollutants is equal to the sum of all partial pollutants, while maximum and ordered weighted sums could be used for problems where influence cannot be added. Such problems could include those involving different types of pollutants and an acceptable level of pollution. As in the case of the classical MinCLP model, the FMinCLP requires an additional variable for the minimal distance between facilities:

$$D - \text{the minimal distance between facilities.} \tag{25}$$

Therefore, the three models of the FMinCLP are as follows.

The FMinCLP with the maximum conorm:

$$\min \quad \sum y_i$$

subject to:

$$\begin{aligned}
\textstyle\sum_{j\in J} x_j &= P,\\
d(x_i, x_j) &\geqslant D, \forall i \neq j,\\
\mathsf{S}_M(x_j, c_{ij}) &\geqslant y_i, \forall i \in I, \forall j \in J,\\
x_i &\in \{0, 1\},\\
y_i &\in [0, 1].
\end{aligned}$$

The FMinCLP with the limited sum conorm:

$$\min \quad \sum y_i$$

subject to:

$$\begin{aligned}
\textstyle\sum_{j\in J} x_j &= P,\\
d(x_i, x_j) &\geqslant D, \forall i \neq j,\\
\mathsf{S}_L(x_j, c_{ij}) &= 1, \forall i \in I, \forall j \in J,\\
x_i &\in \{0, 1\},\\
y_i &\in [0, 1].
\end{aligned}$$

The FMinCLP with the OWS operator:

$$\min \sum y_i$$

subject to:

$$\begin{aligned}
&\textstyle\sum_{j\in J} x_j = P,\\
&d(x_i, x_j) \geqslant D, \quad \forall i \neq j,\\
&F(x_1, x_2, \ldots, x_n) \geqslant y_i, \forall i \in I,\\
&x_i \in \{0, 1\},\\
&y_i \in [0, 1].
\end{aligned}$$

where F is the operator defined in Definition 5 and the weight coefficients $W = [c_{1j}, c_{2j}, \ldots, c_{nj}], \forall j \in J$.

4 Computational Experiments

This chapter presented 3 classical and 7 fuzzy models of covering location problems. A detailed computational analysis, along with the methods for solving these models and as well as the results surpasses the scope of this chapter and can be found in [11,

Table 1 The results of CPLEX and PSO solvers

			CPLEX		PSO		
n	$R + f_r$	f_t	Solution	Time (ms)	Solution	Time (ms)	GAP(%)
70	$5 + 0.5$	0.1	68.018	38794	68.018	259	0
70	$5 + 0.5$	0.2	68.087	34786	68.087	376	0
70	$5 + 1$	0.1	68.416	112369	68.416	202	0
70	$5 + 1$	0.2	68.398	112369	68.398	418	0
80	$5 + 0.5$	0.1	76.811	3415977	76.811	276	0
80	$5 + 0.5$	0.2	76.85	2887284	76.85	329	0
80	$5 + 1$	0.1	77.315	4431462	77.315	269	0
80	$5 + 1$	0.2	77.294	10166557	77.294	269	0
90	$5 + 0.5$	0.1	N/A	3293544	85.506	1414	–
90	$5 + 0.5$	0.2	N/A	4737207	85.343	1614	–
90	$5 + 1$	0.1	N/A	8276707	86.762	2170	–
90	$5 + 1$	0.2	N/A	8956056	86.75	1502	–

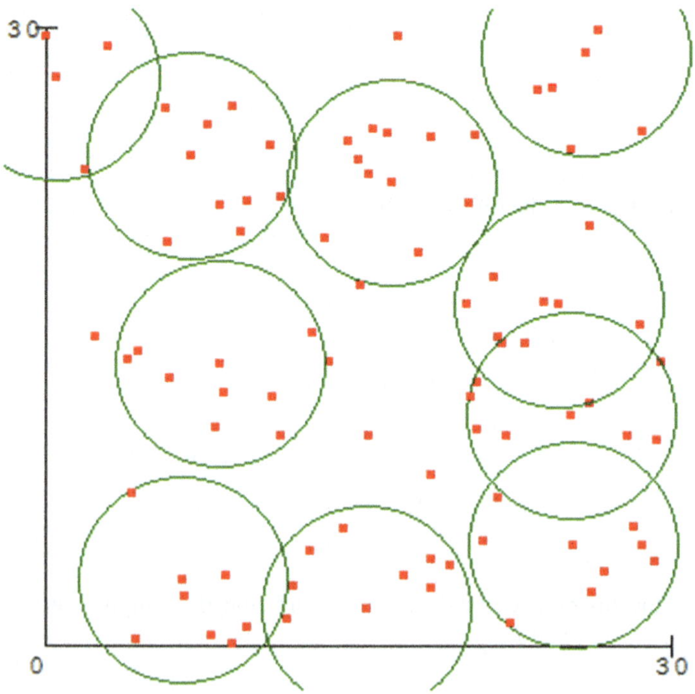

Fig. 4 A solution for problem $n = 90,\ P = 10,\ R + f_r = 5 + 0,\ f_d = 0$

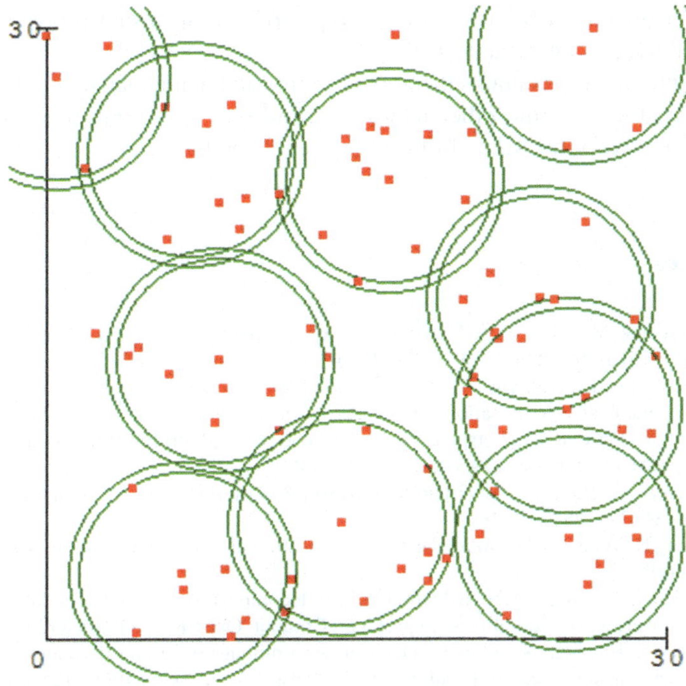

Fig. 5 A solution for problem with input parameters $n = 90$, $P = 10$, $R + f_r = 5 + 0.5$, $f_d = 0.1$

13, 14]. Solutions for a few instances of FMCLP, taken from [11], will be shown here, as an illustration.

Computational experiments were run on generated instances of the problem, the locations are randomly placed in a 30x30 grid. Instances were solved by the exact solver of the IBM CPLEX optimizer and the PSO metaheuristic described in Sect. 1.1. Table 1 presents computational results for the instances of size 70, 80 and 90, the fuzzy covering radius $R + f_r = 5 + 0.5, 5 + 1$, fuzzy travel times $f_t = 0.1, 0.2$ and the number of facilities $P = 10$.

Figures 4 and 5 present graphical representations of the solutions for instances with the parameters $n = 90$, $P = 10$, $R + f_r = 5 + 0$, $f_d = 0$ and $n = 90$, $P = 10$, $R + f_r = 5 + 0.5$, $f_d = 0.1$.

5 Conclusion

This chapter demonstrates how to use fuzzy logic to model uncertainty in various combinatorial optimization problems. Although this chapter only presented covering location problems with two fuzzy variables for covering radius and travel time, a similar mechanism can be used in other types of combinatorial problems, which have some degree of uncertainties.

This chapter mainly contains theoretical results and mathematical models related to fuzzy variables. On the other hand, the results of some instances are briefly described, mainly for the sake of illustrating fuzzy models.

References

1. Arana-Jiménez, M., Blanco, V., Fernández, E.: On the fuzzy maximal covering location problem. Eur. J. Oper. Res. **283**(2), 692–705 (2020)
2. Basar, A., Catay, B., Unluyurt, T.: A multi-period double coverage approach for locating the emergency medical service stations in Istanbul. J. Oper. Res. Soc. **62**, 627–637 (2011)
3. Batanović, V., Petrović, D., Petrović, R.: Fuzzy logic based algorithms for maximum covering location problems. Inf. Sci. **179**(12), 120–129 (2009)
4. Bhattacharya, U., Rao, J.R., Tiwari, R.N.: Fuzzy multi-criteria facility location problem. Fuzzy Sets Syst. **51**, 277–287 (1992)
5. Church, R., ReVelle, C.: The maximal covering location problem. Pap. Region. Sci. Assoc. **32**, 101–118 (1974)
6. Corrêa, F.d.A., Lorena, L.A.N., Ribeiro, G.M.: A decomposition approach for the probabilistic maximal covering location-allocation problem, Comput. Oper. Res. **36**(10), 2729-2739 (2009)
7. Curtin, K.M., Hayslett, K., Qiu, F.: Determining optimal police patrol areas with maximal covering and backup covering location models. Netw. Spatial Econ. **10**(1), 125–145 (2010)
8. Darzentas, J.: A discrete location model with fuzzy accessibility measures. Fuzzy Sets Syst. **23**, 149–154 (1987)
9. Davari, S., Zarandi, M.H.F., Hemmati, A.: Maximal Covering Location Problem (MCLP) with fuzzy travel times. Expert Syst. Appl. **38**(12), 14534–14541 (2011)
10. Davari, S., Zarandi, M.H.F., Turksen, I.B.: A greedy variable neighborhood search heuristic for the maximal covering location problem with fuzzy coverage radii. Knowl. Based Syst. **41**, 68–76 (2013)
11. Drakulić, D., Takači, Marić M.: Fuzzy covering location problems with different aggregation operators. Filomat **31** (2) (2017)
12. Drakulić, D., Takači, A., Marić, M.: New model of maximal covering location problem with fuzzy conditions. Comput. Inf. **35**(3) (2016)
13. Drakulić, D.: Fuzzy sets in modelling location problems of combinatorial optimization (Doctoral dissertation, in Serbian). University of East Sarajevo (2016)
14. Drakulić, D., Takači, Marić M.: Solving minimal covering location problems with single and multiple node coverage, Scientific Bulletin of the Petru Maior University of Targu Mures **1**(2) (2016)
15. Drezner, Z., Hamacher, H.W. (eds.): Facility Location: Applications and Theory. Springer, Berlin, Heidelberg (2004)
16. Eiselt, H.A., Marianov, V.: Foundations of Location Analysis. Springer, US (2011)
17. Guzman, V.C., Pelata, D.A., Verdeay, J.L.: An approach for solving maximal covering location problems with fuzzy constraints. Int. J. Comput. Intell. Syst. **9**(4), 734–744 (2016)
18. Kennedy, J., Eberhart, H.: Swarm Intelligence. Morgan Kaufmann Publishers, San Francisco (2001)
19. Kennedy, J., Eberhart, H.: A discrete binary version of the particle swarm algorithm. IEEE Conf. Syst. Man Cybern. **5**, 1942–1948 (1997)
20. Kennedy, J., Eberhart H.: Particle Swarm Optimization. In: Proceedings of 4th IEEE Conference on Neural Networks, Perth, Australia, 1942–1948 (1995)

21. Klement, E.P., Mesiar, R., Pap, E.: Triangular Norms, Trends in Logics 8. Kluwer Academic Publishers, Dordrecht/Boston/London (2000)
22. Marianov, V., Serra, D.: Probabilistic maximal covering location-allocation models for congested systems. J. Region. Sci. **38**(3), 401–424 (1998)
23. Marianov, V., ReVelle, C.: The queuing probabilistic location set covering problem and some extensions. Socio-Econ. Plan. Sci. **28**(3), 167–178 (1994)
24. Murawski, L., Church, R.L.: Improving accessibility to rural health services: the maximal covering network improvement problem. Socio-Econ. Plan. Sci. **43**, 102–110 (2009)
25. Perez, J.A.M., Vega, J.M.M., Verdegay, J.L.: Fuzzy location problems on networks. Fuzzy Sets Syst. **142**, 393–405 (2004)
26. ReVelle, C.S., Eiselt, H.A.: Location analysis: a synthesis and survey. Eur. J. Oper. Res. **165**(1), 1–19 (2005)
27. ReVelle, C., Hogan, K.: The maximum reliability location problem and α-reliable p-center problem: derivatives of the probabilistic location set covering problem. Ann. Oper. Res. **18**, 155–174 (1989)
28. ReVelle, C., Hogan, K.: A reliability-constrained siting model with local estimates of busy fractions. Environ. Plan. B: Plan. Des. **15**, 143–152 (1988)
29. Toregas, C., Swain, R., ReVelle, C., Bergmann, L.: The location of emergency service facilities. Oper. Res. **19**, 1363–1373 (1971)
30. Wenxian, G., Jianxin, X., Hongmei, G.: A hybrid PSO for optimizing locations of booster chlorination stations in water distribution systems. In: 2010 International Conference on Intelligent Computation Technology and Automation, vol. 1, pp. 126-129 (2010)
31. Zadeh, L.A.: Fuzzy sets. Inf. Control **8**, 338–353 (1965)

An Improved BAT Algorithm for Solving Job Scheduling Problems in Hotels and Restaurants

Tarik A. Rashid, Chra I. Shekho Toghramchi, Heja Sindi, Abeer Alsadoon, Nebojša Bačanin, Shahla U. Umar, A. S. Shamsaldin, and Mokhtar Mohammadi

Abstract One popular example of metaheuristic algorithms from the swarm intelligence family is the Bat algorithm (BA). The algorithm was first presented in 2010 by Yang and quickly demonstrated its efficiency in comparison with other common algorithms. The BA is based on echolocation in bats. The BA uses automatic zooming to strike a balance between exploration and exploitation by imitating the deviations of the bat's pulse emission rate and loudness as it searches for prey. The BA maintains solution diversity using the frequency-tuning technique. In this way, the BA can quickly and efficiently switch from exploration to exploitation. Therefore, it becomes an efficient optimizer for any application when a quick solution is needed. In this

T. A. Rashid (✉) · C. I. Shekho Toghramchi · A. S. Shamsaldin
Department of Computer Science and Engineering, University of Kurdistan of Hewler, Erbil, KRG, Iraq
e-mail: tarik.ahmed@ukh.edu.krd

H. Sindi
Business and Management Department, School of Management and Economics, University of Kurdistan Hewlêr, Erbil, KRG, Iraq

A. Alsadoon
School of Computing and Mathematics, Charles Sturt University (CSU), Wagga Wagga, Australia

School of Computer Data and Mathematical Sciences, University of Western Sydney (UWS), Sydney, Australia

Asia Pacific International College (APIC), Information Technology Department, Sydney, Australia

Kent Institute Australia, Information Technology Department, Sydney, Australia

N. Bačanin
Singidunum University, Belgrade, Serbia

S. U. Umar
Technical College of Informatics, Sulaimani Polytechnic University, Sulaimani, KRG, Iraq

Network Department, College of Computer Science and Information Technology, Kirkuk University, Kirkuk, Iraq

M. Mohammadi
Department of Information Technology, Lebanese French University, Erbil 44001, KRG, Iraq

E. Pap (ed.), *Artificial Intelligence: Theory and Applications*,
Studies in Computational Intelligence 973,
https://doi.org/10.1007/978-3-030-72711-6_9

paper, an improvement on the original BA has been made to speed up convergence and make the method more practical for large applications. To conduct a comprehensive comparative analysis between the original BA, the modified BA proposed in this paper, and other state-of-the-art bio-inspired metaheuristics, the performance of both approaches is evaluated on a standard set of 23 (unimodal, multimodal, and fixed-dimension multimodal) benchmark functions. Afterwards, the modified BA was applied to solve a real-world job scheduling problem in hotels and restaurants. Based on the achieved performance metrics, the proposed MBA establishes better global search ability and convergence than the original BA and other approaches

Keywords Swarm intelligence · Bat algorithm (ba) · MBA · Benchmark test function · Job sheduling · Hotel · Restaurant

1 Introduction

Swarm intelligence (SI) is described as the collective behaviour of self-organized and distributed systems that might be artificial or natural. It was introduced in 1989, by Beny [1]. Since then, a great number of algorithms have been proposed. Swarm intelligence algorithms substantiated their ability to tackle numerous optimization problems in several areas. Self-organization and the separation of work are the essential characteristics of swarm intelligence. Self-organization means the ability to organize any system into a prefered pattern or model without any external support. Self-organization also depends on several characteristics, i.e., multiple interactions, negative feedback, fluctuations, and positive feedback. To achieve amplification and stabilization, positive and negative feedback are implemented, while fluctuations are useful for randomness. Meanwhile, different interactions take place when swarm individuals share information among themselves in the search space [2].

Advanced optimization algorithms based on swarm intelligence are inspired by nature. The main sources are biological systems, such as Ant Colony Optimization(ACO), the Genetic Algorithm (GA), the Cuckoo Search Algorithm (CS), the Bat Algorithm, and many others [2]. Nowadays, nature-inspired meta-heuristic algorithms have become a useful tool for finding the answers to optimization problems that companies encounter. In meta-heuristic algorithms, there are two types of solutions. One is single-solution-based and the other is population-based. In single solution-based algorithms, the search process begins with only a single candidate solution, which is then improved over the iterations of the algorithm. In contrast, agents in the population-based one initialize random population in the first iteration and have better communication. Possible solutions exchange information about the search space, which leads to the search process moving to more feasible areas and avoiding being trapped in local optima, Furthermore, it has better exploration ability [3, 4].

The Bat Algorithm is characterized by its efficiency in the exploitation phase, while it suffers from a weak exploration capacity. Sometimes bats get outside the solution search space. Also, acceptance of the scheme is one of the limitations of the BA's need to reduce the probability of accepting a poor solution. In practice, it shows that the BA converges very quickly in the early stages, while it later slows down. However, the accuracy is sufficient when the number of functions is high. The BA does not save the best loudness in every single iteration, even though a global solution has very little impact on a new solution. The main aim of this work is to present a modified version of the BA, then test it against 23 benchmark functions. The results of the BA are then compared to those of the MBA. Finally, a real-world job scheduling problem is solved using the MBA. The main contribution of the MBA is to increase the convergence level and improve the accuracy of the standard BA.

The rest of the paper is presented as follows: in section two, a literature review is presented. In section three, details of the Bat algorithm are explained. Section four introduces the MBA. Section five presents the implementation of both the BA and the MBA. Finally, in Sect. 6, the results are provided along with an analysis.

2 Literature Review

In 2013, Ramesh, Mohan, and Reddy presented an application which aimed to eliminate the environmental effects caused by fossil-fuel power plant emissions. The application used a Bat algorithm to deal with a multi-objective optimization problem in the power system [5].

Yilmaz, Kucuksille, and Cengiz proposed an improvement to the exploration mechanism of the Bat algorithm in 2014. They modified the loudness and pulse rate emission of the Bat algorithm since these two factors affect all dimensions in the solutions while searching. They checked the performance of the modified Bat algorithm and BA on 15 benchmark test functions. When it comes to solving optimization problems, the MBA outperformed the original BA [6].

In 2015, Fister, Rauter, Yang, Ljubic, and Fister Jr, proposed an MBA as an intelligent planner for sports training sessions. They modified the Bat algorithm to solve the mentioned problem. They applied the MBA algorithm on ten basic training sessions for a specific athlete. The idea was to plan a cycling training session The results showed a plan whose prediction which complied with the high standards required by cycling coaches [7].

In 2016, Kielkowics and Grela proposed two modifications of the bat algorithm. They introduced a different scheme of acceptance for newly found solutions and the velocity equation was modified. Furthermore, they modified the acceptance scheme to reduce the probability of accepting a poor solution, and also changed the velocity update equation by archiving the component and introducing a cognitive coefficient. As the last step, they created additional memory by storing the best solutions that were found during the optimization process. They compared the modified bat algorithm with the existing bat algorithm and tested the modifications in a few simulation experiments [8].

Cai, Gao, and Xue introduced an improved Bat algorithm in 2016. The improvements focus on the exploitation and exploration capabilities of the Bat algorithm, which was limited, and the multi-model numerical problem. A new search mechanism was suggested using an optimal forage approach to improve the bat's movements and directions, as well as to utilize the random disturbance method to enhance the exploration phase. They tested and compared the performance with four other evolutionary algorithms. The simulation results indicated that the improved Bat algorithm was the most effective [9].

In 2019, Osab et al. proposed a discrete and improved Bat algorithm (DaIBA) that concentrated on the problem of real-world drugs spreading with pharmacological waste collection. They modelled the problem as a multi-attribute, or rich vehicle routing problem (RVRP), i.e. as a clustered vehicle routing problem for deliveries and pickups, asymmetric variable costs, and forbidden paths. This is the first time that the problem was addressed by the scientific community. Thereupon, they designed a benchmark consisting of 24 datasets which contained from 60 to 1000 customers. The results concluded that the proposed algorithm was a promising technique for the specified problem [10].

3 The Bat Algorithm

There are some rules which are used to develop Bat-inspired algorithms [11];

1. Distance: In general, bats use echolocation to sense distance. They recognize the difference between food/prey and background barriers in a remarkable way.
2. Frequency: They can fly arbitrarily with a velocity of v_i position x_i with a fixed frequency f_{min}, varying wavelength γ, and loudness A_0 to search for prey. They can automatically adjust the wavelength (or frequency) of their emitted pulses and adjust their rate of pulse emission $r = [0, 1]$, depending on their proximity to their target.
3. Loudness: Although the loudness can vary in many ways, it can be assumed that loudness could diverge from a significant (positive) A_0 to a minimum constant A_{min}.

Generally, (x_i) was used to represent the bat's position in the search space, while the (v_i) and (f_i) used to denote the velocities and the frequency of the pulse respectively. Whereas (β) refers to a vector of random numbers between 0 and 1. As well as, (x_*) indicates the best solution obtained till now.

$$f_i = f_{min} + (f_{max} - f_{min})\beta \quad (1)$$

$$v_i^t = v_i^{(t-1)} + (x_i^t - x_*) f_i \quad (2)$$

$$x_i^t = x_i^{(t-1)} + v_i^t \quad (3)$$

After comparing all the other solutions of n bats. Each bat is assigned a random frequency from $[f_{max}, f_{min}]$. Using a random walk for the local search, only one solution is selected from the best ones at this moment and a new solution for each bat is generated

$$x_{new} = x_{old} + \epsilon A^t \tag{4}$$

A^t, which is equal to A_i^t and is the average loudness at this time step for all of the bats and $\epsilon \in [-1, 1]$ is a random number. As the iteration proceeds, the emission must be updated, as must the loudness A_i and the rate r_i of the pulse. When a bat finds its prey, the loudness decreases and the pulse emission increases.

$$A_i^t = \alpha A_i^t, \quad r_i^{(t+1)} = r_i^0[1 - \exp(-\gamma t)] \tag{5}$$

$$If(rand > A^i \& f(x_i) < f(x_*)), \tag{6}$$

where α and γ are constants, for any $0 < \alpha < 1$ and $\gamma > 0$,

$$A_i^t \rightarrow 0, r_i^t \rightarrow r_i^0, \text{as} \quad t \rightarrow \infty. \tag{7}$$

For the sake of simplicity, $\alpha = \gamma$ is used. Through the randomization process, different values for the loudness and pulse emission rate of each bat [11] is computed. Algorithm 1, shows the pseudo-code of the Bat Algorithm.

Algorithm 1 Pseudo-code of Bat algorithm

Objective function $f(x)$ where $x = (x_1, ..., x_d)^T$
Initialize the bat population $x_i (i = 1, 2, ..., n)$ and v_i
Define pulse frequency f_i at x_i
Initialize pulse rates r_i and the loudness A_i
while ($t <$ max number of iterations) **do**
  Generate new solutions by adjusting, updating velocities and locations use Eqs. (1)–(3)
  if ($rand > r_i$) **then**
    choose a solution among the best solutions
    $\boldsymbol{x_{new} = x_{old} + \epsilon A^t}$
  end if
  Generate a new solution by flying randomly
  if ($rand < A^i$ & $f(x_i) < f(x_*)$) **then**
    Accept the new solutions
    Increase r_i and decrease A_i
  end if
  Rank the bats and find the current best x_*
end while

4 The Modified Bat Algorithm

In a meta-heuristic based population, one of the essential aspects is the balance between the exploration and exploitation phases of the search space. Exploration is sometimes called diversification and is responsible for the global search ability of the algorithm. Exploitation is sometimes called intensification and is responsible for the local search ability of the algorithm. The bat algorithm is more effective at exploitation in a local search; although it may become trapped in local optima which leads to an inability to perform a global search [8]. Generally, the BA depends entirely on random walks, therefore swift convergence cannot be guaranteed [11]. The bat algorithm is modified to mitigate the aforementioned limitation. The main improvement is achieved by replacing the old position as opposed to adding a global position and global loudness for all bats. This is done by modifying Eq. (4) to match Eq. (8)

$$x_{new} = x^* + \epsilon A^t + A^* \tag{8}$$

The improvement aims to speed up convergence and make the method more suitable for a wider range of applications. If a new solution improves the result, then the loudness and pulse rate will be updated. After generating the new solution, the loudness of the best solution until that point is saved as the best loudness. Algorithm 2 shows the pseudo-code of the modified Bat Algorithm.

Algorithm 2 Pseudo-code of Modified Bat algorithm

Objective function $f(x)$ where $x = (x_1, ..., x_d)^T$
Initialize the bat population $x_i (i = 1, 2, ..., n)$ and v_i
Define pulse frequency f_i at x_i
Initialize pulse rates r_i and the loudness A_i
while ($t <$ max number of iterations) **do**
  Generate new solutions by adjusting, updating velocities and locations use equation (1), (2), and (3)
  if ($rand > r_i$) **then**
    choose a solution among the best solutions
    $\boldsymbol{x_{new} = x^* + \epsilon A^t + A^*}$
  end if
  Generate a new solution by flying randomly
  if ($rand < A^i$ & $f(x_i) < f(x_*)$) **then**
    Accept the new solutions
    Store the loudness of the best solution
    Increase r_i and decrease A_i
  end if
  Rank the bats and find the current best x_*
end while

Table 1 Unimodal benchmark test functions

Function name	Function	Range
Sphere function	$f_1(x) = \sum_{i=1}^{n} x_i^2$	[−100, 100]
Schwefel' problem 2.22	$f_2(x) = \sum_{i=1}^{n} \lvert x_i \rvert + \prod_{i=1}^{n} \lvert x_i \rvert$	[−10, 10]
Schwefel problem 1.2	$f_3(x) = \sum_{i=1}^{n} \left(\sum_{j-1}^{i} x_j\right)^2$	[−100, 100]
Schwefel's problem 2.21	$f_4(x) = max_i \{\lvert x_i \rvert, \ 1 \le i \le n\}$	[−100, 100]
Generalized Rosenbrock's function	$f_5(x) = \sum_{i-1}^{n-1} \left[100\left(x_{i+1} - x_i^2\right)^2 + (x_i - 1)^2\right]$	[−30, 30]
Step function	$f_6(x) = \sum_{i=1}^{n} ([x_i + 0.5])^2$	[−100, 100]
Quartic function with noise	$f_7(x) = \sum_{i=1}^{n} ix_i^4 + random[0, \ 1]$	[−1.28, 1.28]

5 Implementation

The BA and MBA are tested on a set of standard well-known benchmark test functions. A simulation experiment was performed on a MacBook Pro computer, running macOS Sierra version 10.12.6. on Intel Core i7 2.2 GHz with 16 GB of memory. The algorithms are implemented in C++ and compiled with MATLAB R2018a. This paper is divided into several parts, as shown below:

5.1 Benchmark Test Functions

To conduct a comprehensive evaluation of the bat algorithm and the modified bat algorithm, and the effects they have on problems, their performance was tested using a large set of standard benchmark functions. These benchmark functions are listed in Tables 1, 2, and 3. In total, there are 23-benchmark functions, which are classified into three categories according to their characteristics, as follows:

5.1.1 Unimodal Benchmark Functions

This group consists of functions from F1–F7 and they are high dimensional. They are relatively easy to solve in comparison to the other groups. Table 1 shows the unimodal benchmark function's name, equation, and the lower and upper bounds used for the function.

5.1.2 Multimodal Benchmark Functions

This group is composed of functions F8–F13. They are considered to be the most challenging problems in the set of benchmark functions. Table 2 shows the Multimodal benchmark function's name, equation, and the lower and upper bounds used for the function.

5.1.3 Fixed-Dimension Multimodal Benchmark Functions

This group includes functions F14-F23 and they have smaller dimensions. Table 3 shows the Fixed-Dimension Multimodal benchmark function's name, equation, and the lower and upper bounds used for the function.

6 Results and Discussion

The bat algorithm and the modified bat algorithm searched for an optimum solution with a population size of 30 and a maximum number of iterations of 500. The goal was to find an optimum solution over 500 iterations after which the average value and standard deviation results of both the BA and the MBA were calculated. Many comparisons were conducted to evaluate the BA and the MBA.

6.1 The Average Value and the Standard Deviation

The averages and standard deviation results of the BA and the MBA compared as follows:

1. Unimodal function comparison: In Table 4, for the average value, the MBA had a better result than the BA in all functions, which means that the MBA is a better optimizer. Also, because the unimodal function is an optimum global value, the MBA performed better. At the same time, in Table 5, the standard deviation results showed that the MBA provides better results in 6 benchmark functions (F1, F2, F3, F5, F6, and F7) while the BA is better in only one (F4). This shows the outperforming of the modified bat algorithm against the standard bat.
2. The multimodal function and fixed-dimension multimodal functions: Table 4 illustrates that the MBA was more efficient on average in 13 of the benchmark functions (from F8 until F20), while the BA produced a better optimum value in 3 functions (F21, F22, and F23). These functions have many local optima that depend on the variable design, the number of optima increased. Also for the standard deviation result of the multimodal function and fixed-dimension multimodal functions, Table 5 shows that the MBA produced a better optimum

Table 2 Multimodal benchmark test functions

Function name	Function	Range
Generalized Schwefel's problem 2.26	$f_8(x) = \sum_{i=1}^{n} -x_i sin\left(\sqrt{\lvert x_i \rvert}\right)$	[−500, 500]
Generalized Rastrigin's function	$f_9(x) = \sum_{i=1}^{n} \left[x_i^2 - 10cos(2\pi x_i) + 10\right]$	[−5.12, 5.12]
Ackley's function	$f_{10}(x) = -20exp\left(-0.2\sqrt{\frac{1}{n}\sum_{i=1}^{n} x_i^2}\right) - exp\left(\frac{1}{n}\sum_{i=1}^{n} cos(2\pi x_i)\right) + 20 + e$	[−32, 32]
Generalized Griewank function	$f_{11}(x) = \frac{1}{4000}\sum_{i=1}^{n} x_i^2 - \prod_{i=1}^{n} cos\left(\frac{x_i}{\sqrt{i}}\right) + 1$	[−600, 600]
Generalized penalized function	$f_{12}(x) = \frac{\pi}{n}\left\{10\, sin(\pi y_1) + \sum_{i=1}^{n-1} (y_i - 1)^2 \left[1 + 10sin^2(\pi y_{i+1})\right] + (y_n - 1)^2\right\} + \sum_{i=1}^{n} u(x_i, 10, 100, 4)$ $y_i = 1 + \frac{x_i + 1}{4}$ $u(x_i, a, k, m) = \begin{cases} k(x_i - a)^m & x_i > a \\ 0 & -a < x_i < a \\ k(-x_i - a)^m & x_i < -a \end{cases}$	[−50, 50]
Generalized penalized function	$f_{13}(x) = 0.1\left\{sin^2(3\pi x_1) + \sum_{i=1}^{n} (x_i - 1)^2 \left[1 + sin^2(3\pi x_i + 1)\right] + (x_n - 1)^2 \left[1 + sin^2(2\pi x_n)\right]\right\} + \sum_{i=1}^{n} u(x_i, 10, 100, 4)$	[−50, 50]

Table 3 Fixed-dimension multimodal benchmark test functions

Function name	Function	Range
Shekel's Foxholes function	$f_{14}(x) = \left(\frac{1}{500} + \sum_{j=1}^{25} \frac{1}{j+\sum_{i=1}^{2}\left(x_i - a_{ij}\right)^6}\right)^{-1}$	[−65.536, 65.536]
Kowalik's function	$f_{15}(x) = \sum_{i=1}^{11}\left[a_i - \frac{x_1\left(b_i^2 + b_i x_2\right)}{b_i^2 + b_i x_3 + x_4}\right]^2$	[-5, 5]
Six-hump camel back function	$f_{16}(x) = 4x_1^2 - 2.1x_1^4 + \frac{1}{3}x_1^6 + x_1 x_2 - 4x_2^2 + 4x_2^4$	[−5, 5]
Brain function	$f_{17}(x) = \left(x_2 - \frac{5.1}{4\pi^2}x_1^2 + \frac{5}{\pi}x_1 - 6\right)^2 + 10\left(1 - \frac{1}{8\pi}\right)\cos x_1 + 10$	[−5,0], [10, 15]
Goldstein-Price function	$f_{18}(x) = \left[1 + (x_1 + x_2 + 1)^2\left(19 - 14x_1 + 3x_1^2 - 14x_2 + 6x_1x_2 + 3x_2^2\right)\right] \times$ $\left[30 + (2x_1 - 3x_2)^2 \times \left(18 - 32x_1 + 12x_1^2 + 48x_2 - 36x_1x_2 + 27x_2^2\right)\right]$	[−2, 2]
Hartman's family	$f_{19}(x) = -\sum_{i=1}^{4} c_i exp\left(-\sum_{j=1}^{3} a_{ij}\left(x_j - p_{ij}\right)^2\right)$	[0, 1]
Hartman's family	$f_{20}(x) = -\sum_{i=1}^{4} c_i exp\left(-\sum_{j=1}^{6} a_{ij}\left(x_j - p_{ij}\right)^2\right)$	[0, 1]
Shekel's family	$f_{21}(x) = -\sum_{i=1}^{5}\left[\left(x - a_i\right)\left(x - a_i\right)^T + c_i\right]^{-1}$	[0, 10]
Shekel's family	$f_{22}(x) = -\sum_{i=1}^{7}\left[\left(x - a_i\right)\left(x - a_i\right)^T + c_i\right]^{-1}$	[0, 10]
Shekel's family	$f_{23}(x) = -\sum_{i=1}^{10}\left[\left(x - a_i\right)\left(x - a_i\right)^T + c_i\right]^{-1}$	[0, 10]

Table 4 Average value comparison

Function	Average Value of BA	Average Value of MBA	Min AVG
F1	1.510E+01	3.406E+00	+
F2	1.898E+01	4.096E+00	+
F3	9.743E+02	7.504E+00	+
F4	9.673E-01	4.172E-01	+
F5	2.638E+03	3.038E+02	+
F6	2.298E+01	3.296E+00	+
F7	1.048E+02	2.323E+01	+
F8	−5.498E+01	−1.164E+02	+
F9	2.298E+01	1.519E+01	+
F10	3.136E+00	8.926E-01	+
F11	6.362E-01	1.307E-01	+
F12	2.549E+00	3.723E-01	+
F13	1.020E+00	9.926E-01	+
F14	1.268E+01	1.267E+01	+
F15	1.620E-02	1.030E-02	+
F16	−4.655E-01	−6.314E-01	+
F17	8.158E+00	7.072E-01	+
F18	2.449E+02	4.845E+01	+
F19	−2.456E+00	−2.643E+00	+
F20	−3.643E-01	−5.514E-01	+
F21	−2.478E+00	−2.111E+00	−
F22	−2.724E+00	−2.100E+00	−
F23	−2.756E+00	−2.494E+00	−

result in 11 of the benchmark functions (F8, F11, F12, F14-F18 and F21-F23), while the BA produced a better optimum value in 5 of the benchmark functions (F9, F10, F13, F19, and F20). The results show that the modified bat algorithm is an improvement on the bat algorithm and that it had better exploration ability

As the results show, the BA is poor at exploration, which is a weak point for most problems. The MBA helps the BA to be more global and converge on a result more quickly. In the unimodal functions, the MBA obtained a significant result that had global values, this proved that the MBA functions better globally.

6.2 Statistical Tests

The BA and the MBA results were tested for statistical significance with the Wilcoxon rank-sum test. The t-test is used to determine that the two samples come from the

Table 5 Standard deviation comparison

Function	Standard Deviation of BA	Standard Deviation of MBA	Min STD
F1	1.139E+01	4.258E+00	+
F2	8.009E+00	5.571E+00	+
F3	1.126E+03	1.987E+01	+
F4	1.555E-01	4.340E-01	–
F5	7.944E+02	4.225E+02	+
F6	1.344E+01	4.059E+00	+
F7	1.227E+02	3.008E+01	+
F8	1.210E+01	2.450E-01	+
F9	1.219E+01	3.140E+01	–
F10	1.179E+00	1.488E+00	–
F11	2.352E-01	2.005E-01	+
F12	1.732E+00	2.302E-01	+
F13	1.537E-01	2.264E-01	–
F14	4.420E-02	3.904E-04	+
F15	1.410E-02	9.600E-03	+
F16	4.749E-01	3.491E-01	+
F17	2.950E+00	3.077E-01	+
F18	2.205E+02	4.170E+01	+
F19	7.736E-01	7.862E-01	–
F20	4.199E-01	5.340E-01	–
F21	1.105E+00	8.086E-01	+
F22	1.068E+00	9.243E-01	+
F23	1.258E+00	9.892E-01	+

same two populations with the same mean. This t-test was used to calculate the P-Value of the BA and the MBA. The P-Value of the t-test must lower than 0.05 for results to be statistically significant. As shown in Table 6 60% of the P-Values of the BA and the MBA are below 0.05, which shows that the results of our algorithms are statistically significant.

6.3 *The MBA Verses the BA*

The results shown in this paper are taken from the original paper published by Yang, in which the new Bat algorithm is compared with the Genetic Algorithm (GA) and Particle Swarm Optimization (PSO). This comparison uses the same type of approaches (Yang 2010). In the simulation of the MBA, six functions being used, i.e. the Generalized Rosenbrock's function, the Spher function, the Ackley's function,

Table 6 The Wilcoxon rank-sum test for benchmark function

Function	BA verses MBA (P-Value)
F1	2.689E-05
F2	5.009E-11
F3	6.028E-05
F4	3.513E-07
F5	6.508E-15
F6	9.137E-09
F7	2.004E-03
F8	1.857E-22
F9	2.384E-01
F10	6.594E-08
F11	4.298E-09
F12	7.831E-08
F13	5.781E-01
F14	1.329E-01
F15	6.671E-02
F16	1.272E-01
F17	1.543E-14
F18	6.831E-05
F19	3.189E-01
F20	1.101E-01
F21	2.001E-01
F22	2.584E-02
F23	4.106E-01

the Schwefel problem 1.2, the Generalized Rastrigin's function, and the Generalized Griewank function. The number of populations is set to 40 and the dimension of the algorithm was also changed as follows:

1. The Generalized Rosenbrock's function changes the dimension to (dim = 16).
2. De Jong's standard sphere function changes the dimension to (dim = 256).
3. The Schwefel problem 1.2 changes the dimension to (dim = 128).
4. The Ackley's function changes the dimension to (dim = 128).

Following the evaluation, which included the MBA, the PSO and the GA, Table 7 shows a comparison with six functions, which indicates that the MBA's results are better than the PSO and the GA, in all six functions. This proves that the MBA is superior to other algorithms in terms of accuracy and efficiency.

Another comparison involves the MBA and the Genetic Algorithm. The GA data are taken from [12], and used 19 benchmark functions. However, 13 of them are from our list of unimodal and multimodal functions. Table 8, shows the performance

Table 7 Comparing MBA with PSO and GA

Name	Function number	MBA AVG	GA AVG	PSO AVG
Generalized Rosenbrock's function	F5	2.685E+02	5.572E+04	3.276E+04
Sphere function	F1	2.150E+00	2.541E+04	1.704E+04
Ackley's function	F10	1.176E+00	3.272E+04	2.341E+04
Schwefel problem 1.2	F3	1.237E+02	2.273E+05	1.452E+04
Generalized Rastrigin's function	F9	2.735E+01	1.105E+05	7.949E+04
Generalized Griewank function	F11	1.077E-01	7.093E+04	5.597E+04

Table 8 Comparing MBA and GA

F	GA AVG	MBA AVG	Min AVG
F1	7.486E+02	3.406E+00	+
F2	5.971E+00	4.096E+00	+
F3	1.949E+03	7.504E+00	+
F4	2.116E+01	4.172E-01	+
F5	1.333E+05	3.038E+02	+
F6	5.639E+02	3.296E+00	+
F7	1.669E-01	2.323E+01	–
F8	-3.407E+03	-1.164E+02	–
F9	2.552E+01	1.519E+01	+
F10	9.499E+00	8.926E-01	+
F11	7.720E+00	1.307E-01	+
F12	1.859E+03	3.723E-01	+
F13	6.805E+04	9.926E-01	+

of the modified bat algorithm compared with the Genetic Algorithm. The results for the modified bat algorithm were better in 11 functions. The GA was better in only two functions,—one of the unimodal functions and one of the multimodal functions.

6.3.1 Solving a Real-World Problem Using the MBA

In general, optimization depends on finding the best solution for a function to minimize or maximize a particular area. Many problems business has been solved using a variety of different approaches. Optimization problems are important with regards

Table 9 The restaurant example

	Worker 1	Worker 2	Worker 3	Worker 4
Job1	216	247	541	222
Job2	437	937	849	543
Job3	82	329	325	289
Job4	578	264	776	158

to business, industry and even our daily lives. Literature has provided different solutions for scheduling problems, job assignments, call centres, task distribution and many other optimization-related problems. This is done by showing how different types of algorithms work. These problems can be solved using linear programming. However, this programming is time-consuming and sometimes requires the use of many different mathematical equations. This is particularly true for exponentially complex problems. When faced with a small problem, it is possible solve it quickly, but whenever there is a problem that grows in complexity exponentially, finding a solution quickly becomes problematic. Here is an example of using the scheduling problem in a restaurant. The problem was solved using a linear programming technique (Table 9).

In this example, if job 1 is assigned to worker 3, then the expected work time will be 541 s. Also, if job 2 is assigned to worker 2, then expected work time will be 937 s. The maximum solution from the matrix above is ($541 + 937 + 289 + 578 = 2345$ s). This will require close to 9 min and 46 s of average work time. Most of the solutions for the above matrix will take 7 min on average. However, our primary goal is to find the minimum amount of time necessary to solve the problem. Using linear programming the minimum solution is ($247 + 437 + 325 + 158 = 1157$ s) or 4 min and 52 s. This result is a significant improvement in the average result of 7 min. The above example is solved with linear programming. A linear program uses a complex mathematical equation to solve the optimization problem and to get the final minimum solution.

This problem is a well-known one. The restaurant scheduling problem aims to allocate or distribute work and jobs in the most efficient way possible. Each worker has specific skills and experience, such as cooking, servicing, cleaning, moving quickly, etc. At the same time, each job needs to be assigned to the right worker so that the best solution is found. For example, if there are four workers and four jobs, the requirement is to allocate the jobs to these workers. Logically, ($4! = 24$) possible combinations exist, one of which is the optimal solution. In much the same way, a restaurant with 5 workers would have ($5! = 120$) potential combinations. If the restaurant were to have ten workers with ten jobs, then ($10! = 3{,}628{,}800$) combinations would exist. Here the problem of using another type of programming, such as linear programming, occurs, because it is impossible to solve this number linearly and look through all of the combinations to find the optimal solution quickly.

The MBA based on the swarm intelligence algorithm is implemented to solve the above problem of restaurant scheduling. Every restaurant has many jobs and workers.

Table 10 Result of application

Dimension	Average value	Standard deviation	Worktime in minutes
4	266.0409	64.2047	4.434
5	304.7108	50.3414	5.0785
6	315.3420	59.0484	5.2557
7	337.8279	47.6988	5.6305
8	344.9468	55.3843	5.7491

The number of workers and jobs should be equal, which represents one dimension of the algorithm. In this application, there are 4 workers and 4 jobs. After running the application, the optimal solution for the restaurant job scheduling problem is 4.3734 min. Table 10 demonstrates the results of the application. The application is implemented with different dimensions starting from four to eight.

7 Conclusion

In this paper, the BA and the proposed MBA were evaluated against 23 benchmark functions. The average value, standard deviation, and P-value results were calculated and assessed. Furthermore, the MBA was compared to the GA and the PSO with six functions. Also, further comparison between MBA and GA was conducted using unimodal and multimodal functions. As a result, the following conclusions were reached:

1. The BA is poor at exploration, whereas the MBA is more global and converges on results more quickly as it performed well on unimodal testing functions. Also, for the multimodal and fixed-dimension multimodal functions, the results of the MBA were significant since their optima values increased.
2. The average values of the MBA were better for 20 functions. The standard deviation results of the MBA were better for 17 functions
3. The comparison between the MBA and the GA and PSO showed that the MBA performed better on all of the functions.
4. The comparison between the MBA and the GA showed that the MBA produced better results on eleven out of thirteen functions.

References

1. Fister Jr, I., Fister, D., Yang, X.-S.: A hybrid bat algorithm (2013). arXiv:1303.6310
2. Ab Wahab, M.N., Nefti-Meziani, S., Atyabi, A.: A comprehensive review of swarm optimization algorithms. PloS One **10**(5), e0122827 (2015)

3. A. S. Shamsaldin, T. A. Rashid, R. A. Al-Rashid Agha, N. K. Al-Salihi, and M. Mohammadi, "Donkey and smuggler optimization algorithm: A collaborative working approach to path finding," *Journal of Computational Design and Engineering*, vol. 6, no. 4, pp. 562–583 (2019)
4. Mirjalili, S., Mirjalili, S.M., Lewis, A.: Grey wolf optimizer. Adv. Eng. Softw. **69**, 46–61 (2014)
5. Ramesh, B., Mohan, V.C.J., Reddy, V.V.: Application of bat algorithm for combined economic load and emission dispatch. Int. J. Electr. Eng. Telecommun. **2**(1), 1–9 (2013)
6. Yılmaz, S., Kucuksille, E.U., Cengiz, Y.: Modified bat algorithm. Elektronika ir Elektrotechnika **20**(2), 71–78 (2014)
7. Fister, I., Rauter, S., Yang, X.-S., Ljubič, K., Fister, I., Jr.: Planning the sports training sessions with the bat algorithm. Neurocomputing **149**, 993–1002 (2015)
8. K. Kiełkowicz and D. Grela, "Modified bat algorithm for nonlinear optimization," *International Journal of Computer Science and Network Security (IJCSNS)*, pp. 46–50 (2016)
9. Cai, X., Gao, X.-Z., Xue, Y.: Improved bat algorithm with optimal forage strategy and random disturbance strategy. Int. J. Bio-Inspired Comput. **8**(4), 205–214 (2016)
10. Osaba, E., Yang, X.-S., Fister, I., Jr., Del Ser, J., Lopez-Garcia, P., Vazquez-Pardavila, A.J.: A discrete and improved bat algorithm for solving a medical goods distribution problem with pharmacological waste collection. Swarm Evolut. Comput. **44**, 273–286 (2019)
11. X.-S. Yang, "A new metaheuristic bat-inspired algorithm," in *Nature inspired cooperative strategies for optimization (NICSO 2010)*, pp. 65–74, Springer (2010)
12. Abdullah, J.M., Ahmed, T.: Fitness dependent optimizer: inspired by the bee swarming reproductive process. IEEE Access **7**, 43473–43486 (2019)

Applications

Patterns of PCB-138 Bioaccumulation in Small Pelagic Fish from the Eastern Mediterranean Sea Using Explainable Machine Learning Prediction

Andreja Stojić, Bosiljka Mustać, Gordana Jovanović, Jasna Đinović Stojanović, Mirjana Perišić, Svetlana Stanišić, and Snježana Herceg Romanić

Abstract Fish consumption, especially consumption of oily marine species, is increasing globally due to its recommendation by dieticians. This is due to high polyunsaturated ω-3 and ω-6 (PUFAs) fatty acid content in the tissue of the fish. The health benefits of PUFA ingestion coincide with the risk of intaking hazardous lipophilic persistent pollutants including organochlorine pesticides (OCPs) and related polychlorinated biphenyls (PCBs). We examined the impact of 17 fatty acids (FAs) and 36 toxic organic and inorganic contaminants on the behavior patterns of the indicator congener PCB-138 in marine fish using eXtreme Gradient Boosting (XGBoost), SHapley Additive exPlanations (SHAP), and SHAP value fuzzy clustering. XGBoost indicated non-linear relationships between PCB-138 and other investigated variables that were explained by SHAP values. The ten obtained fuzzy

A. Stojić (✉) · G. Jovanović · M. Perišić
Institute of Physics Belgrade, National Institute of the Republic of Serbia, University of Belgrade, Belgrade, Serbia
e-mail: andreja@ipb.ac.rs

G. Jovanović
e-mail: gordana.vukovic@ipb.ac.rs

M. Perišić
e-mail: mirjana.perisic@ipb.ac.rs

B. Mustać
Department of ecology, agronomy and aquaculture, University of Zadar, Zadar, Croatia
e-mail: bmustac@unizd.hr

J. Đinović Stojanović
Institute of Meat Hygiene and Technology, Kaćanskog 13, 11 000 Belgrade, Serbia
e-mail: jasna.djinovic@inmes.rs

S. Stanišić
Environment and Sustainable Development, Singidunum University, Belgrade, Serbia
e-mail: sstanisic@singidunum.ac.rs

S. Herceg Romanić
Institute for Medical Research and Occupational Health, Ksaverska cesta 2, PO Box 291, 10001 Zagreb, Croatia
e-mail: sherceg@imi.hr

E. Pap (ed.), *Artificial Intelligence: Theory and Applications*,
Studies in Computational Intelligence 973,
https://doi.org/10.1007/978-3-030-72711-6_10

clusters of SHAP values revealed that a higher intake of saturated myristic-C14:0 and margaric-C17:0 acids followed by the intake of nutritionally beneficial eicosadienoic acid (C20:2n-6) mostly do not increase the bioaccumulation of PCB-138. Important effects on PCB-138 behavior patterns were also recorded for the chemically allied indicator congeners (−153, −180, −118 and−101) and organochlorines' metabolite p,p'-DDE. Associations between the target congener and the toxicologically relevant PCBs (−123 and −170) were less prominent.

Keywords Persistent organic pollutants (pops) · (omega-3-6) fatty acids · Heavy metals · Shapley additive explanations (shap)

1 Introduction

Whether intentionally or unintentionally, anthropogenic activities have led to the release of numerous man-made synthetic chemicals which continuously harm human health. The contaminants have also caused several global, regional, and local environmental problems related to air, water and soil pollution, a gradual decrease of the stratospheric ozone layer, a decrease in biodiversity, etc. Organic and inorganic contaminant emissions and dispersion caused by human activities refer to different classes of polychlorinated biphenyls (PCBs), polycyclic aromatic pollutants, trace metals, and natural radioactivity. Organochlorine pesticides (OCPs) and PCBs are well-known persistent organic pollutants (POPs). They are long-lived contaminants, which resist photolytic, biological, and chemical degradation to differing extents. Because of their resistance to metabolic breakdown and high lipid solubility, POPs are bioaccumulated in the fatty tissues of living organisms and cause numerous adverse effects [20]. Although their production and use are limited or prohibited in most countries, investigating the behavior of POPs is still challenging as long-term monitoring of their impacts was not mandatory during their early use. Recognized as hydrophobic, POPs bind to particles in soil and sediment tightly, which can act as a secondary source of contamination for environmental media (water, air and living organisms).

At a global level, the final sink/destination of POPs are marine environments, which are polluted via coastal outfalls, rivers, and deposition from the atmosphere. Oceans act as a secondary source of contamination because POPs are slowly degraded and adsorbed on suspended particles or bioaccumulated in benthic marine organisms, which are at the bottom of the food chain and represent a source of toxins in human nutrition. The assessment of environmental exposure to marine toxins is based on data on the concentrations of POPs in samples of water, plants, and food, surrounding conditions (e.g., salinity, temperature), the pollutant physico-chemical properties and biotic factors (e.g., organisms' feeding modality, trophic position). Heavy metals can occur naturally in marine environments and some of them are essential to living organisms. However, levels are increased by anthropogenic activities and all of them can be toxic above threshold levels.

Small pelagic oily fish are a highly recommended source of nutrients worldwide due to their content of protein, minerals and healthy fats, including long-chain omega-3 (ω-3) and 6 (ω-6) polyunsaturated fatty acids (PUFAs), such as eicosapentaenoic (EPA) and docosahexaenoic acid (DHA) [8]. According to FAO [2018a] small pelagic fish are considered low-priced fish, which means they are an important food source in many developing countries, whereas in other countries they are largely processed into fishmeal and fish oil. The Mediterranean Sea is an important commercial fishing ground whose catches are primarily small pelagic fish, accounting for around 50% of all catches [9]. However, it is a semi-enclosed sea, which is particularly vulnerable to chemical contamination from the surrounding heavily industrialized and agricultural countries. The benefits of fatty acid consumption are associated with normal growth and development, the prevention of cardiovascular and inflammatory diseases, as well as the prevention of cognitive decline and dementia. However, these FAs represent a suitable matrix for the bioaccumulation of highly lipophilic xenobiotics such as POPs. More than 90% of human exposure to POPs is through food; mainly meat, dairy products, fish, and shellfish [24].

Although the profiles and organochlorine content of FAs have been evaluated in numerous marine fish species worldwide, there are few data on their interrelations. In our previous studies, a preliminarily investigation was conducted on the presence of OCPs and PCBs in different edible fish species [12], followed by elaboration on the joint use of advanced algorithms (Kohonen self-organizing maps and Decision Tree Learning) to study the spatio-temporal distribution of POPs in fish from the Croatian Adriatic [23]. In this study, the presence of OCPs and PCBs was examined in small pelagic edible fish species: sardine *Sardina pilchardus* (Walbaum, 1792), anchovy *Engraulis encrasicolus* (Linnaeus, 1758), round sardinella *Sardinella aurita* (Valenciennes, 1847), chub mackerel *Scomber japonicus* (Houttuyn, 1782) and horse mackerel *Trachurus trachurus* (Linnaeus, 1758). We applied eXtreme Gradient Boosting (XGBoost), SHapley Additive exPlanations (SHAP), and SHAP value fuzzy clustering aiming to obtain a detailed insight into the distribution of the indicator congener PCB-138 in the fish species. The impacts of the following factors including the level of OCPs, PCBs, saturated fatty acids (SFAs), monounsaturated fatty acids (MUFAs), PUFAs and heavy metals were evaluated using SHAP since the method offers uniquely consistent and locally accurate solutions that have been confirmed in previous studies of environmental phenomena [22]. An earlier version of this paper was presented at the International Scientific Conference on Information Technology and Data Related Research (Sinteza 2020), Singidunum University [21].

2 Materials and Methods

2.1 Sampling

Fish samples were collected along the Croatian part of the Eastern Adriatic Sea in 2014 and 2016. Details about sampling were described previously [23]. A total of 107 fish samples were collected in various fisheries in coastal (A, E, and F) and off-coast (B and C) zones. Approximately 50 specimens were sampled randomly using purse seine catches (mesh size: 8 mm/bar length) totaling 107 pooled samples from the fillet of the specimens.

2.2 Chemical Analyses of POPs

The chemical analysis of POPs was described previously elsewhere [12, 23]. In brief, from each pool approximately 5 g of two aliquots was ground with 2 g of sodium sulphate in a glass mortar and extracted with 40 mL of n-hexane. After passing through filter paper (Whatman No. 1) into a pre-weighted test tube, the extract was reduced under a nitrogen steam. The lipid extracts were dissolved in 5 mL of n-hexane and purified three times with 4 mL of 96% sulphuric acid; the solvent was evaporated under a gentle nitrogen stream. Finally, the tissue lipid content was determined gravimetrically.

Seven OCPs (HCBα-, β-, and γ-HCH, *p,p'*-DDT, *p,p'*-DDE, and *p,p'*-DDD), six indicator PCB congeners (PCB-28, PCB-52, PCB-101, PCB-138, PCB-153, and PCB-180), eight mono ortho congeners (PCB-105, PCB-114, PCB-118, PCB-123, PCB-156, PCB-157, PCB-167, PCB-189) and PCB-60, PCB-74 and PCB-170 were analyzed. The extracted residues were dissolved in 1.0 mL of n-hexane and 5 μ L of the sample was injected. The pollutant content was determined by high-resolution gas chromatography with an electron capture detector (s) performed on a CLARUS 500 chromatograph [23]. Two capillary columns (Restek, Bellefonte, PA, USA) were used simultaneously: (1) 60 m $\times$ 0.25 mm, Rtx-5 film thickness of 0.25 μ m, and (2) 30 m $\times$ 0.25 mm, Rtx-1701 film thickness of 0.25 μ m. The column temperature was attained after the following steps: (1) initial heating from 100 °C to 110 °C at 4 °C min^{-1}, (2) isothermal heating for 5 min at 110 °C, (3) heating from 240 °C at 15°C min^{-1}, and (4) 50 min of isothermal heating at 240 °C. Nitrogen was used as the carrier gas. The injector and detector temperatures were 250 and 270 °C, respectively. Only compounds identified on both columns were evaluated.

The certified reference material CRM 430 (Community Bureau of Reference, Commission of the European Communities, Brussels, Belgium) was used to determine the relative accuracy of the method. The results agreed with the reference range. The concentrations of analytes in blanks were below the detection limit of the instrument.

To test the recovery and reproducibility of the method, standard addition was used. A known amount of all of the analyzed compounds (between 0.61 and 0.68 ng g^{-1} of fresh weight for OCPs and between 0.42 and 0.7 ng g^{-1} of fresh weight for PCBs) were added to the five aliquots of homogenized samples before extraction. The recoveries for the PCBs ranged between 75% and 89% while a relative standard deviation (RSD) of 1% to 11% was obtained. The recoveries for OCPs ranged from 76% to 86%, with RSD from 1% to 11% for both PCBs and OCPs, while the determination limits were 0.01 ng g^{-1} of fresh weight. The presented data were recalculated based on the recovery values.

2.3 Chemical Analyses of Elements

Sixteen macro- and micro-elements including toxic metals (Na, Mg, K, Ca, As, Cd, Co, Cr, Cu, Mn, Fe, Hg, Ni, Zn, Pb and Se) were analyzed using the previously described procedure [5]. Prior to analysis, the frozen fish samples were thawed at 4 °C. The homogenized fish samples (0.5 g) were basted with 5 mL of nitric acid (67% TraceMetal grade, Fisher Scientific, Bishop, UK) and 1.5 mL of hydrogen peroxide (30% analytical grade, Sigma-Aldrich, St. Louis, MA, USA). Subsequently, microwave digestion (Start D Microwave Digestion System, Milestone, Sorisole, Italy) was performed under the following conditions: maximum power (1000W); heating ramp to 180 °C in 5 min; hold at 180 °C for 15 min; cooling in an oven for 20 min and then at room temperature for 15 min. After cooling at room temperature, the digests were diluted to 100 mL with deionized water into polypropylene volumetric flasks. Analysis of the elements was performed by inductively coupled plasma mass spectrometry (ICP-MS), (iCap Q mass spectrometer, Thermo Scientific, Bremen, Germany). The operating conditions were: RF power (1550 W); cooling gas flow (14 L min^{-1}); nebulizer flow (1 L min^{-1}); collision gas flow (1 mL min^{-1}); operating mode (Kinetic Energy Discrimination-KED); and dwell time (10 ms). The most abundant isotopes were used for quantification.

For five-point calibration, solutions of Fe, Zn, Cu, Mn, Se, Cr, Co, Ni, Na, K, Mg, Ca, Cd, Pb, Hg and As were prepared from standard stock solutions (VGH labs, Manchester, UK) containing 1000 mg L^{-1} of each element. The concentration of the calibration-solution for Fe, Zn, Cu, Mn, Se, Cr, Co, Ni, Na, K, Mg and Ca was in a concentration range of 0.2–2.0 mg L^{-1}, for Cd, Hg and As, it ranged from 0.2 to 2.0 $\mu g\ L^{-1}$ and for Pb it ranged between 2.0 and 20.0 $\mu g\ L^{-1}$.

To check the accuracy of the analysis, the certified reference material NIST SRM 1577c (bovine liver, Gaithersburg, MD, USA) was analyzed in the same manner as the fish samples. For all elements, the obtained results were within a satisfactory range of the certified values.

2.4 Chemical Analyses of Fatty Acids

The concentrations of 6 SFAs (myristic, pentadecylic, palmitic, palmitoleic, margaric, and stearic acid), 3 MUFAs (oleic, paullinic and arachidonic acid) and 8 PUFAs, ω-6 and ω-3 families (linoleic, α-linolenic, eicosadienoic, dihomo-γ-linolenic, eicosatrienoic, eicosapentaenoic, docosapentaenoic and docosahexaenoic) were determined according to the previously described procedure [18]. Prior to analysis, the samples were partially thawed at +4 °C. To determine the total lipids for fatty acid content, accelerated solvent extraction (ASE 200, Dionex, Sunnyvale, CA), using a mixture of n-hexane and iso-propanol (60:40 v/v) was applied to an extraction cell heated at 100 °C and under nitrogen pressure of 10.3 MPa. The samples were analyzed as fatty acid methyl esters (FAME) obtained by transesterification using trimethyl sulfonium hydroxide (ISO 2000 standard). Afterwards, gas-liquid chromatography (Shimdzu, Japan) was performed. A fused silica cianopropyl HP-88 column (length 100 m, i.d. 0.25 mm, film thickness 0.20 μ m, J&W Scientific, USA) and a flame ionization detector (GC/FID) were used. Hydrogen was used as a carrier gas at a 1.3 mL flow rate and an injector split ratio of 1:50. The following temperature program was used: (1) heating from 125 °C to 175°C at a rate of 10 °C min^{-1}, (2) isothermal heating at 175 °C for 10 min, (3) a temperature increase from 175 °C to 210°C (rate 5 °C min^{-1}), (4) isothermal heating at 210 °C for 5 min, and (5) final temperature incline to 230 °C at a rate of 2 °C min^{-1}. The injector and detector temperatures were 250 °C and 280 °C, respectively. The chromatographic peaks were identified and quantified using Supelco 37 Component mix standard and internal standard (heneicosanoic acid methyl ester), respectively.

The chromatographic peaks were identified and quantified using Supelco 37 Component mix standard and internal standard (heneicosanoic acid methyl ester), respectively.

2.5 Data Analysis

The relationships between PCB-138 and all the other measured parameters were modeled through XGBoost regression. Details on the method were given elsewhere [23]. Briefly, it is a decision-tree-based ensemble machine learning algorithm that applies a gradient boosting framework. The method boosts weak learners by sequentially correcting the errors made by existing models. XGBoost is based on a gradient descent algorithm, used to minimize loss when adding new models. The algorithm includes system optimization and algorithmic enhancements through parallelized sequential tree building, tree pruning, regularization, weighted quantile sketch algorithm implementation, cross-validation, etc. In this study, we used Python XGBoost implementation. The dataset was split into training (80%) and validation (20%) sets. Hyperparameter tuning was implemented using a brute-force grid search and 10-fold stratified cross-validation. The best performing hyperparameter values were used for the final model.

The explainability of the produced XGBoost model, that operates with high-dimensional input data in a non-linear fashion, was obtained using the explainable artificial intelligence method SHapley Additive exPlanations (SHAP) [2, 13]. Based on game theory, Shapley explanations represent the only possible locally accurate and globally consistent feature attribution values. In this study, we applied fuzzy clustering of absolute SHAP attributions to identify and characterize the relations among the measured parameters responsible for PCB-138 behavior.

3 Results and Discussion

3.1 Fatty Acid Content

Descriptive statistics of the content of the FAs in the examined fish species is presented in Fig. 1. The molecules of FAs are classified according to the presence and sum of double bonds, which determine their physico-chemical properties, functional characteristics, and metabolic fate in living organisms. The most prevalent and equally abundant FAs in sardine, anchovy and mackerel species were palmitic acid (C16:0), which is an SFA with no double bond, and oleic acid (C18:1n-9), a MUFA, which

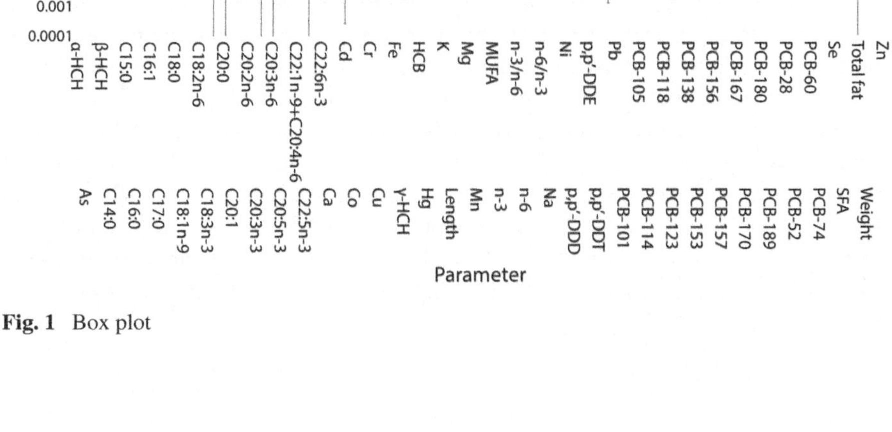

Fig. 1 Box plot

has a single double bond. In the studied fish, the highest percentage of these acids are usually found in the form of free FAs in which they are transported to the large intestine and extracted. In addition, the absorption of palmitic acid by human enterocytes is limited since the best absorption rate of SFAs is achieved when they are presented as sn-2 monoacylglycerols [16, 17].

Although some SFAs and MUFAs can be produced in the human body from carbon groups, which are constituents of carbohydrate and protein macromolecules, humans cannot synthesize some ω-3 and ω-6 PUFAs since they lack the delta 12 and delta 15 desaturase enzymes, which arrange a double bond at the n-3 and n-6 positions of FAs. These PUFAs are known as essential FAs and their uptake must be provided via food intake including fish and seafood. From the parent acids, ω-3 α-linoleic (C18:3n-3) and ω-6 linoleic (C18:2n-6) acid, 20 carbon long-chain (or more) FAs could be synthesized through successive reactions of desaturation and elongation [11]. The possible products are dihomo-γ-linolenic acid (20:3n-6), arachidonic acid (20:4n-6), EPA (20:5n-3) and DHA (22:6n-3). However, the metabolite-conversion efficiency is low and the consumption of EPA- and DHA-enriched food is recommended to ensure an optimal level of FAs in human blood, which reflects both dietary intake and metabolic production [4].

Eicosapentaenoic acid and DHA are the most recognized n-3 PUFAs with two or more double bonds because of their nutritionally and physiologically beneficial effects for healthy adults, pregnant women, and children. As presented in Fig. 1, the content of DHA was significant in the studied marine fish and varied from 10.4% to 21.3% while EPA was less abundant, from 3.3% to 4.6%. The accumulation pattern of both FAs was as follows: sardines>anchovy>chub mackerel>horse mackerel while the DHA:EPA ratio was approximately 2.5:1 implying that a variety of different fish species should be consumed to gain a sufficient supply of nutritionally relevant FAs. A considerably higher amount of EPA and DHA has been reported in the class of phospholipids compared to other classes including monoacylglycerols, diacylglycerols, free fatty acids, triacyl-glycerols, sterol esters, neutral lipids, and glycolipids. In this way, EPA and DHA are more bioavailable and less susceptible to oxidative degradation whereas EPA and DHA free forms of FA are less efficiently absorbed than other lipid classes and are easily eliminated from human intestines where they are dispersed into mixed micelles and bound to soluble lipid-binding proteins [3, 17]. Among the other essential FAs, the content of linoleic acid (1.5%–2.2%) was three times higher than the levels (0.4%–0.7%) of α-linoleic acid while the metabolite of linoleic acid, dihomo-γ-linoleic acid, where levels of up to 1% were found (Fig. 1).

Many previous studies have indicated that the ω-6 to ω-3 PUFA ratio in the human diet evolved from 1:1 to the current 20:1. While increased amounts of ω-6 PUFAs can cause the pathogenesis of many diseases, high levels of ω-3 PUFAs show suppressive effects [15]. In this study, the ω-6 to ω-3 ratio ranged from between 0.025 to 22.63 (average value = 1.80) and showed pronounced maximums in sardines and chub mackerel. As the results demonstrated, small pelagic fish species are a good source of FAs and their nutritional benefits are comparable to those of Atlantic salmon [14] or Atlantic Bluefin tuna [19].

3.2 Pollutant Toxicological Profile

Small edible pelagic fish live briefly at the bottom of the marine food chain (plankton<sardine species, anchovy<mackerel species), and are expected to bioaccumulate low levels of environmental contaminants. The transfer of accumulated toxins to the human body via fish intake is generally considered to be low. As the results of the study indicated, inorganic compounds, macro-elements, and toxic heavy metals, were notably more abundant in fish tissue than organic xenobiotics, OCPs and PCBs (Fig. 1).

The biological effects of inorganic elements depend on processes which include absorption, accumulation, elimination, and biotransformation into less or more toxic metabolites. Several minutes after absorption in the gut, the elements reach the internal organs such as the heart, liver, kidney, and the brain, while their penetration into the muscles and adipose tissue may take up to several hours [1]. However, when they are present in low concentrations in aquatic environments, their accumulation is also less common. In the studied fish species, the concentrations of macro- (K > Na > Ca > Mg) and micro-elements (Fe > Zn > Cu) were within the normal physiological range, below levels that could potentially cause pathological changes in tissues and organs, because of the absence of leading sources of pollution. In addition, levels of the following elements: Hg, Pb and Cd, were lower than the maximum thresholds prescribed by existing EU regulations [6].

Among POPs, *p,p'*-DDE, PCB-153, PCB-138 and PCB-180 were the most prevalent. Although it is expected that larger, longer-lived species that are at higher trophic levels, such as chub mackerel and horse mackerel, will uptake more organic contaminants than sardine species and/or anchovy. Toxicological parameter values, such as the total quantity of indicator PCBs and WHO-dioxin-like PCBs toxic equivalents, were below the threshold concentrations of 75 ng g^{-1} w.w. and 6.5 pg g^{-1} [7]. Therefore, it can be concluded that the examined fish species appear to be safe for human consumption.

3.3 PCB-138 Patterns

As demonstrated by the Pearson's correlation analysis (Fig. 2), significant linear correlation coefficients (r>0.90) were found between the following pairs of investigated variables: *p,p'*-DDE-PCB-118; *p,p'*-DDE-PCB-138; *p,p'*-DDD-PCB-105; *p,p'*-DDD-PCB-180; PCB-138-PCB-118; PCB-138-PCB-153; PCB-153-PCB-118; PCB-153-PCB-170 and PCB-156-PCB-180.

The correlations between pairs of POPs: γ-HCH-PCB-170; γ-HCH-PCB-156; γ-HCH-PCB-118; γ-HCH-PCB-114; γ-HCH-PCB-180; *p,p'*-DDE-PCB-153; *p,p'*-DDT-PCB-170; *p,p'*-DDT-PCB-156; *p,p'*-DDT-PCB-118; *p,p'*-DDT-PCB-153; PCB-138-PCB-170; PCB-180-PCB-170; PCB-180-PCB-156; PCB-105-PCB-180; PCB-105-PCB-156 and PCB-105-PCB-170 ranged from 0.80 to 0.90. The listed

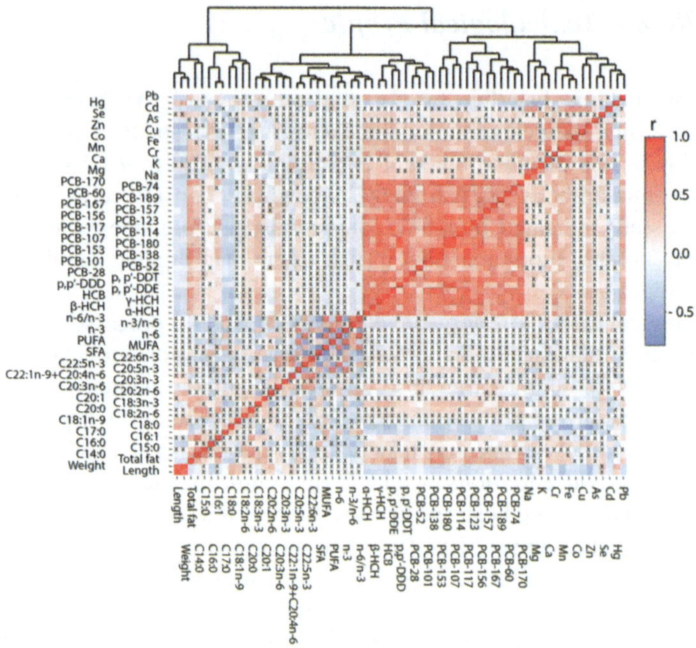

Fig. 2 Parameter value correlations matrix

species with similar chemical structures and common origins display similar behavior patterns, as discussed below. Since OCPs and PCBs are highly liphophilic chemicals, their toxicodynamics are considered to be dependent on the amount of adipose tissue and various fatty compounds in living organisms. No significant Pearson's correlation coefficients were observed between the 18 FAs and the studied POPs(Fig. 2). Therefore, we assume that methods other than the typical correlation matrices could be utilized to describe the associations between POPs and FAs or inorganic contaminants in more detail.

In this study, XGBoost was utilized for the investigation of non-linear relationships between PCB-138 levels and key variables that shape its behavior patterns in marine fish tissue. The predicted/observed calculated relative error was 11.2%, while the r2 was 0.99 (Fig. 3).

The fuzzy clustering of SHAP values extracted ten groups (relative error $< 10\%$) of similar variables that shape the dynamics of PCB-138 (Fig. 4). The constituents of six clusters, C2 and C6-C10, had a highly positive impact on PCB-138 concentrations from the mean to maximum values (0.6–1.9 ng g^{-1}). Prominent impacts, described by relative SHAP values (%), are attributed to the following compounds: *p,p'*-DDE (C2: 23, C6: −10, C7: 33, C8: 25, C9: 31, and C10: 23), PCB-153 (C2: 39, C6: 36, C7: 13, C8: 34, C9: 25, and C10: 28), PCB-118 (C2: 8, C6: 13, C7: 10, C8: 7.8, C9: 10, and C10: 12), PCB-180 (C2: 6.3, C6: 5.3, C9: 3.9, and C10: 5.0), and PCB-170 (C2: 6.0, C6: 4.3, C8: 6.6, C9: 5.0, and C10: 6.2). Less significant effects

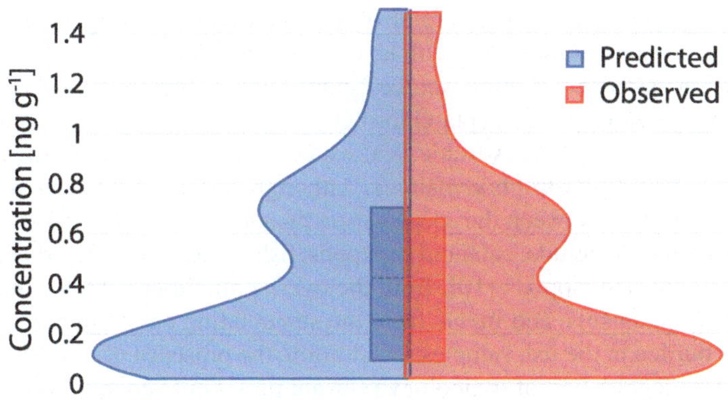

Fig. 3 Model evaluation

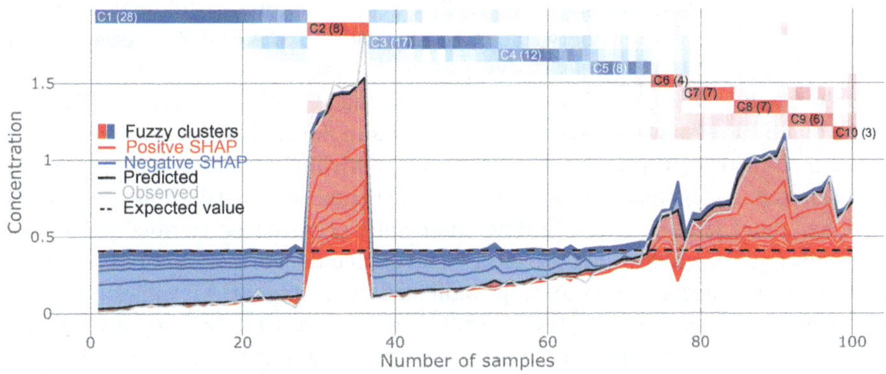

Fig. 4 SHAP value force plot

were recorded for PCB-101 (C2: 1.2 and C9: 2.1) and PCB-123 (C2: 1.8, C8: 1.8, and C9: 2.0). The average and high PCB-138 concentrations were accompanied by levels (ng g^{-1}) of listed pollutants, which ranged from the median and 75th percentile values to the upper fence values: *p,p'*-DDE (C2: 1.3, C6: 0.3, C7: 0.7, C8: 0.9, C9: 0.8, and C10: 0.5), PCB-153 (C2: 2.5, C6: 1.4, C7: 1.0, C8: 1.7, C9: 1.4, and C10: 1.2), PCB-118 (C2: 0.4, C6: 0.2, C7: 0.2, C8: 0.2, C9: 0.2, and C10: 0.1), PCB-180 (C2: 0.8, C6: 1.0, C9: 0.4, and C10: 0.3), PCB-170 (C2: 0.4, C6: 0.3, C8: 0.1, C9: 0.2, and C10: 0.2), PCB-101 (C2: 0.2 and C9: 0.1), and PCB-123 (C2: 0.4, C8: 0.2, and C9: 0.2). Although indicator congeners (−28, −52, −101, −138, −153, and −180), which are often classified as non-dioxin-like PCBs, overshadowed dioxin-like PCBs, both classes were significant in the evaluation of PCB-138 patterns. The relationships between PCB congeners and their presence at higher concentration levels are due to their rigid molecular structure and number of attached halogen atoms, and consequently, their persistence and prolonged half-lives in marine environments.

The more chlorinated congeners are also more resistant to endogenous metabolic degradation in fish in comparison with less chlorinated congeners.

In addition, similar pollution sources contribute to the relationships between "heavy" long-lived penta- to hepta-chloro congeners. The studied fish species live in the epipelagic zone of the Adriatic Sea, which extends from the surface to 200 m depth and shows similar trophodynamics. Although certain differences in nutrition cycles were reported between the species investigated in this study, the main prey of all small pelagic fish include: calanoid copepods, fish larvae, decapod larvae, copepod eggs, amphipods and isopods [10]. Both the overlapping diet, which contains low-niche marine organisms, and inputs from the surrounding environment contributed to the POP burden in the fish samples. In addition, the presence of *p,p'*-DDE reflects the metabolic degradation of toxicology relevant persistent xenobiotics.

Of the 17 examined FAs, saturated myristic (C14:0) acid impacts the PCB-138 bioaccumulation patterns represented in C2 and C6-C10 the most, followed by the nutritionally beneficial eicosadienoic (C20:2n-6) and dihomo-γ-linolenic (C20:2n-3) acids. The interrelations between PCB-138, and the SFA and PUFAs were indicated by relative SHAP values (%): C14:0 (C7: −0.6, C9: −0.4, and C10: −0.5), C20:2n-6 (C6: −0.8, C8: −1.5, and C10: −1.2), and C20:3n-6 (C10: 1.3). We have noted that these FAs were not the most abundant in the studied fish species, and the prevalence of other FAs was recorded in the following order: palmitic > oleic > docosahexaenoic > stearic > myristic > eicosapentaenoic > palmitoleic > linoleic acid. Saturated acids such as C14:0 predominantly occurred as esterified polar phospholipids and therefore, SFAs are more bioavailable than FAs in non-esterified free form. However, negative relationships indicated by relative SHAP values imply that low content (%) of FAs (C14:0-C7: 6.5, C9: 6.1, and C10: 8.0; C20:2n-6-C6: 0.4, C8: 0.7, and C10: 0.8; and C20:3n-6-C10: 0.07) do not increase PCB-138 uptake.

Eicosapentaenoic acid and DHA are widely known as the most nutritionally relevant ω-3 PUFAs in oily blue fish such as mackerel, sardines, and anchovies, but they appear to have no impact on PCB-138 accumulation. Chub mackerel has been known to contain high percentages of PUFAs in the bioavailable phospholipid form, which is of nutritional and physiological relevance [17]. Therefore, consumption of the studied fish species, which ensure a continuous supply of dietarily beneficial levels of EPA and DHA, does not coincide with POPs burden in humans.

Fuzzy clustering identified four clusters (C1, C3–C5) representing groups of the variable interrelations associated with low to minimal levels of PCB-138, from 0.013 to 0.1965 ng g^{-1} (Fig. 4). The same parameters as discussed above shaped the clusters, but their negative influences on PCB-138 distribution were recorded: *p,p'*-DDE (C1: −22, C3: -25, C4: −25, and C5: −27), PCB-153 (C1: −42, C3: −36, C4: −30, and C5: −8.5), PCB-118 (C1: −7.8, C3: −8.4, C4: −8.7, and C5: −5.4), PCB-180 (C1: −6.2 and C3: −5.2), PCB-170 (C1: −7.7, C3: −6.7, and C4: −7.8), PCB-101 (C3: −1.0 and C5: −1.6), and PCB-123 (C1: −1.8, C3: −2.0, and C4: −2.3). Low concentrations of PCB-138 were related with POPs levels, which ranged from the mean to minimum values (ng g^{-1}): *p,p'*-DDE (C1: 0.056, C3: 0.117, C4: 0.023, and C5: 0.212), PCB-153 (C1: 0.125, C3: 0.279, C4: 0.408, and C5: 0.660), PCB-118 (C1: 0.023, C3: 0.041, C4: 0.057, and C5: 0.079), PCB-180 (C1: 0.031 and C3:

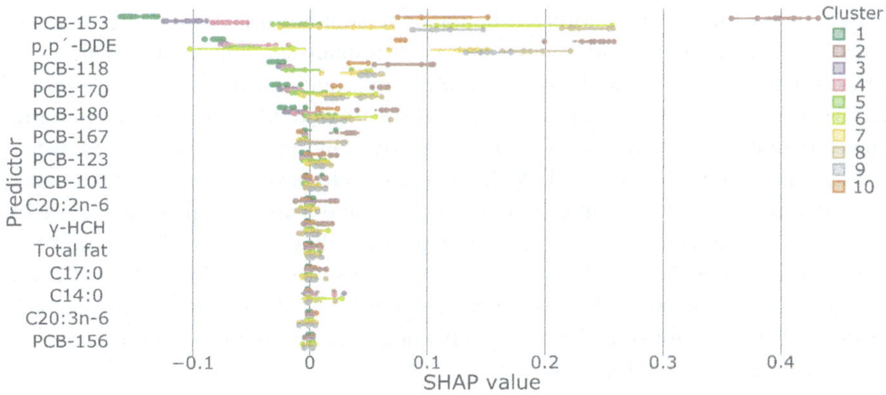

Fig. 5 SHAP value top ten clusters

0.055), PCB-170 (C1: 0.028, C3: 0.047, and C4: 0.071), PCB-101 (C3: 0.026 and C5: 0.030), and PCB-123 (C1: 0.023, C3: 0.034, and C4: 0.048). The results indicate that the uptake of POPs from different sources is found to have variable impacts depending on the sampling time, season, and other fishery zone-related factors.

As discussed above, the negative impact of myristic (C14:0) and margaric (C17:0) acids was also evident in clusters C1, C3–C5. The negative relative SHAP values, %, (C14:0-C1: −0.54 and C5: −0.85; C17:0-C4: −0.76 and C5: −1.7) were accompanied by the following concentrations (%): C14:0 (C1: 3.8 and C5: 5.5) and C17:0 (C4: 1.3 and C5: 1.3).

The plot in Fig. 5 shows the impact each cluster has on the output of the model. High levels of the constituents of clusters (C2, C6–C10) primarily PCB-153, *p,p'*-DDE, PCB-118, PCB-170, PCB-180 and myristic acid, had a high and positive impact on PCB-138 patterns, as shown by the right-oriented long distribution tail. Lower concentrations of these variables, grouped in C1 and C3–C5 clusters, are negatively correlated with the target variable, as indicated by the negative relative SHAP value. To a lesser extent, the influences are also of importance for bioaccumulation fate of *p,p'*-DDE, PCB-118, PCB-170, and PCB-180, while their relevance is weaker for describing environments which shape the occurrence of volatile POPs such as γ-HCH and FAs.

4 Conclusion

Small edible fish species can be considered both a nutritionally valuable food source and a source of hazardous organochlorine pollutants which negatively affect human health. Significant indications of the relationships between the uptake of POPs and fatty acid in fish tissue have been reported worldwide. In this study, we presented a promising methodology, explainable artificial intelligence methods (XGBoost and

SHAP), with the aim of gaining a better understanding of the specific interrelations between the fatty acid content and contaminants in consumable marine fish species. Out of 17 fatty acids, two saturated (myristic and margaric) and two ω-3 and 6 (eicosadienoic and dihomo-γ-linolenic) acids were identified as crucial for the bioaccumulation of PCB-138 in sardine, anchovy, and mackerel species. However, nutritionally beneficial EPA and DHA are assumed to have no impact on the uptake of contaminants. The content of macro-elements and heavy metals is not related to PCB-138 chemodynamics in fish tissue, while the influences of *p,p'*-DDE and both indicator and toxicological congeners (−101, −118, −123, −153, and −180) were evident. Finally, the methods have been successfully verified as a reliable means for examining the relationships between POPs and FAs that improve upon commonly employed statistical approaches.

Acknowledgements The authors acknowledge the funding provided by the Institute of Physics, Belgrade, through research supported by the Ministry of Education, Science and Technological Development and the Science Fund of the Republic of Serbia, #GRANT No. 6524105, AI - ATLAS. In addition, this study was supported by the institutional financing of scientific activity 2018-2020 Project "Persistent Organic Pollutants-Environmental Impact Assessment and Stability of Human Genetic Material" (IMROH, 2018–2021, Zagreb, Croatia).

References

1. Brucka-Jastrzêbska, E., Kawczuga, D., Rajkowska, M., Protasowick, M.: Levels of microelements (Cu, Zn, Fe) and macroelements (Mg, Ca) in freshwater fish. J. Elementol. **14**, 437–447 (2009)
2. Chen, T., Guestrin, C.: Xgboost: A scalable tree boosting system. In: Proceedings of the 22nd Acm sigkdd International Conference on Knowledge Discovery and Data Mining, pp. 85–794 (2016)
3. Cook, C.M., Hallaråker, H., Sæbø, P.C., et al.: Bioavailability of long chain omega-3 polyunsaturated fatty acids from phospholipid-rich herring roe oil in men and women with mildly elevated triacylglycerols. Prostaglandins Leukot Essent Fatty Acids **111**, 17–24 (2016)
4. Davidson, M.H.: Omega-3 fatty acids: new insights into the pharmacology and biology of docosahexaenoic acid, docosapentaenoic acid, and eicosapentaenoic acid. Curr. Opin. Lipidol. **24**, 467–474 (2013)
5. Đinović-Stojanović, J., Nikolić, D., Vranić, D., Babić, J., Milijašević, M., Pezo, L., Janković, S.: Zinc and magnesium in different types of meat and meat products from the Serbian market. J. Food Compost. Anal. **59**, 50–54 (2017)
6. European Commission: Commission Regulation (EC) No 1881/2006 of 19 December 2006 setting maximum levels for certain contaminants in foodstuffs 20.12.2006 Off J Eur Union L 364, 5-24 (2006)
7. Commission, European: Commission Regulation (EU) No 1259/2011 of 2 December 2011 amending Regulation (EC) No 1881/2006 as regards maximum levels for dioxins, dioxin-like PCBs and non-dioxin-like PCBs in foodstuffs. Off. J. Eur. Union L **320**, 18–23 (2011)
8. Food and Agriculture Organization of the United Nations, FAO: The state of world fisheries and aquaculture worldwide 2018-Meeting the sustainable development goals. 1-227 (2018)
9. Food and Agriculture Organization of the United Nations, FAO: The State of Mediterranean and Black Sea Fisheries. General Fisheries Commission for the Mediterranean 172, Licence: CC BY-NC-SA 3.0 IGO (2018)

10. Hure, M., Mustać, B.: Feeding ecology of Sardina pilchardus considering co-occurring small pelagic fish in the eastern Adriatic Sea. Mar. Biodivers. **50**, 40 (2020)
11. Jump, D.B., Depner, C.M., Tripathy, S.: Omega-3 fatty acid supplementation and cardiovascular disease. J. Lipid Res. **53**, 2525–2545 (2012)
12. Kljaković-Gašpić, Z., Herceg Romanić, S., Klinić, D., Tiina, V.: Chlorinated compounds in the muscle tissue of fish from the Croatian Adriatic: preliminary data on contamination. Arh. Hig. Rada Toksikol. **66**, 299–308 (2015)
13. Lundberg, S., Lee, S.: A unified approach to interpreting model predictions. Adv. Neural Inf .Process Syst. **4765–4774** (2017)
14. Lundebye, A.K., Lock, E.J., Rasinger, J.D., et al.: Lower levels of Persistent Organic Pollutants, metals and the marine omega 3-fatty acid DHA in farmed compared to wild Atlantic salmon (Salmo salar). Environ. Res. **155**, 49–59 (2017)
15. Nasir, M., Bloch, M.H.: Trim the fat: the role of omega-3 fatty acids in psychopharmacology. Ther. Adv. Psychopharmacol. **9**, 2045125319869791 (2019)
16. Ramírez, M., Amate, L., Gil, A.: Absorption and distribution of dietary fatty acids from different sources. Early Hum. Dev. **65**, S95–S101 (2001)
17. Rincón-Cervera, M.Á., González-Barriga, V., Valenzuela, R., López-Arana, S., Romero, J., Valenzuela, A.: Profile and distribution of fatty acids in edible parts of commonly consumed marine fishes in Chile. Food Chem. **274**, 123–129 (2019)
18. Špirić, A., Trbović, D., Vranić, D., Đinović, J., Petronijević, R., Matekalo-Sverak, V.: Statistical evaluation of fatty acid profile and cholesterol content in fish (common carp) lipids obtained by different sample preparation procedures. Anal. Chim. Acta **672**, 66–71 (2010)
19. Sprague, M., Dick, J. R., Medina, A., Tocher, D. R., Bell, J. G., Mourente, G.: Lipid and fatty acid composition, and persistent organic pollutant levels in tissues of migrating Atlantic bluefin tuna (Thunnus thynnus, L.) broodstock. Environ. Pollut. **171**, 61–71 (2012)
20. Stockholm Convention: Stockholm Convention for Persistent Organic Pollutants (2001). http://chm.pops.int/. Cited November 2020
21. Stojić, A., Mustać, B., Jovanović, G.: Explainable machine learning prediction of PCB-138 behavior patterns in edible fish from Croatian Adriatic. In: Book of Proceedings of International Scientific Conference on Information Technology and Data Related Research (Sinteza 2020), Singidunum University, pp. 23–28 (2020)
22. Stojić, A., Stanić, N., Vuković, G., Stanišić, S., Perišić, M., Šoštarić, A., Lazić, L.: Explainable extreme gradient boosting tree-based prediction of toluene, ethylbenzene and xylene wet deposition. Sci. Total Environ. **653**, 140–147 (2019)
23. Vuković, G., Herceg Romanić, S., Babić, Ž., Mustać, B., Štrbac, M., Deljanin, I., Antanasijević, D.: Persistent organic pollutants (POPs) in edible fish species from different fishing zones of Croatian Adriatic. Marine Pollut. Bull. **137**, 71–80 (2018)
24. World Health Organisation, WHO: Dioxins and their effects on human health. In: World Health Organization Fact Sheet, No. 225 (2014)

Patterns of PCB-138 Occurrence in the Breast Milk of Primiparae and Multiparae Using SHapley Additive exPlanations Analysis

Gordana Jovanović, Marijana Matek Sarić, Snježana Herceg Romanić, Svetlana Stanišić, Marija Mitrović Dankulov, Aleksandar Popović, and Mirjana Perišić

Abstract Breastfeeding provides numerous health benefits for newborns by meeting infants' nutritional needs and supporting associated immunological protection. Maternal milk is high in fat, and therefore, represents a very suitable medium for the bioaccumulation of lipophilic pollutants, such as organochlorine pesticides (OCPs) and polychlorinated biphenyls (PCBs). This makes breast milk the infant's primary source of postnatal exposure to persistent toxic xenobiotics. In this study, we applied a novel SHapley Additive exPlanations (SHAP) method to examine the key parameters that govern the distribution of PCB-138, an indicator of non-dioxin congeners, in the mother's milk. According to the accuracy metrics, the eXtreme Gradient Boost-

G. Jovanović (✉) · M. Perišić
Institute of Physics Belgrade, National Institute of the Republic of Serbia, Environment and Sustainable Development, Singidunum University, Belgrade, Serbia
e-mail: gordana.vukovic@ipb.ac.rs

M. Perišić
e-mail: mirjana.perisic@ipb.ac.rs

M. Matek Sarić
Department of Health Studies, University of Zadar, Zadar, Croatia
e-mail: marsaric@unizd.hr

S. Herceg Romanić
Institute for Medical Research and Occupational Health, Ksaverska cesta 2, PO Box 291, 10001 Zagreb, Croatia
e-mail: sherceg@imi.hr

S. Stanišić
Environment and Sustainable Development, Singidunum University, Belgrade, Serbia
e-mail: sstanisic@singidunum.ac.rs

M. Mitrović Dankulov
Institute of Physics Belgrade, National Institute of the Republic of Serbia, University of Belgrade, Belgrade, Serbia
e-mail: mitrovic@ipb.ac.rs

A. Popović
Faculty of Chemistry, University of Belgrade, Belgrade, Serbia
e-mail: apopovic@chem.bg.ac.rs

E. Pap (ed.), *Artificial Intelligence: Theory and Applications*,
Studies in Computational Intelligence 973,
https://doi.org/10.1007/978-3-030-72711-6_11

ing regression was employed successfully, with a predicted/observed relative error below 20% and a high correlation coefficient (r = 0.97), for modeling the relationships between PCB-138 and other non-dioxin congeners, the mother's age, and the number of births. According to the results, PCB-156, PCB-180, HCB, HCH and PCB-118 have a major impact, while PCB-28, PCB-52 and PCB-189 have a minor impact on PCB-138 distribution in breast milk. Similar contaminant behaviors, which belong to both the indicator congener group (−28, −52, −180) and the toxicologically relevant PCBs (−118, −189), were also noted. The SHAP conclusions were only partially consistent with the results of the correlation analysis suggesting that POPs exhibit non-linear dynamics and interrelations. Therefore, current knowledge on the contamination of complex biomatrices would benefit from further detailed analyses of pollutant intermittent relationships.

Keywords Human biomonitoring · Persistent organic pollutants (POPs) · Organochlorine pesticides (OCPs) · Polychlorinated biphenyls (PCBs) · SHapley Additive exPlanations (SHAP)

1 Introduction

Persistent organic pollutants (POPs), including organochlorine pesticides (OCPs), polychlorinated biphenyls (PCBs), polychlorinated dibenzodioxins (PCDDs) and polychlorinated dibenzofurans (PCDFs), have posed an environmental and health threat for over a century. Due to their common properties such as persistence, bioaccumulation, toxicity, and their ability to be transported through the air over long distances, the production and use of POPs was prohibited and/or restricted by the Stockholm Convention from 2001. Because of their persistence and lipophilic properties, they accumulate and biomagnify in the food chain, mainly in the lipid-rich tissues of humans and animals.

Research has shown that organochlorine compounds, such as endocrine modulators, affect human health. The toxic potential of individual compounds is different, but the overall toxic effect is additive. According to epidemiological and laboratory studies, depending on the compound, these toxins can increase mortality and the risk of cancer, cause allergies and hypersensitivity, damage the central and peripheral nervous system, cause reproductive disorders, and immune system disorders [5, 7]. Humans and wildlife are exposed to PCBs and OCPs in the polluted environment via air, water, sediment soil and food; though ingestion, inhalation, or skin contact. Previous studies suggest that more of these compounds are introduced into the body through food intake than by inhalation or through the skin. It is also known that children are prenatally exposed to organochlorine compounds that pass from the mother to the fetus through the placenta. Prenatal exposure to organochlorines has been linked to impaired fetal development, reduced cognitive and neurological progress and negative effects on the function of respiratory organs [3, 8].

Biological monitoring of POPs in the body requires invasive techniques involving the surgical extraction of adipose tissue. The alternative monitoring approach, and one of the most commonly used non-invasive methods for the assessment of human exposure, is the analysis of human milk's POPs content. Human milk, as the natural and superior food for newborns, meets infants' necessary nutritional requirements and provides numerous health benefits associated with immunological and psychological advancement. However, the high fat content of human milk makes it a suitable matrix for the accumulation of toxic lipophilic chemicals such as POPs, which is a result of the dietary habits and occupational exposure of the breastfeeding mother. Since 1987, the World Health Organization (WHO) has carried out six global surveys on POPs in human milk to evaluate in utero and lactational exposure to these compounds and identify the measures which should be taken to limit or minimize the impact of organochlorines on humans and the environment. Although the studies indicated that the presence of POPs in the environment has decreased, and consequently that their intake by humans has declined, evaluation of the risk factors and determinants of breastfed infants' exposure to POPs is still a challenge [26].

A number of literature sources have reported on POPs levels in breast milk regarding regional and global quantitative differences and time-trends and/or their dependence on parameters like the mother's age and weight loss, childbirth, dietary habits and occupation [6, 14, 20, 25]. Maternal age and parity are among the most frequently considered parameters. The concentration of organochlorines has been anticipated to increase in the fatty tissues of older mothers, whereas parity is expected to reduce their levels since lactation is recognized as the main pathway for the elimination of POPs from the body. Our previous research aimed to investigate the presence of OCPs and PCBs in human milk as well as their mutual interrelations and associations with the mother's age and parity using machine learning (ML) algorithms [10, 12, 13]. In this study, we applied the additive feature attribution method—SHapley Additive exPlanations (SHAP), which offers uniquely consistent and locally accurate attribution values in comparison with the conventional attribution methods and has become an increasingly popular tool for predicting environmental phenomena [21, 24]. We aimed to obtain a detailed insight into the parameters, including organochlorine pollutants, the mother's age, and parity, that shape the PCB-138 distribution in the mother's milk and to offer a promising methodology for the evaluation of the observed interrelations. We chose PCB-138 since among the PCB6 group, it has been considered as an indicator PCB behavior patterns in various samples and has been also taken as the most suitable target compound for evaluating non-dioxin-like PCBs. The classification of non-dioxin- and dioxin-like congeners is used in European food and feed regulations where dioxin-like PCBs resemble biochemical and toxicological characteristics of 2,3,7,8-tetrachlorodibenzo-p-dioxin (TCDD). The present chapter is an extended version of the study presented at the International Scientific Conference on Information Technology and Data Related Research (Sinteza 2020), Singidunum University [23].

2 Materials and Methods

2.1 Sampling

Breast milk samples were collected from 150 healthy primiparae, secundiparae, and multiparae (having delivered three children), aged between 19 and 45 living in Zadar (Croatia). The sampling was performed three times between 2011 and 2016 and the second and third campaigns were part of the WHO survey on POP levels in breast milk. The samples were gathered when mature lactation was achieved, from 2 to 38 weeks postnatally. Milk was manually expressed into pre-cleaned glass bottles and stored at $-20\,°C$ until analysis.

All of the mothers were aware of the aim and relevance of the study and signed informed written consent forms to confirm their participation. The mothers completed an enclosed questionnaire and according to the results, the participants had no history of accidental or occupational exposure to the analyzed POPs. The collected samples and the personal and clinical data were only used for the purposes of the research. The study and the sampling were approved by the local Zadar County Health Centre Ethics Committee (01-745/2011 and 01-405/2014). More details about the sampling were given previously [10].

2.2 Chemical Analysis of PCBs and OCPs

The analytical procedure has been described in detail previously [13]. Briefly, two subsamples of each unfrozen milk sample (5 g) were extracted twice with a mixture of chloroform and methanol and dried under a nitrogen flow. Subsequently, the milk fat was dissolved in n-hexane and purified with sulfuric acid. Adsorption chromatography was additionally performed on a multilayer silica column using 4% diethyl-ether in n-hexane as a solvent. As a final step prior to the chromatography analysis, the organochlorine compounds were fractionated on carbon commercial tubes (ENVI-Carb SPE tubes, Supelco, the USA), eluted with n-hexane/toluene (99:1) mixture, reduced to dryness and dissolved in n-hexane.

Seven OCPs [hexachlorobenzene (HCB), hexachlorocyclohexane isomers (α-, β-, and γ-HCH), 1,1,1-trichloro-2,2-di(4-chlorophenyl)ethane (*p,p'*-DDT), 1,1-dichloro-2,2-di(4-chlorophenyl) ethylene(*p,p'*-DDE), 1,1-dichloro-2,2-di(4-chlorophenyl)ethane (*p,p'*-DDD)], 17 PCB congeners [PCB-28, PCB-52, PCB-101, PCB-138, PCB-153, PCB-180 (six indicator congeners), PCB-105, PCB-114, PCB-118, PCB-123, PCB-156, PCB-157, PCB-167, PCB-189 (eight mono ortho congeners), PCB-60, PCB-74 and PCB-170] were analyzed. High-resolution gas chromatography with an electron capture detector was performed on a CLARUS 500 chromatograph using two capillary columns (Restek, Bellefonte, PA, USA) simultaneously: (1) $60\,m \times 0.25\,mm$, Rtx-5 film thickness of $0.25\,\mu m$, and (2) $30\,m \times 0.25\,mm$, Rtx-1701 film thickness of $0.25\,\mu m$. The column temperature was programmed as follows: (1) a gradient increase from 100 °C to 110 °C with incremental changes of 4 °C min^{-1}, (2) isothermally heating at 110 °C for 5 min, (3) a gradient increase to 240 °C with incremental changes of 15 °C min^{-1}, and (4) isothermally heating at 240 °C for 50 min. Nitrogen carrier gas was used with injector and detector temperatures of 250 °C and 270 °C, respectively.

The LODs for the analyzed compounds were 0.5 $ng\,g^{-1}$ milk fat for PCB congeners, 0.1 $ng\,g^{-1}$ milk fat for α-HCH and HCB, 0.2 $ng\,g^{-1}$ milk fat for *p,p'*-DDE, 0.3 $ng\,g^{-1}$ milk fat for β-HCH, γ-HCH and *p,p'*-DDD, and 0.6 $ng\,g^{-1}$ milk fat for *p,p'*-DDT. The average recoveries for PCBs ranged between 58 and 86% and for organochlorine pesticides between 59 and 92%. The reproducibility of the method expressed as a relative standard deviation was between 6% and 22%, and 7% and 24% for PCBs and OCPs respectively.

2.3 Data Analysis

The relationships between PCB-138 in breast milk and all other measured parameters were modeled using XGBoost regression. Details on the method are available elsewhere [4]. Briefly, XGBoost is a highly effective machine learning method based on a sequential tree-growing algorithm, where each decision tree tries to complement all others and correct for residuals in the predictions made by the previous trees. Due to its core advantages concerning computational efficiency, excellent predictive performance, competitive accuracy, highly optimized multicore, and tree complexity penalization that was not widely used in previous additive tree models, XGBoost has been successfully applied in various domains [9, 18, 19]. In this study, we used Python XGBoost implementation. The dataset was split into training (80%) and validation (20%) sets. Hyperparameter tuning was implemented using a brute-force grid search and a 10-fold stratified cross-validation. The best performing hyperparameter values were used for the final model.

The explainability of the produced XGBoost model, that operates with high-dimensional input data in a non-linear fashion, was obtained by using the explainable artificial intelligence method SHapley Additive exPlanations. This method is capable of providing a straightforward and meaningful interpretation of machine learning model-derived decisions, which is now focused on user-readable logic rules [15]. Based on game theory and Shapley values that provide distributed payouts among the cooperating players (features) depending on their contribution to the joint payout (prediction), the SHAP values, as a measure of feature importance, represent the only possible locally accurate and globally consistent feature attribution values. SHAP overcomes the main drawback of attribution inconsistency and minimizes the possibility of underestimating the importance of a feature with a certain attribution value. In this study we used Python SHAP implementation (SHAP Python package). The captured attributed importance of a feature is visually presented as a SHAP summary plot.

To clearly indicate a change in the absolute SHAP value relative to other feature attributions, we obtained relative SHAP values [22]. Relative SHAP values, defined as the share of absolute SHAP in the total attributed importance of all features for the case, show the relative influence of a feature on the prediction. To test the stability of the obtained absolute and relative SHAP values, they were evaluated with a 50 times-replicated bootstrap method.

3 Results and Discussion

3.1 Pollutant Profiles

The concentrations of organochlorines found in the milk samples of primiparae, secundiparae and multiparae aged between 19 and 45 (with an average age of 30 years) are given in Fig. 1.

The most abundant pollutants were from the DDT group, exhibiting concentrations in the following ranges: 0.15–77.75 ng g^{-1} (average: 8.34 ng g^{-1}) for *p,p'*-DDE, 0.10–49.48 ng g^{-1} (average: 2.67 ng g^{-1}) for *p,p'*-DDD and 0.30–92.64 ng g^{-1} (average: 5.58 ng g^{-1}) for *p,p'*-DDT. The levels are comparable or even lower than the contamination reported in the investigations from Europe and the rest of the world [1, 2, 6]. The occurrence of *p,p'*-DDT in breast milk primarily depends on the mother's diet or direct occupational contact with this compound and therefore, high levels of *p,p'*-DDT may be associated with recent exposure to the xenobiotic. However, during active exposure, humans metabolically dechlorinate *p,p'*-DDT to *p,p'*-DDE or *p,p'*-DDD and increased values of *p,p'*-DDE indicate enhanced metabolic conversion. The *p,p'*-DDE compound is the most persistent metabolite with the longest half-life in the environment in comparison with its counterparts, but individual lev-

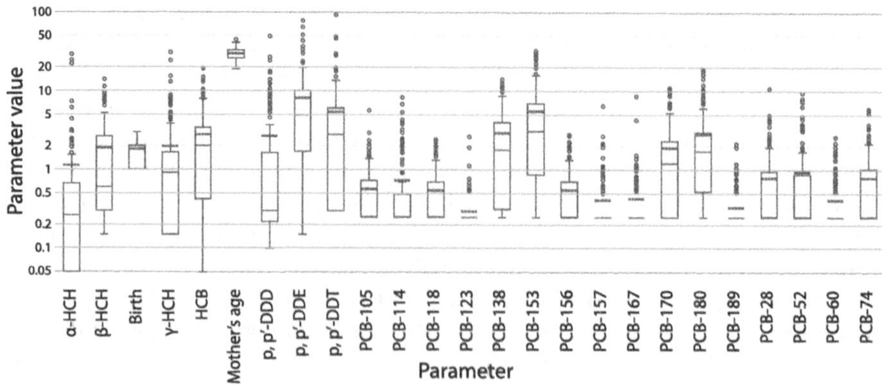

Fig. 1 Obtained parameter box-plot

els of *p,p'*-DDE in the mothers' body, which are the result of *p,p'*-DDT breakdown, are difficult to distinguish from direct ingestion of the already presented *p,p'*-DDE metabolite in the environment. To imply fresh inputs of *p,p'*-DDT in the organism as well as the rate of its transformation or excretion, some researchers suggest using the ratio *p,p'*-DDT/*p,p'*-DDE as an indicator [11]. The values of this marker were higher than 1 in approximately 30% of the samples collected in this study, indicating that one third of the mothers were recently exposed to *p,p'*-DDT via diet including home-produced free-range eggs [1] or that their metabolism is limited at reducing levels of this pollutant. In addition, since *p,p'*-DDT and *p,p'*-DDE bind to different hormone receptors, dissimilar biological effects are expected to occur in the mothers' body with different *p,p'*-DDT/*p,p'*-DDE ratios and consequently, to affect breastfed infants in another manner.

Among the other POPs, the most prevalent were by-products of lindane, γ-HCH and HCB, occurring in ranges from 0.10 to 19.31 $ng\,g^{-1}$ (average: 2.86 $ng\,g^{-1}$) and from 0.31 to 30.85 $ng\,g^{-1}$ (average: 2.01 $ng\,g^{-1}$), respectively. Among potential non-dietary (e.g., mothers age, nursing history, parity, residence) and dietary (e.g., fish consumption and its frequency, inclusion of milk products, intake of meat and eggs) predictors that could affect accumulation of HCH compounds and HCB in the mothers' organism, fatty fish and the reception of breastfeeding were suggested to be the most influential determinants [1].

The PCB contamination profile was dominated by PCB-153 (average: 5.76 $ng\,g^{-1}$) followed by other non-dioxin like congeners: PCB-138 (average: 3.04 $ng\,g^{-1}$), PCB-180 (average: 3.06 $ng\,g^{-1}$), and PCB-170 (average: 2.06 $ng\,g^{-1}$). This distribution represents a likely pattern of congener accumulation in the milk samples around the world reflecting the stronger resistance of heavy hexa- and hepta-chlorinated PCBs to biological degradation. However, individual concentrations are higher than those reported in the region [29], but lower than the congener levels around the world [1, 6, 16].

Significant linear correlation coefficients ($r \geq 0.90$) were found between the following pairs of investigated variables: PCB-170-PCB-138; PCB-170-PCB-153; PCB-170-PCB-180; PCB-153-PCB-180 and PCB-153-PCB-138 (Fig. 2). The results indicate that the listed pollutants have similar molecular structures and metabolic pathways, which is discussed below. However, we assumed that advanced methods, other than commonly applied correlation matrices, could be employed to deepen current understanding of PCB-138 behavior patterns in breast milk.

3.2 Interrelations of Pollutant Patterns

For investigating the non-linear relationships between PCB-138 and other congeners, the mother's age, and number of births, XGBoost regression analysis was used, with a predicted/observed relative error below 20% and a high correlation coefficient ($r = 0.97$) (Fig. 3). As shown by the highest positive (up to 4) and negative (up to -1.5) SHAP values, the most important variables that shaped PCB-138 behavior patterns

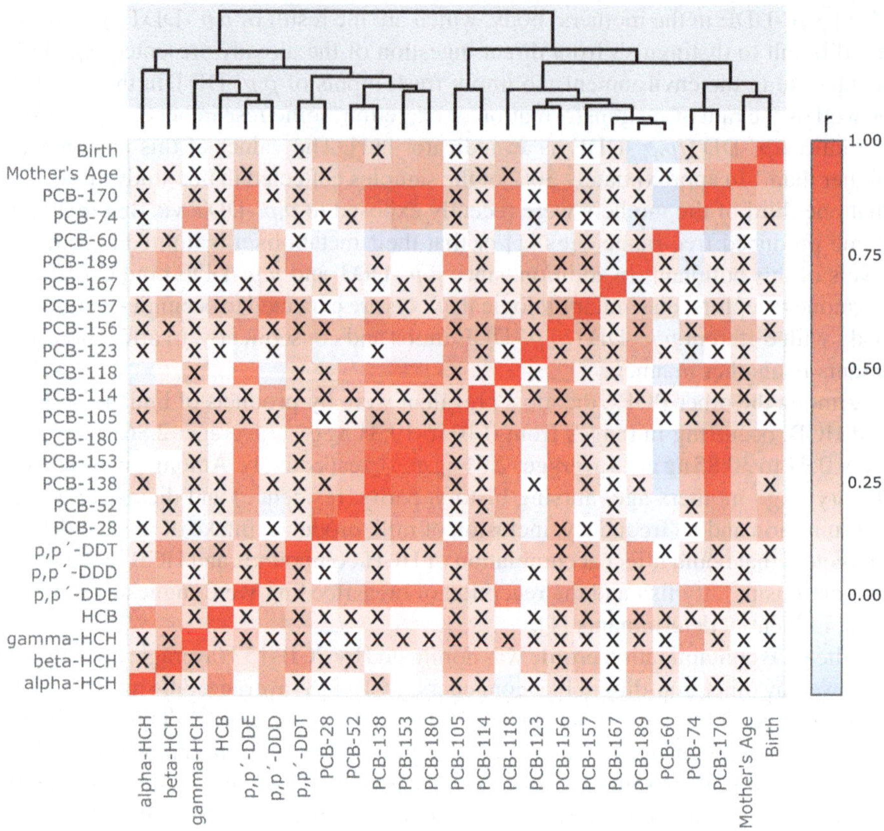

Fig. 2 Obtained parameter correlation matrix ('x'—p-value > 0.1)

in the examined milk samples were PCB-170 and PCB-153 (Fig. 4). The strongest influences were observed to be related to elevated concentrations of the listed pollutants, suggesting that "heavy" congeners substituted with chlorine atoms at 2' and 6 ortho positions are more prone to bioaccumulation in human milk compared to other PCBs. The position of the halogen substituting biphenyl ring of PCB-153 and PCB-170 provides the molecules with a structure and rigidity such that PCBs are able to pass from the blood to the breast milk. Other molecular descriptors including lipophilicity (octanol-water partitioning coefficients-logKow), molecular diameter and weight, the number of hydrogen-bond acceptors and the number of attached halogens appeared to be less determinative for PCB partitioning between blood and milk [27].

Several studies reported on the high serum/milk ratio of congeners with six or more chlorine substitutes suggesting that an increase in molar size, polarizability, and hydrophobicity cause a low rate of partition between blood serum and milk [17]. When considering compounds with the same number of attached chlorines, higher

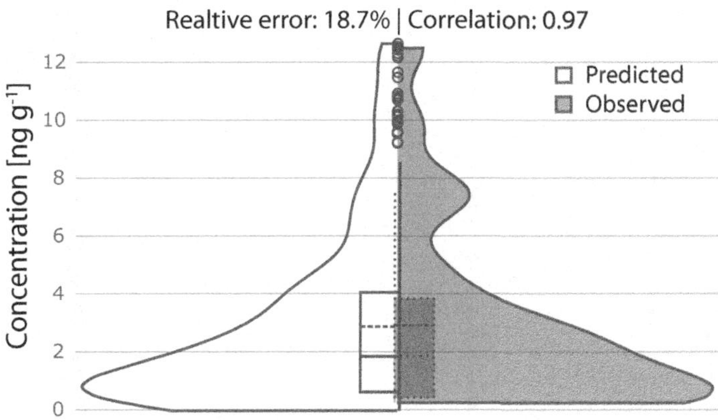

Fig. 3 XGBoost evaluation

mean serum/milk ratios were recorded for the toxicologically relevant dioxin-like PCBs compared to the indicator non-dioxin-like PCB, although a twofold difference between the lowest and highest serum/milk ratio of an individual congener was reported. The observations suggested a significant role is had by other chemical properties, but which one directly determines the serum/milk ratio is difficult to establish because of the collinearity of PCB molecules. The PCDD/Fs, with a molar volume comparable to PCBs and polybrominated diphenyl ethers, exhibited a higher serum/milk ratio that confirms a reduced transfer of molecules with more rigid structures and higher hydrophobicity from serum to milk [17]. The comparison of the PCB serum/milk ratio from different studies is limited by the different methodologies used for the ratio calculation, an inconsistent number of PCB serum-milk pairs included in the studies, and analytical uncertainties associated with the determination of congeners. Further research is needed to support the conclusion of our study and provide valid information on the complex characteristics that influence PCB partition kinetics between human serum and milk. In addition, data on the chemical characteristics alone are insufficient to predict the serum/milk partition of PCBs and descriptors such as metabolic breakdown through dechlorination or binding to different protein sites in serum and milk should be considered.

The SHAP analyses revealed less significant impacts of PCB-156, PCB-180 and PCB-118 on PCB-138 behavior patterns in milk samples. The pollutants belong to a non-dioxin-like/indicator congener group (−138, −153 and −180) and the toxicologically relevant PCBs (−118 and −156), which elicit aryl hydrocarbon receptor-mediated biochemical and toxic responses and resistance in the food chain. A minor negative dependency was observed between PCB-138 and elevated levels of low-chlorinated congeners (PCB-52 and PCB-28), which are more volatile and susceptible to metabolic breakdown and excretion, and organochlorine *p,p'*-DDT.

Numerous studies reported that POPs are eliminated from the body during breastfeeding, which results in the lower level of pollutants in the breast milk of multipara,

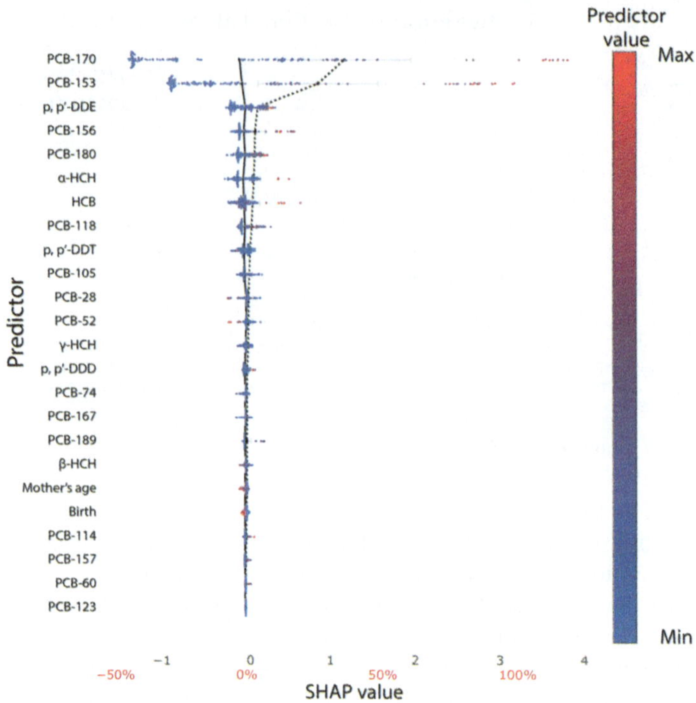

Fig. 4 PCB-138 SHAP summary plot

while the review of PCB changes during lactation emphasized non-consistent elimination rates of pollutants [16]. In addition, it is usually reported that the milk of older mothers contained higher concentrations of organochlorine xenobiotics. The positive associations may be related with the higher cumulative exposure of older mothers to POPs, and possibly a higher body mass index associated with increased fat content [1]. In the present study, the following age ranges were considered: 19–24 (N = 22), 25–30 (N = 53), 31–36 (N = 52) and 37–45 (N = 13; three mothers were above 40 years old). The pollutant levels were largely unaltered by the age of first-time mothers whereas their content increased in the breast milk of older multiparae. A slight decline in OCPs, the indicator and toxicologically relevant congeners, was shown in the milk of 29- and 30-year-old secundiparae, but the contaminant maximums were achieved in the samples of 35-year-old mothers. For the mothers with third child delivery, similar patterns were confirmed only for the toxicologically relevant congeners and PCB-74. However, as indicated by low SHAP values approaching zero, the mother's age and parity had no significant influences on the PCB-138 patterns (Fig. 4). Although the dataset is limited, it is comparable to the number of samples in other studies where a positive association between age, parity and POP levels was reported. For predicting the bioaccumulation of POPs in breast milk, information on exposure routes including dietary habits, residence, working environment, and the mother's childhood nursing is required.

Ten clusters, representing groups of variable interrelations associated with certain PCB-138 levels, were identified through fuzzy clustering of the SHAP values (Fig. 5). Two of the clusters (C7 and C9) with the highest PCB-138 concentrations (up to 12 ng g^{-1}) were the most differentiated ones. Congeners PCB-153 and PCB-170 explain the higher PCB-138 concentrations in breast milk and their impact, which is characterized by relative SHAP values, and accounts for 69% and 75% in total of C7 and C9 respectively (relative error < 7.7%). Positive interrelations between PCB-138, and PCB-153 and PCB-170 were indicated by the SHAP (PCB-153: 0.4-2.8; PCB-170: 0.5–3.5 ng g^{-1}) and relative SHAP values (PCB-153: 19.8–44.7%; PCB-170: 23.8–44.7%). The higher and extreme values of PCB-138 are related to the higher and extreme values of PCB-153 (3.6–8.2 ng g^{-1}) and PCB-170 (7.4–22.9 ng g^{-1}), which are mostly above upper fence values. The impact of the congeners −170 and −153 increase as their concentration increases as well.

Fuzzy clustering identified three clusters (C1–C3) representing groups of variable interrelations associated with medium PCB-138 levels (relative error < 12%). Medium PCB-138 levels were associated with PCB-153 and PCB-170 concentrations above their mean values, 5.6 and 1.9 ng g^{-1} respectively. Apart from C2 and C3, the negative impact of PCB-153 on the PCB-138 pattern is evident in C1. Since the concentration of both congeners is low compared to other clusters, the outcome could be the result of differing intakes of PCB contaminants through differences in the mothers' diets. The C1 cluster is also characterized by the higher concentrations and influence of *p,p'*-DDT, α-HCH, and HCB as well as by the lower levels and impact *p,p'*-DDE. In some participants, fresh inputs of pollutants from the OCP group are evident, but some of the mothers could possess a limited capacity to metabolically transform and eliminate organochlorine xenobiotics as well. Overall, the result indicates a similar origin of OCPs and PCBs in the maternal body and/or similar metabolic pathways as discussed in detail in previous paragraphs.

The overall implication is that compounds such as DDT and lindane, which were previously thought to have exclusively agricultural uses, may have also been used in cities for insect control and these cities may be significant sources of the compounds in the atmosphere. As indicated by the relative SHAP values given in brackets, the occurrence of high (up to 7.7 ng g^{-1}) and medium (up to 5.3 ng $^{-1}$) PCB-138 concentrations are explained by the following pollutants: *p,p'*-DDE (C1: 1.9%, C2: 4.4%, C3: 7.5%, C7: 3.2%, and C9: 2.8%), PCB-118 (C2: 2.5%, C3: 5.3%, and C7: 1.9%), PCB-156 (C9: 4.8%), PCB-180 (C1: 3.8%, C2: 4.8%, C3: 3.9%, C7: 3%, and C9: 2.6%) and α-HCH (C1: 3.4%). In the clusters, the listed compounds exhibited the following concentrations (ng g^{-1}): *p,p'*-DDE (C1: 7.5, C2: 16.3, C3: 12.0, C7: 14.7, and C9: 30.4, which ranged from mean to above upper fence values), PCB-105 (C1: 0.7, 75th percentile value), PCB-118 (C2: 0.8, C3: 0.9, and C7: 1.3, which ranged from 75th percentile to upper fence values), PCB-156 (C1: 0.5, and C9: 1.8 as referred to above mean and upper fence values), PCB-180 (C1: 2.6, C2: 5.1, C3: 2.9, C7: 7.5 and C9: 12.8, above mean and upper fence values). The listed clusters are characterized by excessive concentrations of POPs which imply similar

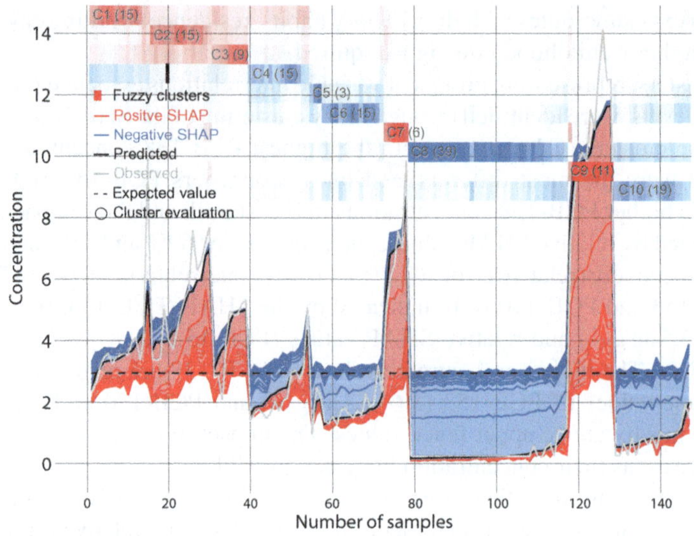

Fig. 5 PCB-138 Force plot

exposure to the pollutants because of fresh inputs or metabolic transformation and degradation.

Low PCB-138 concentrations (<4 ng g^{-1}) are distributed in five clusters (C4–C6, C8 and C10) (Fig. 5) and the same constituents as discussed above determine the clusters (relative error < 20%). The low concentrations of PCB-138 are related with the occurrence (ng g^{-1}) of: PCB-170 (C4: 1.7, C5: 1.2, C6: 1.0, C8: 0.3, and C10: 0.45), PCB-153 (C4: 3.6, C5: 3.1, C6: 3.0, C8: 0.4, and C10: 1.7), PCB-118 (C5: 0.3), PCB-180 (C5: 2.1 and C8: 0.4), PCB-156 (C5: 0.3 and C8: 0.3), *p,p'*-DDE (C5: 6.0, C6: 5.5, and C8: 1.2), α-HCH (C4: 2.0 and C6: 0.2), PCB-52 (C4: 0.56), and *p,p'*-DDD (C5: 4.8). The strength of the effects is described by the following relative SHAP values (%): PCB-170 (C4: 9.5, C5: −33, C6: −42, C8: −44, and C10: −44), PCB-153 (C4: −25, C5: −27, C6: −20, C8: −29, and C10: −25), PCB-118 (C5: −3.1), PCB-180 (C5: −3.3 and C8: −3.9), PCB-156 (C5: −4.1 and C8: −2.5), *p,p'*-DDE (C5: 2.0, C6: 1.3, and C8: −5.1), α-HCH (C4: 1.6 and C6: 1.8), PCB-52 (C4: 1.6), and *p,p'*-DDD (C5: 2.1). The results imply the congener uptake from different sources related to variations in the mothers' occupational exposure and dietary habits. The low positive SHAP relative values of *p,p'*-DDD and *p,p'*-DDE suggests intensified metabolic degradation of the POPs. Out of the extracted groups, the clusters characterized by excessive concentrations of POPs could be considered the most influential for the observed high levels of PCB-138 as indicated by the absolute (Fig. 6) and relative SHAP values (Fig. 7), which indicate the impact of each cluster on the model's output.

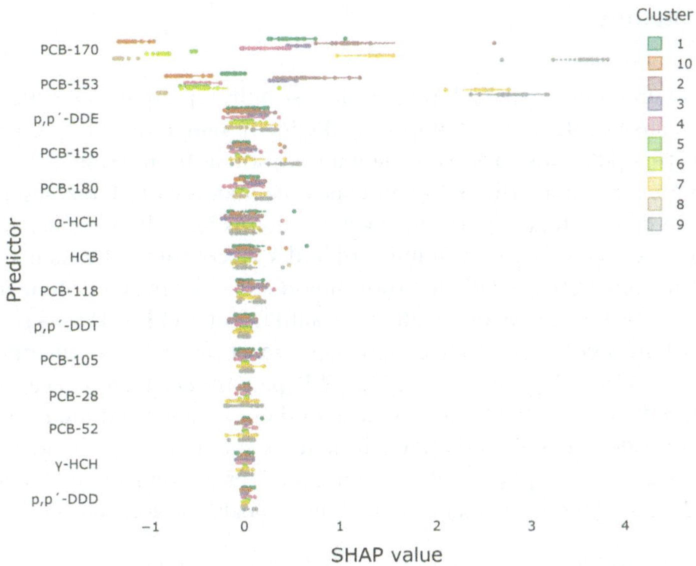

Fig. 6 Distribution of SHAP values within cluster for predictors explaining 90% of PCB-138 concentrations

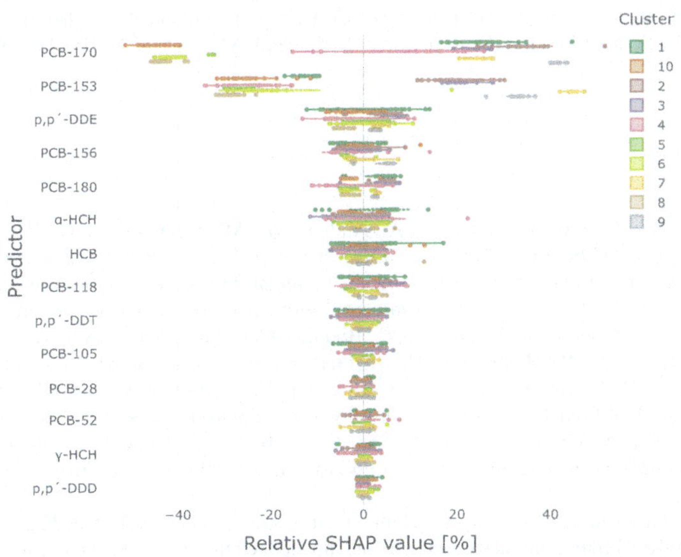

Fig. 7 Distribution of relative SHAP values within cluster for predictors explaining 90% of PCB-138 concentrations

4 Conclusion

The health burden of organochlorines in breast milk represents an issue of global concern because of the adverse impact of POPs on human health, particularly on sensitive sub-populations such as women and children. In this study, we presented a promising explainable artificial intelligence methodology (XGBoost and SHAP) with the aim of investigating organochlorine patterns as well as the dependence of PCB-138 on the mother's age and number of births. According to the results, similarly structured molecules, which belong to both non-dioxin-like/indicator congener group (−28, −52, −180) and toxicologically relevant PCBs (−118, −189, −156) as well as *p,p'*-DDT metabolite (*p,p'*-DDE) have an impact on PCB-138 distribution. No significant functional dependencies of PCB-138 patterns and maternal age and parity were observed, suggesting that the impacts of dietary habits and the health burden of POPs in residential and working environments should be investigated further in the future. The herein applied analyses could serve as a promising methodology for future epidemiological investigations and human health protection.

Acknowledgements The authors acknowledge the funding provided by the Institute of Physics, Belgrade, through a grant by the Ministry of Education, Science, and Technological Development of the Republic of Serbia and the bilateral scientific research project between the Republic of Serbia and Croatia 2019–2021 (No. 337-00-205/2019-09/22) "Persistent Organic Compounds in breast milk and their effects on the level of primary DNA damage in human cells". Also, this paper was financed by the Science Fund of the Republic of Serbia #GRANT No. 6524105, AI—ATLAS.

References

1. Aerts, R., Van Overmeirea, I., Colles, A., Andjelković, M., Malarvannan, G., Poma, G., Den Hond, E., Van de Mieroop, E., Dewolf, M.C., Charlet, F., Van Nieuwenhuyse, A., Van Loco, J., Covaci, A.: Determinants of persistent organic pollutant (POP) concentrations in human breast milk of a cross-sectional sample of primiparous mothers in Belgium. Environ. Int. **131** (2019)
2. Antignac, J.P., Main, K.M., Virtanen, H.E., Boquien, C.Y., Marchand, P., Venissieux, A., Guiffard, I., Bichon, E., Wohlfahrt-Veje, C., Legrand, A., Boscher, C., Skakkebæk, N.E., Toppari, J., Le Bizec, B.: Country-specific chemical signatures of persistent organic pollutants (POPs) in breast milk of French, Danish and Finnish women. Environ. Pollut. **218**, 728–738 (2016)
3. Berghuis, S.A., Van Braeckel, K.N.J.A., Sauer, P.J.J., Bos, A.F.: Prenatal exposure to persistent organic pollutants and cognition and motor performance in adolescence. Environ. Int. **121**, 13–22 (2018)
4. Chen, T., Guestrin, C.: Xgboost: a scalable tree boosting system. In: Proceedings of the 22nd ACM SIGKDD International Conference on Knowledge Discovery and Data Mining, pp. 785–794 (2016)
5. De Cock, M., De Boer, M.R., Govarts, E., Iszatt, N., Palkovicova, L., Lamoree, M.H., Schoeters, G., Eggesbø, M., Trnovec, T., Legler, J., Van de Bor, M.: Thyroid-stimulating hormone levels in newborns and early life exposure to endocrine-disrupting chemicals: analysis of three European mother-child cohorts. Pediatr. Res. **82**, 429–437 (2017)
6. Fång, J., Nyberg, E., Winnberg, U., Bignert, A., Bergman, Å.: Spatial and temporal trends of the Stockholm convention POPs in mothers' milk—a global review. Environ. Sci. Pollut. Res. **22**, 8989–9041 (2015)

7. Fry, K., Power, M.C.: Persistent organic pollutants and mortality in the United States, NHANES 1999–2011. Environ. Health **16**, 105 (2017)
8. Govaerts, A., Verhaert, V., Covaci, A., Jaspers, V.L.B., Berg, O.K., Addo-Bediako, A., Jooste, A., Bervoets, L.: Distribution and bioaccumulation of POPs and mercury in the Ga-Selati River (South Africa) and the rivers Gudbrandsdalslågen and Rena (Norway). Environ. Int. **121**, 1319–1330 (2018)
9. Hartmann, J.: Classification using decision tree ensembles. Available at SSRN 3484009 (2019)
10. Jovanović, G., Herceg Romanić, S., Stoji'c, A., Klini'c, D., Matek Sari'c, M., Grzunov Letinić, J., Popović, A.: Introducing of modeling techniques in the research of POPs in breast milk—a pilot study. Ecotoxicol. Environ. Saf. **172**, 341–347 (2019)
11. Kezios, L.K., Xinhua, L., Cirillo, M.P., Cohn, A.B., Kalantzi, I.O., Yunzhu, W., Petreas, X.M., Park, J.S., Factor-Litvak, P.: Dichlorodiphenyltrichloroethane (DDT), DDT metabolites and pregnancy outcomes. Reprod. Toxicol. **35**, 156–164 (2013)
12. Klinić, D., Herceg Romanić, S., Brić Karaonji, I., Matek Sarić, M., Grzunov Letinić, J., Brajenović, N.: Organochlorine pesticides and PCBs (including dl-PCBs) in human milk samples collected from multiparae from Croatia and comparison with primiparae. Environ. Toxicol. Pharmacol. **45**, 74–79 (2016)
13. Klinić, D., Herceg Romanić, S., Matek Sarić, M., Grzunov, J., Dukić, B.: Polychlorinated biphenyls and organochlorine pesticides in human milk samples from two regions in Croatia. Environ. Toxicol. Pharmacol. **37**, 543–552 (2014)
14. Lignell, S., Winkvist, A., Bertz, F., Rasmussen, K.M., Glynn, A., Aune, M., Brekke, H.K.: Environmental organic pollutants in human milk before and after weight loss. Chemosphere **159**, 96–102 (2016)
15. Lundberg, S., Lee, S.: A unified approach to interpreting model predictions. Adv. Neural Inf. Process Syst. 4765–4774 (2017)
16. Man, Y.B., Chow, K.L., Guan, H.X., Janet, K.Y.C., Sheng, C.W., Wong, M.H.: A pilot study on health risk assessment based on body loadings of PCBs of lactating mothers at Taizhou, China, the world's major site for recycling transformers. Environ. Pollut. **227**, 364–371 (2017)
17. Mannetje, A., Coakley, J., Mueller, F.J., Harden, F., Toms, L.-M., Douwes, J.: Partitioning of persistent organic pollutants (POPs) between human serum and breast milk: a literature review. Chemosphere **89**, 911–918 (2012)
18. Mitchell, R., Fank, E.: Accelerating the XGBoost algorithm using GPU computing. PeerJ Comput. Sci. **3**, e127 (2017)
19. Nielsen, D.: Tree boosting with XGBoost. NTNU Norwegian University of Science and Technology (2016)
20. Polder, A., Skaare, J.U., Skjerve, E., Løken, K.B., Eggesbø, M.: Levels of chlorinated pesticides and polychlorinated biphenyls in Norwegian breast milk (2002–2006), and factors that may predict the level of contamination. Sci. Total Environ. **407**, 4584–4590 (2009)
21. Stanišić, S., Perišić, M., Jovanović, G., Milićević, T., Herceg Romanić, S., Jovanović, A., Šoštarić, A., Udoviić, V., Stojić, A.: The PM2.5-bound polycyclic aromatic hydrocarbon behavior in indoor and outdoor environments, part I: emission sources. Environ. Res. **193**, 110520 (2021)
22. Stojić, A., Jovanović, G., Stanišić, S., Herceg Romanić, S., Šoštarić, A., Udoviić, V., Perišić, M.: The PM2.5-bound polycyclic aromatic hydrocarbon behavior in indoor and outdoor environments, part II: explainable prediction of benzo[a]pyrene levels. Submitted for publication (2021)
23. Stojić, A., Matek Sarić, M., Herceg Romanić, S.: SHapley additive explanations of indicator PCB-138 distribution in breast milk. In: Book of Proceedings of International Scientific Conference on Information Technology and Data Related Research (Sinteza 2020), Singidunum University, pp. 35–40 (2020)
24. Stojić, A., Stanić, N., Vuković, G., Stanišić, S., Perišić, M., Šoštarić, A., Lazić, L.: Explainable extreme gradient boosting tree-based prediction of toluene, ethylbenzene and xylene wet deposition. Sci. Total Environ. **653**, 140–147 (2019)
25. United Nations Environment Programme.: Results of the global survey on concentrations in human milk of persistent organic pollutants by the United Nations Environment Programme

and the World Health Organization. In: Conference of the Parties to the Stockholm Convention on Persistent Organic Pollutants, Sixth meeting, Geneva, 28 April 2010 (2013)
26. Van den Berg, M., Kypke, K., Kotz, A., Tritscher, A., Yong Lee, S., Magulova, K., Fiedler, H., Malisch, R.: WHO/UNEP global surveys of PCDDs, PCDFs, PCBs and DDTs in human milk and benefit-risk evaluation of breastfeeding. Arch. Toxicol. **91**, 83–96 (2017)
27. Vasios, G., Kosmidi, A., Kalantzi, O.I., Tsantili-Kakolidou, A., Kavantzas, N., Theocharis, S., Giaginis, C.: Simple physicochemical properties related with lipophilicity, polarity, molecular size and ionization status exert significant impact on the transfer of drugs and chemicals into human breast milk. Expert Opin. Drug Metab. Toxicol. **12**, 1273–1278 (2016)
28. Vigh, É., Colombo, A., Benfenati, E., Håkansson, H., Berglund, M., Bódis, J., Garai, J.: Individual breast milk consumption and exposure to PCBs and PCDD/Fs in Hungarian infants: a time-course analysis of the first three months of lactation. Sci. Total Environ. **449**, 336–344 (2013)
29. Vukavić, T., Vojinović Miloradov, M., Mihajlović, I., Ristivojević, A.: Human milk POPs and neonatal risk trend from 1982 to 2009 in the same geographic region in Serbia. Environ. Int. **54**, 45–49 (2013)

What Information on Volatile Organic Compounds Can Be Obtained from the Data of a Single Measurement Site Through the Use of Artificial Intelligence?

Svetlana Stanišić, Mirjana Perišić, Gordana Jovanović, Dimitrije Maletić, Dušan Vudragović, Ana Vranić, and Andreja Stojić

Abstract Increasing air pollutant concentrations over the last few decades have been a focus of contemporary scientific research due to adverse effects on public health, the environment and climate change. In this chapter, we used an innovative integrated methodology for spatio-temporal characterization of sources and concentration forecasts of toxic, mutagenic and carcinogenic representatives of volatile organic species—benzene, toluene, ethylbenzene and xylene, commonly referred to as BTEX. The methodology is based on receptor-oriented air circulation modeling and artificial intelligence implemented through machine learning and explainable artificial intelligence methods. The study covered two years of data obtained from a single monitoring station located at 54a Despota Stefana Boulevard (44°49'68" N, 20°28'04" E). This station was selected from the local and state network for air quality monitoring in the territory of Belgrade. The receptor-oriented modeling was effective for classifying sources of BTEX and the assessment of BTEX concentra-

S. Stanišić (✉)
Environment and Sustainable Development, Singidunum University, Belgrade, Serbia
e-mail: sstanisic@singidunum.ac.rs

M. Perišić · G. Jovanović · D. Maletić · D. Vudragović · A. Vranić · A. Stojić
Institute of Physics Belgrade, National Institute of the Republic of Serbia, Environment and Sustainable Development, Singidunum University, Belgrade, Serbia
e-mail: mirjana.perisic@ipb.ac.rs

G. Jovanović
e-mail: gordana.vukovic@ipb.ac.rs

D. Maletić
e-mail: dimitrije.maletic@ipb.ac.rs

D. Vudragović
e-mail: dusan.vudragovic@ipb.ac.rs

A. Vranić
e-mail: ana.vranic@ipb.ac.rs

A. Stojić
e-mail: andreja.stojic@ipb.ac.rs

E. Pap (ed.), *Artificial Intelligence: Theory and Applications*,
Studies in Computational Intelligence 973,
https://doi.org/10.1007/978-3-030-72711-6_12

tions in the Belgrade urban area surrounding the receptor site that was not regularly monitored. The correlations and ratios between BTEX compounds were used for estimating their interrelationships and presence in the air, which contributed to the identification of their origin. Also, this study evaluated the possibilities of BTEX spatio-temporal forecasts based on the integrated methodology. For this purpose, XGBoost was efficient at forecasting BTEX levels, with estimated errors (6–15%) significantly below the uncertainty obtained by conventional models for the evaluation of average annual pollutant concentrations. The results suggest that temperature, wind speed and wind direction represented the main parameters which explain the spatio-temporal distribution of BTEX, while the impact of other factors showed significant variations depending on the locations of the receptor and the compound.

Keywords BTEX · Volatile organic compounds · Machine learning · Explainable artificial intelligence (xai) · XGBoost

1 Introduction

Growing urban populations, economic development, and transport have a significant impact on environmental pollution. Air pollution constitutes a major underestimated cause of non-communicable diseases, being responsible for 19% of all cardiovascular deaths and 23% of lung cancer deaths globally [19]. Around 91% of the world's population lives in areas where air pollution levels exceed the World Health Organization's recommended values [43]. People often perceive air pollution as an issue that affects people living in middle- and low- income countries or people living in megacities. However, it has been estimated that out of 8.8 million global deaths associated with air pollution in 2015, 8% were citizens of high-income countries, mostly those living in small urban areas where the topography and climate contribute to high air pollutant concentrations [3, 7]. Holgate [17] emphasizes that annually, 40,000 excess deaths in the UK can be attributed to low air quality. Society might be much more aware of this issue if the mortality was a result of drinking polluted water. Air pollution in households has decreased since the 1990's as new fuels such as petroleum gas and renewable sources of energy have replaced biomass. In developing countries, residential combustion of solid fuels for cooking and heating remains a significant source of air pollution and a major concern due to its detrimental health effects, particularly in rural areas [9]. In contrast, in developed countries, indoor air pollution is less important because adverse health effects are associated with exposure to outdoor pollutant concentrations that have decreased over previous decades. However, it has also been estimated that the rapid expansion of metropolises, industrial production, increasing pesticide use, toxic chemicals and motor vehicles will offset the effects of air pollution mitigation measures [13].

Among the air pollutants that are of interest for current and future research due to their detrimental effects on both human health and the environment are volatile organic compounds (VOCs). VOCs are a heterogeneous group of organic species with boiling points <250 °C. Their representatives are benzene and its alkylated derivatives toluene, ethylbenzene and xylene, commonly referred to as BTEX. Over the last few decades in developed countries, reducing BTEX levels has been a challenge [26]. This is due to their abundance, their numerous emission sources, their complex atmospheric chemistry, insufficient funds for the establishment and maintenance of monitoring networks, and the fact that abatement programs might have negative impacts on economic output because they are among the most abundantly produced compounds worldwide. They are used as feedstock for several materials and products upon which modern society has become dependent. These compounds are naturally found in crude oil, while in urban and suburban areas they originate from traffic emissions, commercial and industrial uses of petroleum, gasoline, adhesives, coatings, degreasers, solvents, detergents, explosives, pesticides, resins, ink, paints, and varnishes [30, 33, 35]. With the exception of m-/p-xylene, concentrations of BTEX were mostly reported to be higher in indoor air than outdoor air [4].

The health effects of BTEX are wide-ranging. Studies have shown that BTEX can be found in cord blood and the blood of children and adults, particularly adults that are occupationally exposed [18]. Research has shown that long-term exposure to high benzene concentrations increases the risk of developing malignant blood disorders, while long-term exposure to high toluene concentrations causes renal tubular acidosis [42]. Some studies have shown that after benzene concentrations in industrial and urban areas are reduced, lifetime cancer risk decreases by one order of magnitude [20]. In addition, exposure to ambient levels of BTEX, which in many cases were orders of magnitude below the reference concentrations, can be dangerous. This is particularly true during the susceptibility period when exposure can lead to the disruption of endocrine signaling (which is essential for the growth and development), immune responses, reproduction, cardiovascular function and aging. Therefore, populations in highly industrialized areas, the socioeconomically deprived, as well as children, pregnant women and elderly people, appear to be more susceptible to pollution-related morbidity and mortality [5, 32]. While traffic emissions are known to be the main source of outdoor BTEX, converting to alternative renewable energy sources will not reduce demand for consumer products, and the replacement of benzene with safer alternatives, such as toluene and xylene, might ultimately be a poor solution once long-term toluene and xylene exposure scenarios are considered [4]. According to a United States Environmental Protection Agency [40] report, ethylbenzene is sixth on a list of the top 20 chemicals used in children's products, primarily food packaging, toys, sport equipment, arts, crafts and hobby materials. Toluene, on the other hand, is the seventh most used chemical in consumer products, primarily fuels, paints, and coatings. Although toluene and xylenes are less harmful than benzene, it should be kept in mind that the products of photochemical reactions in which BTEX are involved often have more harmful effects on human health than their precursors [14]. It should be also mentioned that limited ventilation in closed premises can often be the reason why indoor BTEX levels are higher than outdoor levels [29].

Apart from their impacts on human health, BTEX and other VOCs are associated with increases in the oxidation capacity of the atmosphere, as well as with the generation of secondary pollutants, such as tropospheric ozone, polycyclic aromatic hydrocarbons and secondary aerosol through photochemical reactions [8, 36]. As regards the impact on global warming, not only do volatile species directly and indirectly contribute to climate change, but their emission and fate are expected to be influenced and increased by global warming.

Despite vast changes in the development and integration of different approaches in the field of environmental science, spatio-temporal air pollution modeling remains a challenge. Two main approaches are typically utilized to forecast air quality and to identify the factors that govern the concentrations of certain pollutants. The first approach relies on atmospheric diffusion models, while the second refers to statistical models that capture the essential relationships between the variables [24]. Multidimensionality and the size of data sets, as well as the complexity of air pollutant processes and interactions, set requirements that exceed the capabilities of conventional statistical methods. For this reason, machine learning (ML) methods and explainable artificial intelligence (XAI), subfields of artificial intelligence (AI) that enable automatized big data analysis and the development of learning algorithms, have been introduced into environmental science research. In this paper, we used an innovative and integrated methodology based on artificial intelligence and implemented through ML ana XAI methods for the modeling of spatio-temporal air pollution and the characterization of BTEX sources in a wider region surrounding the receptor site that was not covered by regular monitoring. The obtained BTEX correlations and ratios were used for estimating the interrelationships between the species in the air, while XGBoost was utilized for efficient spatio-temporal BTEX forecasting. The present paper is an extended version of the study presented at the International Scientific Conference on Information Technology and Data Related Research (Sinteza 2020), Singidunum University [38].

2 Materials and Methods

Nowadays, there are a large number of libraries implemented in different programming languages (R, Python, JS) that deal with interactive display (plotly, leaflet) and spatial data analysis. In addition to various spatial autocorrelation possibilities, the analysis of spatial data patterns, the interpolation of data by statistical methods such as Kriging, Spline or Inverse Distance Weighting, and machine learning methods such as Random Forests are increasingly used. Furthermore, spatial distributions of air pollutant sources can be estimated using general and local hybrid receptors, as well as by analyzing clustered data [9, 33, 34].

Significant improvements to the general hybrid receptor modeling approach have been made in recent decades. Receptor-oriented methods, based on conditional probability and analyzing the residence time of pollutants in an area, have become widely accepted both for studying the dynamic processes and circulation patterns of pollutants in the atmosphere and for investigating the spatial distribution of potential emission sources and their impact on the receptor site without relying on an emission inventory.

The identification of potential emission sources at a local level, as well as their contribution to the measured concentrations of pollutants at the receptor site, can be determined by models analogous to the general hybrid receptor model such as the Potential Source Contribution Function (PSCF), Concentration-Weighted Trajectories (CWT), sQTBA (Simplified Quantitative Transport Bias Analysis) and RTWC (Residence Time Weighted Concentration). The general models, which are based on the analysis of trajectory end points, can be improved by using local wind parameters. This way it is possible to determine pollution circulation patterns very accurately in the area around a given measuring point. A local three-dimensional hybrid receptor model similar to the 3D CWT model was developed for the purposes of this paper [34]. The study involved two-years of regular measurements of BTEX, suspended particles, inorganic gaseous oxides, and meteorological parameters within the Belgrade City Institute for Public Health's automatic monitoring network (44°49'68" N, 20°28'04" E).

Machine learning is an area of artificial intelligence that involves the development of algorithms that can learn based on input data and can thus be trained to predict value variations. Machine learning algorithms are based on the extraction of patterns and the selection of specific attributes from a large number of data, while eliminating irrelevant information. Through identifying most important prediction attributes, machine learning methods acquire knowledge and define the substantial relationships that exist between input and output parameters by focusing on the aspect of the data that is the most useful for efficient forecasting.

Forecasting the concentration dynamics of BTEX in the air was done using the Xtreme Gradient Boosting (XGBoost) method [11], with meteorological data used as predictors. XGBoost is a general-purpose ensemble method of supervised machine learning which combines the results of many decision trees and achieves high accuracy in a wide range of practical applications. It usually outperforms support vector machines, random forests, and deep learning neural networks [31]. The main advantage of the XGBoost method is its obtainment of more precise predictions than those provided by single constitutive decision algorithms. XGBoost is based on a boosting technique that sequentially defines a series of decision trees for classifying input data into two or more attribute-defined classes. Each consecutive decision tree is trained through iterations, taking into account the previously registered classification errors. The datasets from each of the grid cells used for spatio-temporal BTEX forecasting was split into training (80%) and validation (20%) sets. Hyperparameter tuning was implemented using a brute-force grid search and a 10-fold stratified cross-validation. The best performing hyperparameter values were used in the final model.

Methods based on decision trees, such as Gradient Boosting and Random Forests, have been shown to provide inconsistent attribute contributions. This has led to the development of SHAP (SHapley Additive exPlanation), a method that estimates the contribution of each instance of an attribute, which enables interpretation of the model's outputs [15, 21, 31]. To explain the contribution of each feature to the individual XGBoost prediction, the SHAP method was utilized [21]. It is based on coalitional game theory and provides a distribution of each prediction among the features represented as additive attributions. In this study, we used Python SHAP implementation (SHAP Python package). The captured importance of a feature is visually presented as a SHAP summary plot.

3 Results and Discussion

Previous studies have demonstrated strong gradients and pronounced intra-urban spatial variability of pollutant levels. This depends on traffic density, street configuration and prevailing wind direction [41]. In addition to this, Ning et al. [27] emphasized the importance of terrain complexity for air quality because topographic features can, to a certain extent, limit pollutant dispersion under different weather conditions. The results indicate that BTEX concentrations were significantly higher in areas with dense traffic, in the vicinity of busy streets and intersections, where traffic emission sources have a high impact, where local topography limits natural ventilation and vertical dispersion of pollution, and in areas with narrow streets lined by tall buildings and trees.

3.1 BTEX Levels Surrounding the Receptor Site

Benzene concentrations were estimated to range from 1.2 to 2.6 $\mu g\ m^{-3}$, with an average level of 2.6 $\mu g\ m^{-3}$. In a few locations, estimates of pollutant concentrations were extremely high, exceeding the recommended limit of 5 $\mu g\ m^{-3}$. Extreme benzene levels in areas distant from the receptor site should be taken with caution. This is due to a relatively small number of events and therefore limited data produces the calculated values. It has been estimated that the area surrounding the receptor site was mostly influenced by traffic emissions. Other studies have reported that similar average benzene concentrations are sometimes discovered in urban areas where traffic is the predominant source of pollutant emissions [2]. Concentrations of toluene were estimated to range from 0.9 to 8.9 $\mu g\ m^{-3}$ with an average value of 8.7 $\mu g\ m^{-3}$, while the average m/p-xylene concentrations were assessed to be several times higher (8.4 $\mu g\ m^{-3}$) than the concentrations of o-xylene (1.8 $\mu g\ m^{-3}$). Concentrations of ethylbenzene were estimated to range from 0.2 to 2.1 $\mu g\ m^{-3}$ with an average value of 1.8 $\mu g\ m^{-3}$. As predicted, TEX levels were highest in the city center and the north of the city. The assessed TEX levels were also consistent with the reported values for outdoor TEX concentrations in various studies published over the last two decades [1, 10, 16, 23, 25], with somewhat lower levels of toluene than typically expected in an urban area.

3.2 Seasonal and Daily BTEX Variations

While the intensity of BTEX emission sources tend to be higher in winter, evaporations in warmer parts of the year make an important contribution to total pollutant concentrations. On the other hand, the stability of atmospheric conditions and decrease in chemical reactivity of BTEX in winter results in longer retention in ambient air [29]. Furthermore, BTEX are known to exhibit variations on a daily basis. For instance, a previous study that contained a risk assessment of an accidental benzene release in an urban area using an atmospheric dispersion model, showed that benzene spreads over a much larger area during the nighttime due to a stable boundary layer. In contrast, enhanced vertical mixing results in limited dispersion of the pollutant over the study area during the daytime [39]. Similar results were obtained in our study.

3.3 BTEX Forecasting Based on Meteorological Variables as Predictors

As can be seen in Figs. 1 and 2, high correlation coefficients between predicted and observed values (>0,80) were obtained for most of the analyzed data, it can therefore be concluded that XGBoost was a successful and efficient method for forecasting air pollution in an urban area. It should be emphasized that the estimated method errors (6–15%) were significantly lower than the uncertainty (50%) which is required for the evaluation of average annual pollutant concentrations using conventional modeling.

It should be also emphasized that due to the relatively short atmospheric lifetimes of BTEX, the area that can be affected by BTEX emissions is approx. 15–20 km [29], but the consistency of meteorological conditions significantly affects the extent of volatile pollutant dispersion.

The results of this study suggest that low temperatures and weak to moderate wind represent the main parameters which govern the spatio-temporal distribution of benzene in a majority of locations. The impacts of other factors display significant variation depending on the characteristics of the receptor's location (Fig. 1). Namely, one can note that the horizontal axis marked with wind speed is the longest, which means that benzene concentrations were mostly affected by this parameter. Each axis is composed of a series of points that represent the measured values of the predictor. As can be seen in the example for wind speed, extremely high wind speeds, represented by points located on the far right of the axis, have relatively little impact on benzene concentrations in the air. This suggests that benzene concentrations were mostly influenced by weak to moderate wind, which further suggests that emission sources located in the vicinity of the selected receptor site have the highest impact.

Furthermore, the importance of a certain wind direction for pollutant level prediction is related to the position of the emission source which affects the receptor location the most. Namely, in cases where a single pollutant emission source has a

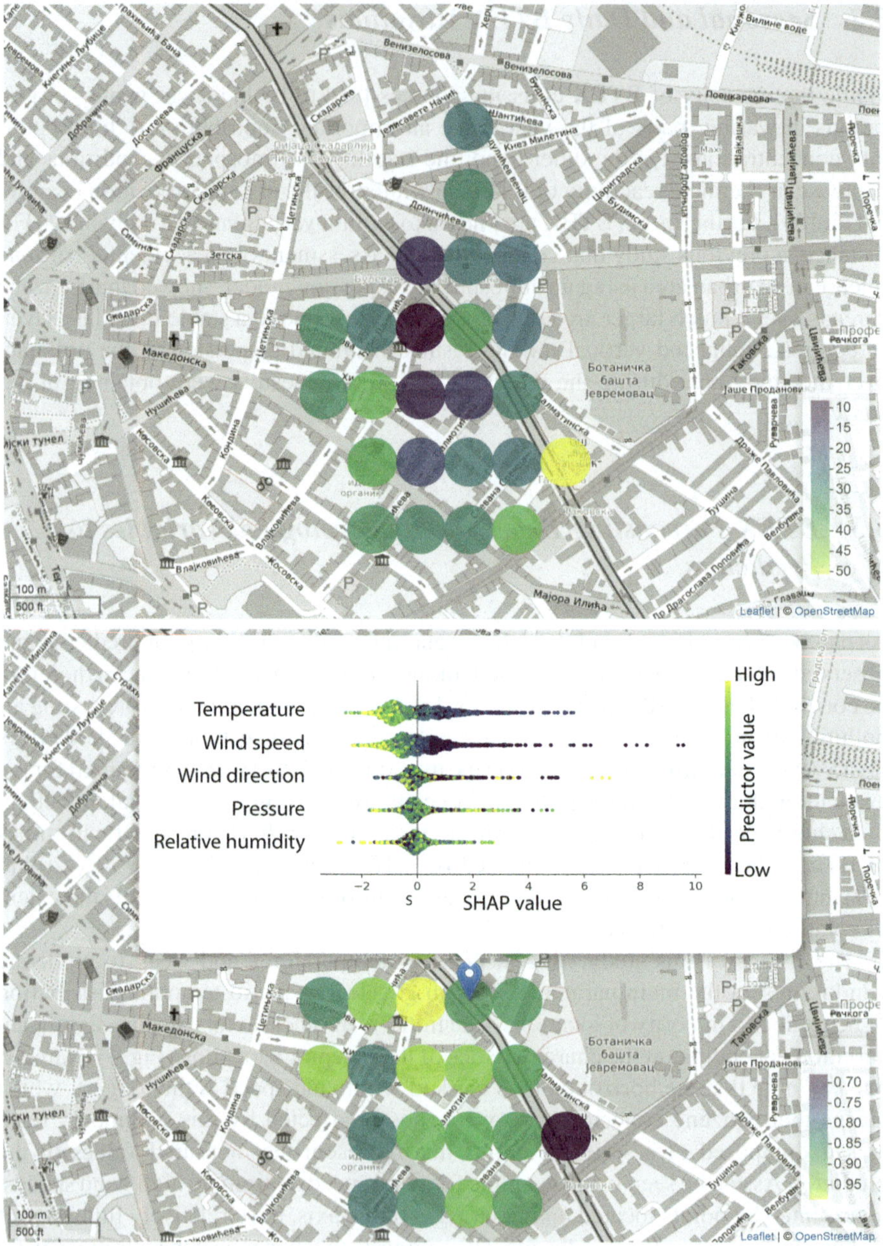

Fig. 1 Benzene forecast based on meteorological parameters—a relative error [%] (top) and SHAP values and predicted/observed correlation coefficients (bottom)

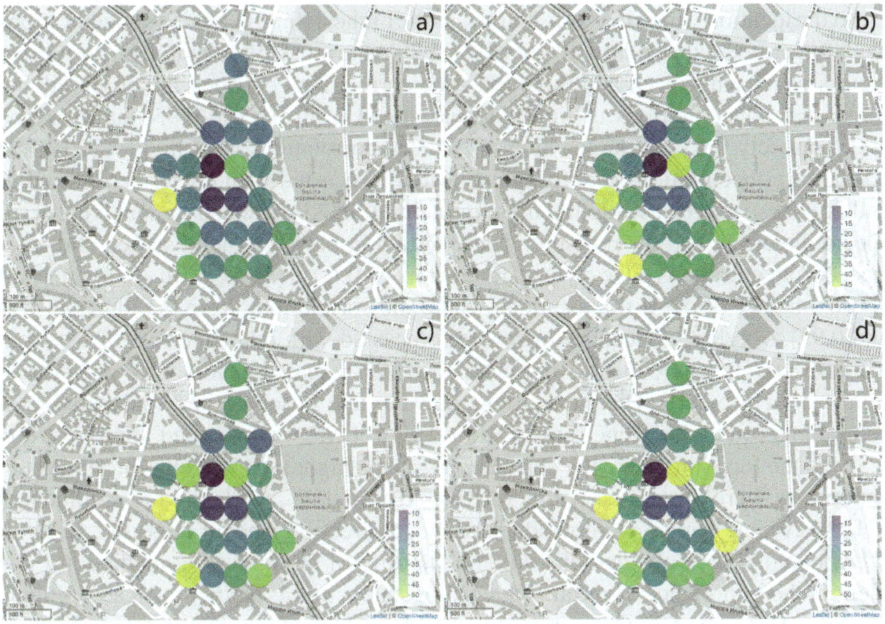

Fig. 2 Toluene (**a**), m,p-xylene (**b**), o-xylene (**c**), and ethylbenzene (**d**) relative error [%] forecasts based on meteorological parameters

major effect on the receptor area, the importance of wind direction as a predicting factor is particularly accentuated. Additionally, the importance of wind direction is evident in situations where tall buildings and trees along the roadside form a corridor that either hinders or assists polluted air masses from the surrounding emission sources, which was the case in this study as well. The importance of relative humidity as predictor for benzene concentrations was predicted to be high in the event of extreme benzene levels. This is because benzene and TEX are partially water-soluble compounds, which suggests that both high and low levels of relative humidity can impact their ambient air levels [28].

The presented figures displaying TEX distributions can be interpreted in the following manner. Atmospheric pressure and wind parameters appear to be the major predictors of TEX behavior in general. The impacts of air temperature and pressure were significantly lower for forecasting toluene levels (Fig. 2) than for forecasting benzene levels, which can be explained by the fact that benzene evaporates at significantly lower temperatures than toluene.

The correlations between the observed and forecasted concentrations of benzene ranged from 0.70 to 0.95, but most often ranged from 0.80 to 0.85. At the location where the correlation was the highest (0.95), benzene concentrations were forecasted with a relative error of 8%, while the most influential SHAP-revealed factor was wind direction. The correlations between the observed and forecasted concentrations of toluene ranged from 0.65 to 0.95, but mostly around 0.80.

At the location where the correlation was the highest (0.95), toluene concentrations were forecasted with a relative error of 9%, while the most influential SHAP-revealed factors were wind speed and direction. The correlations between the observed and forecasted concentrations of m/p-xylenes ranged from 0.65 to 0.95, but averaged approximately 0.75. At the locations where the correlations were highest (0.97 and 0.98), the concentrations of m/p-xylenes were forecasted with relative errors of 6% and 15% respectively, while the most influential SHAP-revealed factors were wind speed, direction, and air pressure. The correlation between the observed and forecasted concentrations of o-xylene ranged from 0.60 to 0.90, with an average of approximately 0.85. At the locations where the correlations were the highest (0.94 and 0.95), o-xylene concentrations were forecasted with relative errors of 12% and 8% respectively, while the most influential SHAP-revealed factors were wind direction and air pressure. The correlations between the observed and forecasted concentrations of ethylbenzene ranged from 0.60 to 0.95, but averaged approximately 0.80. At the location where the correlation was the highest (0.95), ethylbenzene concentrations were forecasted with a relative error of 11%, while the most influential SHAP-revealed factors were wind speed, wind direction, and air pressure.

Although the results demonstrate the capacity of an innovative methodology to identify the importance of certain meteorological factors as predictors of air pollutant concentrations, the fact that variations of meteorological parameters cause changes in other related parameters makes it difficult to distinguish their actual impact on air pollution phenomena. Thus, the impact of meteorological factors should not be observed as the isolated effect of a single parameter and its variations, but as the effect of a certain type of weather. For instance, Liao et al. [22] identified ten typical air circulation types within one of the most polluted areas of China and explored how their synergetic relationship with topography affected the local air quality.

3.4 The Importance of Other Pollutants as Predictors for BTEX Levels

Apart from meteorological parameters, other analyzed factors can be considered important for predicting BTEX concentrations. Namely, for forecasting benzene levels, high CO concentrations appear to be the most important (Fig. 3), with the importance of the other predictors stated here in decreasing order: toluene> ethylbenzene> m/p-xyneles> o-xylene> $PM_{10} > NO_x > NO_2 >$ NO> SO_2.

As regards forecasting toluene levels, high m/p-xylenes, o-xylene and CO concentrations appear be the most important, with the importance of the other predictors stated here in decreasing order: ethylbenzene> benzene> $PM_{10} >$ NO> $NO_x > SO_2 > NO_2$. As regards forecasting m/p-xylenes levels, high toluene, o-xylene, ethylbenzene and CO concentrations appear to be the most important, with the importance of the other predictors stated here in decreasing order: benzene> $NO_2 > NO_x > PM_{10} > SO_2 >$ NO. As regards forecasting o-xylene levels, high m/p-

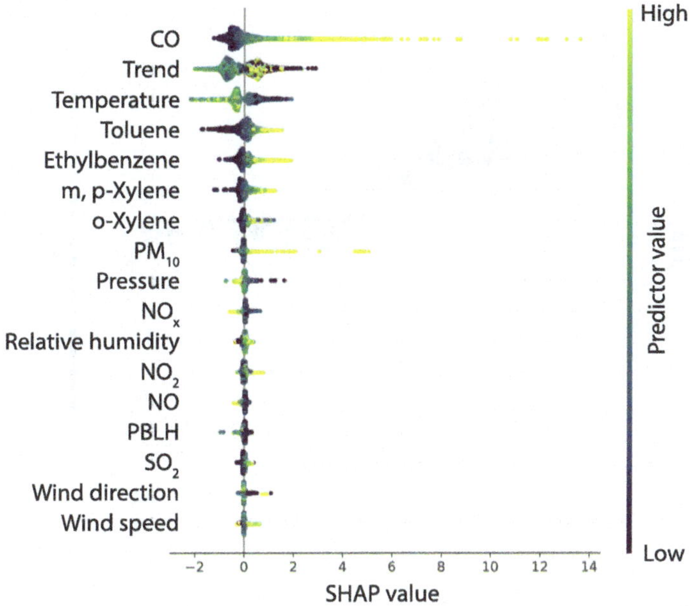

Fig. 3 SHAP summary plot for benzene

xylenes, toluene, ethylbenzene and NO concentrations appear to be the most important, with the importance of the other predictors stated here in decreasing order: NO_x > benzene> CO> PM_{10} > NO_2 > SO_2. As regards forecasting ethylbenzene levels, high m/p-xylenes, o-xylene and benzene concentrations appear to be the most important, with the importance of the other predictors stated here in decreasing order: toluene> NO_2 > SO_2 > CO> PM_{10} > NO> NO_x.

3.5 The Interdependence of BTEX Level Predictors

As part of forecasting pollutant concentrations, an examination of the interdependence between individual predicting factors and their combined effect on BTEX concentrations in the air was performed. As shown in the Fig. 4, during the cold part of the year, when temperatures were below 14 °C, concentrations of benzene are better predicted by toluene levels. Conversely, on days when the temperature exceeded 14 °C, benzene and toluene didn't share the same emission sources. Furthermore, during the cold part of the year, benzene concentrations are either significantly higher or lower than average, depending on whether the location is affected by the burning of fossil fuels for heating or not, while during the warmer part of the year benzene concentrations at different locations tended to be more uniform. This is because higher temperatures cause benzene to evaporate at all locations.

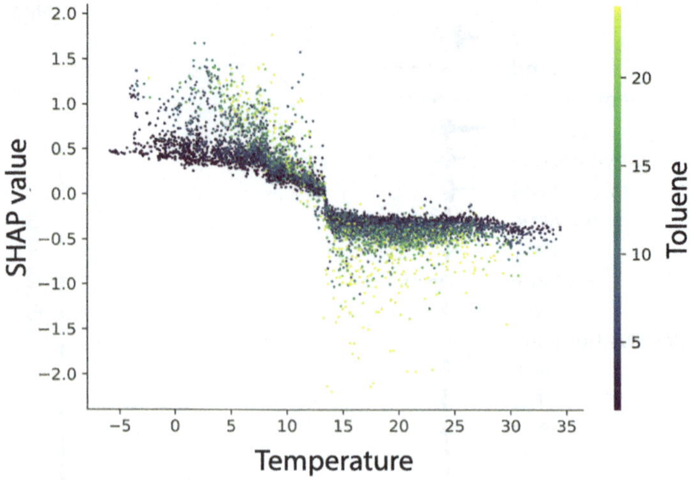

Fig. 4 Benzene SHAP dependency on temperature and toluene

In addition to temperature-toluene interactions, the results suggest that the concentrations of benzene in the ambient air depended on interactions between temperature and carbon monoxide as well. While the effects of interactions between ethylbenzene-m/p-xylenes, ethylbenzene-toluene, ethylbenzene-carbon monoxide and m/p-xylene-carbon monoxide were found to be of minor importance as regards shaping benzene concentrations (Fig. 5).

3.6 Pollutant Correlations and BTEX Origin

The results show moderate correlations between toluene and NO_x and CO concentrations throughout the city. The locations with toluene-NO_x correlations exceeding 0.7 are considered to be affected by burning-related emissions and not by toluene evaporations (Fig. 6).

Relatively high correlations between benzene and carbon monoxide were registered throughout the city, with the notable exception of the south (Fig. 7). The correlations between benzene and SO_2, NO_x and PM_{10} were lower in the eastern, and higher in the western, part of the urban region. These results suggest that benzene in the eastern area of the city is associated with evaporations and emissions from the petrochemical industry such as the Pančevo Oil refinery and the Petrohemija chemical plant. Furthermore, high correlations ($r > 0{,}70$) between benzene and inorganic oxides in the western region of the city may suggest that traffic emissions, as well as remote air pollution sources such as TPP Nikola Tesla A and B in Obrenovac, have a detrimental impact (Fig. 7). The significantly lower correlations between benzene and inorganic oxides, that can be considered indicators of combustion processes in

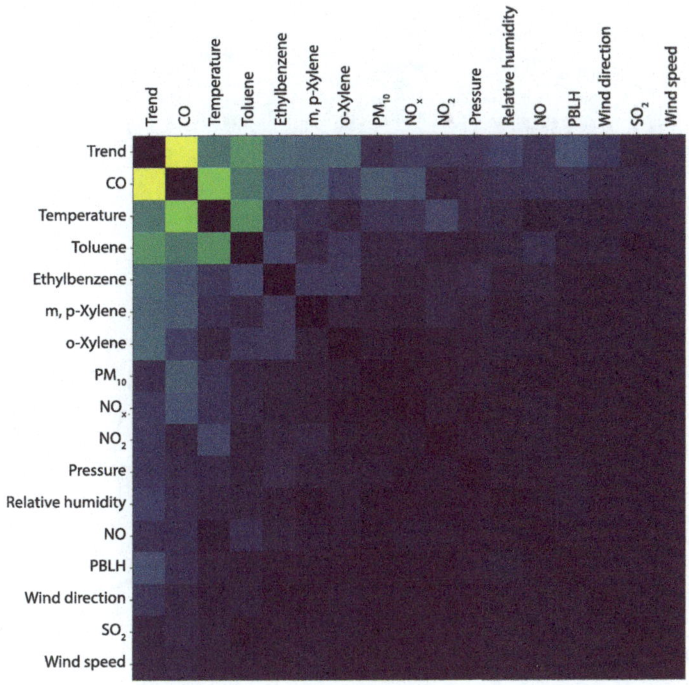

Fig. 5 SHAP interaction plot

the southern part of the city center, probably reflect the impact of gasoline evaporation. This conclusion appears more likely when we take into account that benzene concentrations were assessed to be the lowest in the southern area.

The m/p-xylenes exhibited the highest correlations with NO_x, somewhat lower correlations with CO, while the correlations with SO_2 and PM_{10} tended to be significantly lower throughout the city. A similar calculation was obtained for o-xylene and ethylbenzene.

3.7 BTEX Ratios

The benzene-to-toluene (B/T) ratio is often used as an index for identifying emission sources, while xylenes-to-benzene (X/B) and ethylbenzene-to-benzene (E/B) ratios are generally applied as indices of photochemical reactivity [12]. According to the literature, a B/T ratio below 0.5 suggests that vehicle emissions are the predominant source of BTEX, while X/B and E/B ratios below 1 suggest that sampled air masses are photochemically aged. As the results show, the B/T ratios were approx. 0.3, while the X/B and E/B ratios ranged from 0.7 to 4 with lower ratio values calculated for

Fig. 6 Toluene and NO_x correlations

the central area of the city where narrow streets and tall buildings restrict pollutant dispersion and support the photochemical aging of the emissions. Buczynska et al. [6] showed that B/T ratios ranging from 0.22 to 0.26 for locations that were affected by traffic emissions, while the authors mentioned that ratios below 0.1 reported in previous research suggest additional emission sources of toluene, like industrial emissions, that contribute to high values. It could be concluded from the results that in most locations, BTEX concentrations in ambient air are a result of recent emissions, while traffic emissions might be the major contributor to BTEX concentrations in the analyzed area. Furthermore, BTEX have a similar chemical structure, but there is a difference in the reactivity of these typically unreactive compounds. Namely, benzene is more stable and has a longer atmospheric lifetime [14] which can significantly modify the starting BTEX concentration ratios, and thus another possibility indicated by the B/T ratio is related to aging air masses. Namely, after being emitted from a common source, toluene is about 5 times more reactive than benzene, which is why a high B/T ratio is an indicator of aged air masses. The results suggest that the examined area can be divided diagonally into two segments:

- The north-western and north-eastern area which covers the old city center where the B/T ratio is relatively high. The north-western part is subject to emissions from the industrial area, which includes the petrochemical industry comprised of the Pančevo Oil refinery and the Petrohemija chemical plant. The north-western part is subject to the retention of aged air masses in narrow streets which form an urban

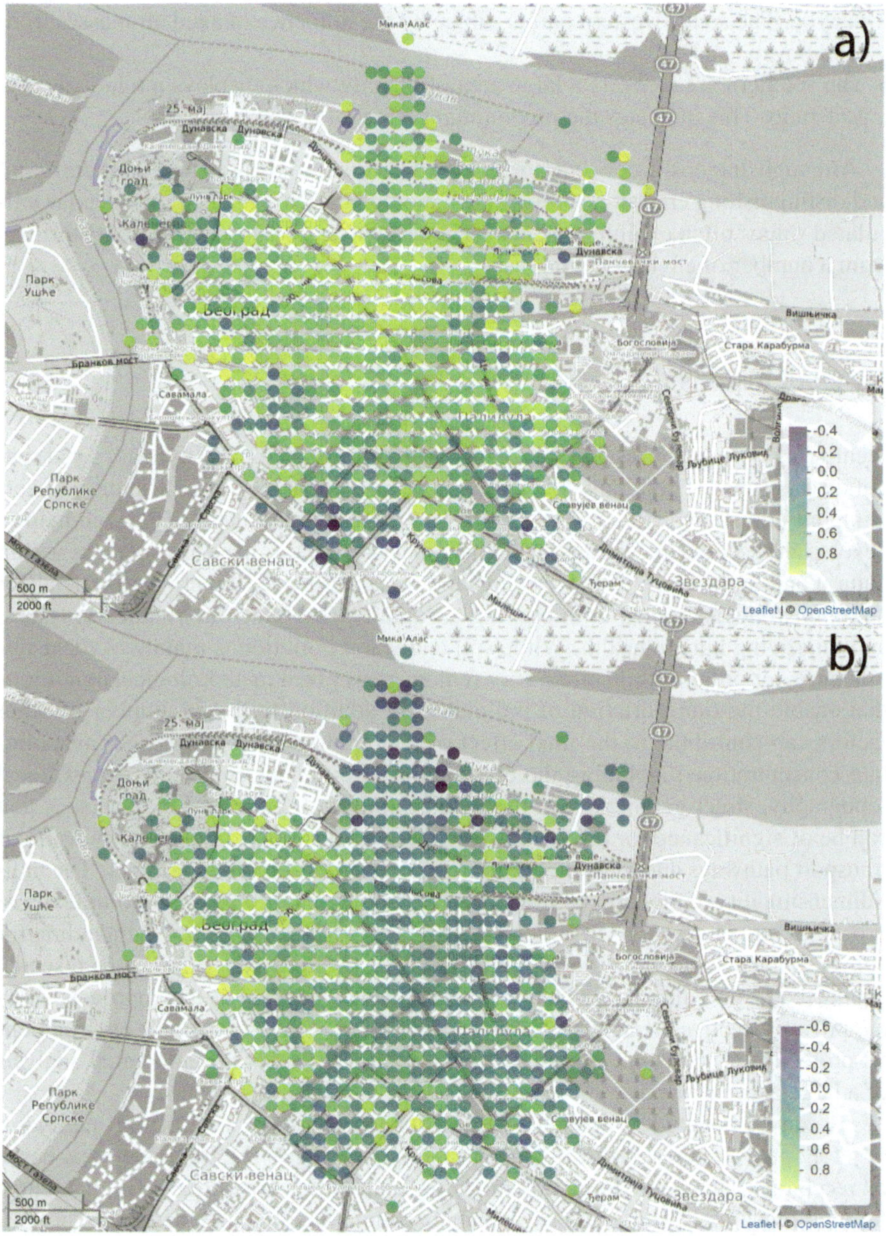

Fig. 7 Benzene correlation coefficients with CO (**a**) and NO_x (**b**)

canyon. The south-western part is subject to the influence of aged air masses from Obrenovac.
- The southern area, covers a large area of the city which displays a relatively low B/T ratio. This indicates the predominant effect of traffic emissions.

Although diagnostic concentration ratios are a commonly used tool for identifying and distinguishing emission sources, they should be used with caution since the calculated values often exhibit seasonal variations and can be susceptible to distortions from a number of environmental factors [37].

4 Conclusions

Identifying pollutant emission sources, investigating the temporal dynamics and spatial distributions of pollutant concentrations, and estimating the contribution of certain emission sources to local air quality are of great importance. They can help to develop an essential understanding of the impact key factors have on processes within natural and anthropogenic ecosystems and their resilience, but are also crucial for defining strategies aimed at improving air quality, resolving environmental issues, and improving human health. While soil and water pollution is mainly of local significance, air pollution can affect very remote areas given meteorological conditions that enable the transportation of pollutants. In addition to this, a number of other factors can contribute to the final effect of meteorological conditions on air pollutant concentrations, including the distribution of pollutant emission sources, local topography, street geometry and the distribution of all elements and surfaces that can be of significance for the air flow regime, pollutant dispersion conditions, their transport pathways and thus, the spatio-temporal variability of their levels. The multidimensionality of modern data, the scope of time series, as well as the complexity of processes and interactions in which air pollutants participate, are too demanding for conventional statistical methods. For this reason, recent decades have seen research conducted to find alternative methods for data analysis. One of the approaches that has proven effective in many professional and scientific fields, including environmental protection, is machine learning which provides the tools for automated analysis of a large amount of data.

As can be concluded, we have demonstrated the use of an efficient methodology for spatio-temporal BTEX concentration modeling in the Belgrade area, based on receptor-oriented air circulation modeling and artificial intelligence implemented through machine learning and explainable artificial intelligence methods. The estimated method errors ranged from 6 to 15%, which is significantly lower than the required uncertainty for conventional models. The presented methodology has the potential to provide a basis for the establishment of a unique and sustainable system for identifying sources of air pollution and enhanced air pollution data coverage that does not require additional investments in monitoring equipment. In the long term, the results of such an approach would provide a solid basis for establishing a sustainable system for improving the management and control of air pollution. As the results show, temperature, pressure, wind speed and wind direction were the main parameters which governed the spatio-temporal distribution of BTEX, while the impact of other factors showed significant variations depending on the characteristics of receptor's location and the compound. In addition, spatial correlations and ratios between different air pollutant concentrations were considered for determining their origin in all the locations covered by our analysis. All examples illustrate how the application of SHAP and other supplemental methods can provide a systematic insight into the impact of emission sources and environmental factors on the presence of BTEX in the air.

Acknowledgements Funding: The authors acknowledge funding provided by the Institute of Physics Belgrade, through a grant by the Ministry of Education, Science and Technological Development of the Republic of Serbia, the Science Fund of the Republic of Serbia #GRANT No. 6524105, AI - ATLAS, and Green Fund of the Ministry of Environmental Protection of the Republic of Serbia (No. 401-00-1219/2018-05).

References

1. Abd Hamid, H., Latif, M.T., Nadzir, M.S.M., Uning, R., Khan, M.F., Kannan, N.: Ambient BTEX levels over urban, suburban and rural areas in Malaysia. Air Qual. Atmos. Health **12**, 341–351 (2019)
2. Baltrėnas, P., Baltrėnaitė, E., Šerevičienė, V., Pereira, P.: Atmospheric BTEX concentrations in the vicinity of the crude oil refinery of the Baltic region. Environ. Monit. Assess. **182**, 115–127 (2011)
3. Behrens, D.A., Koland, O., Leopold-Wildburger, U.: Why local air pollution is more than daily peaks: modelling policies in a city in order to avoid premature deaths. Cent. Eur. J. Oper. Res. **26**, 265–286 (2018)
4. Bolden, A.L., Kwiatkowski, C.F., Colborn, T.: New look at BTEX: are ambient levels a problem? Environ. Sci. Technol. **49**, 5261–5276 (2015)
5. Bose, S., Diette, G.B.: Health disparities related to environmental air quality. In: Health Disparities in Respiratory Medicine, pp. 41–58. Humana Press, Cham (2016)
6. Buczynska, A.J., Krata, A., Stranger, M., Godoi, A.F.L., Kontozova-Deutsch, V., Bencs, L., Naveau, I., Roekens, E., Van Grieken, R.: Atmospheric BTEX-concentrations in an area with intensive street traffic. Atmos. Environ. **43**, 311–318 (2009)
7. Burnett, R., Chen, H., Szyszkowicz, M., Fann, N., Hubbell, B., Pope, C.A., Apte, J.S., Brauer, M., Cohen, A., Weichenthal, S., Coggins, J.: Global estimates of mortality associated with long-term exposure to outdoor fine particulate matter. PNAS USA **115**, 9592–9597 (2018)
8. Campbell, P., Zhang, Y., Yan, F., Lu, Z., Streets, D.: Impacts of transportation sector emissions on future US air quality in a changing climate. Part II: Air quality projections and the interplay between emissions and climate change. Environ. Pollut. **238**, 918–930 (2018)
9. Carter, E., Archer-Nicholls, S., Ni, K., Lai, A.M., Niu, H., Secrest, M.H., Sauer, S.M., Schauer, J.J., Ezzati, M., Wiedinmyer, C., Yang, X.: Seasonal and diurnal air pollution from residential cooking and space heating in the Eastern Tibetan Plateau. Environ. Sci. Technol. **50**, 8353–8361 (2016)
10. Cerón Bretón, J.G., Cerón Bretón, R.M., Martínez Morales, S., Kahl, J.D., Guarnaccia, C., Lara Severino, R.D.C., Rangel Marrón, M., Ramírez Lara, E., Espinosa Fuentes, M.D.L.L., Uc Chi, M.P., Sánchez, G.L.: Health risk assessment of the levels of BTEX in ambient air of one

urban site located in Leon, Guanajuato, Mexico during two climatic seasons. Atmosphere **11**, 165 (2020)
11. Chen, T., Guestrin, C.: Xgboost: a scalable tree boosting system. In: Proceedings of the 22nd ACM SIGKDD International Conference on Knowledge Discovery and Data Mining, pp. 785–794 (2016)
12. Dehghani, M., Fazlzadeh, M., Sorooshian, A., Tabatabaee, H.R., Miri, M., Baghani, A.N., Delikhoon, M., Mahvi, A.H., Rashidi, M.: Characteristics and health effects of BTEX in a hot spot for urban pollution. Ecotoxicol. Environ. Saf. **155**, 133–143 (2018)
13. Fowler, D., Brimblecombe, P., Burrows, J., Heal, M.R., Grennfelt, P., Stevenson, D.S., Jowett, A., Nemitz, E., Coyle, M., Lui, X., Chang, Y.: A chronology of global air quality. Philos. Trans. R. Soc. A **378**, 20190314 (2020)
14. Gallego, E., Roca, F.X., Guardino, X., Rosell, M.G.: Indoor and outdoor BTX levels in Barcelona city metropolitan area and Catalan rural areas. J. Environ. Sci. **20**, 1063–1069 (2008)
15. García, M.V., Aznarte, J.L.: Shapley additive explanations for NO2 forecasting. Ecol. Inform. **56**, (2020)
16. Hajizadeh, Y., Mokhtari, M., Faraji, M., Mohammadi, A., Nemati, S., Ghanbari, R., Abdolahnejad, A., Fard, R.F., Nikoonahad, A., Jafari, N., Miri, M.: Trends of BTEX in the central urban area of Iran: a preliminary study of photochemical ozone pollution and health risk assessment. Atmos. Pollut. Res. **9**, 220–229 (2018)
17. Holgate, S.T.: Every breath we take: the lifelong impact of air pollution-a call for action. Clin. Med. **17**, 8 (2017)
18. Kirman, C.R., Aylward, L.L., Blount, B.C., Pyatt, D.W., Hays, S.M.: Evaluation of NHANES biomonitoring data for volatile organic chemicals in blood: application of chemical-specific screening criteria. J. Expo. Sci. Environ. Epidemiol. **22**, 24–34 (2012)
19. Landrigan, P.J.: Air pollution and health. Lancet Public Health **2**, e4–e5 (2017)
20. Lerner, J.E.C., Kohajda, T., Aguilar, M.E., Massolo, L.A., Sánchez, E.Y., Porta, A.A., Opitz, P., Wichmann, G., Herbarth, O., Mueller, A.: Improvement of health risk factors after reduction of VOC concentrations in industrial and urban area. Environ. Sci. Pollut. Res. **21**, 9676–9688 (2014)
21. Lundberg S., Lee, S.: A unified approach to interpreting model predictions. In: Advances in Neural Information Processing Systems, pp. 4765–4774 (2017)
22. Liao, Z., Gao, M., Sun, J., Fan, S.: The impact of synoptic circulation on air quality and pollution-related human health in the Yangtze River Delta region. Sci. Total Environ. **607**, 838–846 (2017)
23. Liu, A., Hong, N., Zhu, P., Guan, Y.: Understanding benzene series (BTEX) pollutant load characteristics in the urban environment. Sci. Total Environ. **619**, 938–945 (2018)
24. Ly, H.B., Le, L.M., Phi, L.V., Phan, V.H., Tran, V.Q., Pham, B.T., Le, T.T., Derrible, S.: Development of an AI model to measure traffic air pollution from multisensor and weather data. Sensors **19**, 4941 (2019)
25. Marć, M., Namieśnik, J., Zabiegała, B.: BTEX concentration levels in urban air in the area of the Tri-City agglomeration (Gdansk, Gdynia, Sopot), Poland. Air Qual. Atmos. Health **7**, 489–504 (2014)
26. Milazzo, M.J., Gohlke, J.M., Gallagher, D.L., Scott, A.A., Zaitchik, B.F., Marr, L.C.: Potential for city parks to reduce exposure to BTEX in air. Environ. Sci. Process. Impacts **21**, 40–50 (2019)
27. Ning, G., Yim, S.H.L., Wang, S., Duan, B., Nie, C., Yang, X., Wang, J., Shang, K.: Synergistic effects of synoptic weather patterns and topography on air quality: a case of the Sichuan Basin of China. Clim. Dyn. **53**, 6729–6744 (2019)
28. Rattanajongjitrakorn, P., Prueksasit, T.: Temporal variation of BTEX at the area of petrol station in Bangkok, Thailand. APCBEE Procedia **10**, 37–41 (2014)
29. Słomińska, M., Konieczka, P., Namieśnik, J.: The fate of BTEX compounds in ambient air. Crit. Rev. Environ. Sci. Technol. **44**, 455–472 (2014)
30. Stingone, J.A., McVeigh, K.H., Claudio, L.: Early-life exposure to air pollution and greater use of academic support services in childhood: a population-based cohort study of urban children. Environ. Health **16**, 2 (2017)

31. Stojić, A., Stanić, N., Vuković, G., Stanišić, S., Perišić, M., Šoštarić, A., Lazić, L.: Explainable extreme gradient boosting tree-based prediction of toluene, ethylbenzene and xylene wet deposition. Sci. Total Environ. **653**, 140–147 (2019)
32. Stojić, A., Stojić, S.S., Mijić, Z., Šoštarić, A., Rajšić, S.: Spatio-temporal distribution of VOC emissions in urban area based on receptor modelling. Atmos. Environ. **106**, 71–79 (2015b)
33. Stojić, A., Stojić, S.S., Šoštarić, A., Ilić, L., Mijić, Z., Rajšić, S.: Characterization of VOC sources in urban area based on PTR-MS measurements and receptor modelling. Environ. Sci. Pollut. Res. **22**, 13137–13152 (2015a)
34. Stojić, A., Stojić, S.S.: The innovative concept of three-dimensional hybrid receptor modeling. Atmos. Environ. **164**, 216–223 (2017)
35. Šoštarić, A., Stojić, S.S., Vuković, G., Mijić, Z., Stojić, A., Gržetić, I.: Rainwater capacities for BTEX scavenging from ambient air. Atmos. Environ. **168**, 46–54 (2017)
36. Šoštarić, A., Stojić, A., Stojić, S.S., Gržetić, I.: Quantification and mechanisms of BTEX distribution between aqueous and gaseous phase in a dynamic system. Chemosphere **144**, 721–727 (2016)
37. Stanišić, S., Perišić, M., Jovanović, G., Milićević, T., Romanić, S.H., Jovanović, A., Šoštarić, A., Udovičić, V., Stojić, A.: The PM2.5-bound polycyclic aromatic hydrocarbon behavior in indoor and outdoor environments, Part I: emission sources. Environ. Res. 110520 (2020b)
38. Stanišić, S., Perišić, M., Stojić, A.: The use of innovative methodology for the characterization of benzene, toluene, ethylbenzene and xylene sources in the Belgrade area. In: Sinteza International Scientific Conference on Information Technology and Data Related Research, Belgrade, Serbia (2020a)
39. Truong, S.C., Lee, M.I., Kim, G., Kim, D., Park, J.H., Choi, S.D., Cho, G.H.: Accidental benzene release risk assessment in an urban area using an atmospheric dispersion model. Atmos. Environ. **144**, 146–159 (2016)
40. U.S. EPA.: Chemical Data Reporting. United States Environmental Protection Agency, Washington, DC, USA (2016). https://www.epa.gov/sites/production/files/2014-11/documents/2nd_cdr_snapshot_5_19_14.pdf
41. Vardoulakis, S., Solazzo, E., Lumbreras, J.: Intra-urban and street scale variability of BTEX, NO2 and O3 in Birmingham, UK: implications for exposure assessment. Atmos. Environ. **45**, 5069–5078 (2011)
42. Werder, E.J., Engel, L.S., Blair, A., Kwok, R.K., McGrath, J.A., Sandler, D.P.: Blood BTEX levels and neurologic symptoms in Gulf states residents. Environ. Res. **175**, 100–107 (2019)
43. Wright, C.Y., Millar, D.A.: A global statement for air pollution and health. Clean Air J. **29**, 1–2 (2019)

The Linear Fuzzy Space: Theory and Applications

Đorđe Obradović, Zora Konjović, Endre Pap, and Andrej Šoštarić

Abstract The basic properties of dynamic systems include imprecisions and uncertainties in both spatial and temporal data which require adequate representations and further manipulations. Existing literature covers various approaches that address the representations and manipulations of such data, with fuzzy theory being one of the most widely and most frequently used. The linear fuzzy space is an original fuzzy-based approach that provides an effective and compact model of imprecise data. In this chapter we present the main theoretical results of the linear fuzzy space and analyze its capacity to facilitate the modelling of spatial and spatio-temporal systems through representative examples of its applications in selected real-world problems. The chapter consists of five sections. The first section is an introduction which states the problem and briefly presents relevant research. The second section contains a brief, yet comprehensive, overview of linear fuzzy space theory. The third section presents an analysis of fuzzy linear space theory concerning the modelling of spatial and/or temporal systems. The fourth section presents two examples which show its applications in image analysis and time series modelling. The fifth section concludes the chapter and recommends some directions for future research.

Keywords Fuzzy set · Linear fuzzy space. planar spatial objects · Fuzzy point · Fuzzy metric · Spatial modelling · Temporal modelling · 2D image segmentation · GDAS · CAQI index

---

Đ. Obradović (✉) · Z. Konjović · E. Pap
Singidunum University, Belgrade, Serbia
e-mail: djobradovic@singidunum.ac.rs

Z. Konjović
e-mail: zkonjovic@singidunum.ac.rs

E. Pap
e-mail: epap@singidunum.ac.rs

A. Šoštarić
Belgrade City Institute for Public Health, Belgrade, Serbia
e-mail: andrej.sostaric@zdravlje.org.rs

E. Pap (ed.), *Artificial Intelligence: Theory and Applications*,
Studies in Computational Intelligence 973,
https://doi.org/10.1007/978-3-030-72711-6_13

1 Introduction

Models of spatio-temporally determined dynamic systems aim to represent the behavior of spatially determined entities in time. The most general example deals with entities whose spatial properties change over time, especially spatial relations with other entities in the space. An additional property of such systems is the frequent presence of imprecise and uncertain data that describe the spatial entities and temporal aspects of such systems. Namely, data is mainly acquired by various sensors which have limited accuracy and precision. The data is then stored in different formats (images, numerical values, etc.) which results in additional data imprecision and uncertainty.

A digital raster image is a commonly used format which appears often as a data source for modelling spatially determined systems, and therefore deserves special attention. The extraction of basic geometric features from a digital raster image is one of the fundamental processes in image analysis. The main problems with image analysis originate from the fact that a discrete space (a digital raster image) is used for the representation of spatial elements, while spatial relations assume continual space (i.e., points belonging to a line should be collinear by definition). Due to imperfections in either the image data or the edge detector, there may be missing points or pixels on the lines, as well as spatial deviations between the ideal line and the set of imprecise points obtained from the edge detector. This may result in geometric distortion of the image.

Therefore, a topic which is relevant for this chapter is the modelling of basic planar imprecise geometry objects and their relations. Thereupon, we will provide a list of results which represent typical approaches to this problem and give a brief overview of the ones that influenced our linear fuzzy space geometry model.

In general, there are three basic approaches to modelling uncertainty and imprecision in spatial data: exact models [7, 16, 17, 33, 34], probabilistic models [6, 15, 18, 39, 43], and fuzzy models [4, 5, 20–22, 24, 29, 35, 41].

Löffler [16, 17] proposed models for imprecise lines as a set of possible precise planar lines, while Guibas et al. in [7] proposed basic spatial relations (coincidence, between and collinear) between imprecise points modelled as circles with equal diameters. The advantages of these models are their ease of description and efficient computation. A notable disadvantage is the inability to describe a line whose points are known with different levels of accuracy.

The subject of paper [33] is a problem related to spatial objects which do not have homogeneous interiors and sharply defined boundaries. For spatial vagueness called fuzziness, this paper provides a conceptual model of fuzzy spatial objects, together with the following operations: fuzzy geometric union, fuzzy intersection, fuzzy difference, and fuzzy topological predicates. The model relies on a finite, discrete geometric domain called a grid partition, which makes it suitable for the finite-precision number systems of digital computers.

An abstract conceptual model of fuzzy spatial data types (such as fuzzy points, fuzzy lines and fuzzy regions) is proposed in paper [34] and focuses on defining their structure and semantics. The theory of fuzzy sets and fuzzy topologies is a formal framework for the conceptual model.

Finally, Buckley and Eslami in [18] present fuzzy plane geometry. The fuzzy point and fuzzy line are defined as fuzzy sets with membership functions that are convex and upper semi-continuous.

We adopted a fuzzy approach for modelling the uncertainty and imprecision of spatial data. Based on our results presented in [8, 21, 22, 24], in the paper [26] we introduced a new mathematical model for basic objects (a fuzzy line, a fuzzy triangle, and a fuzzy circle) as well as basic spatial relations: coincidence, between, and collinear. These models are the foundation for modelling spatial objects that were used to solve several real-world problems, either alone or in combination with machine learning techniques [11–13, 23, 25, 27].

Modelling the behavior of spatio-temporally determined systems that include imprecise and uncertain data generally requires modelling imprecision/uncertainty in time itself, as well as in time-dependent parameters and variables.

A time series is a series of data with natural temporal ordering that is often used to describe the behavior of dynamic systems. Different models of time series exist, and very vivid research of this topic strives to combine time series models with techniques supporting the manipulation of imprecise and uncertain data. In literature, these are known under the umbrella term Fuzzy Time Series (FTS). The seminal paper [38] was published in 1993. It was followed by many papers which dealt with both the theoretical foundations and applications of Fuzzy Time Series in diverse domains such as economics, technical systems, meteorology, and seismography. An extensive review is provided in sources [30, 37]. Looking at the results, one could say that existing achievements relate mostly to modelling imprecision and uncertainty in time dependent parameters and variables, while modelling time imprecision and uncertainty is less common [3, 19, 31, 36]. Nevertheless, FTS provide acceptable possibilities when it comes to practical tasks like forecasting, classification, etc. [2, 28, 42].

2 Linear Fuzzy Space-Based Models of Planar Spatial Objects, Their Relations, and Topology

This section briefly, but comprehensively, presents mathematical models of basic planar imprecise geometrical objects, basic spatial measurement functions, basic spatial relations, and the topology of linear fuzzy space. First, the fundamental concepts of the linear fuzzy space are shown. Afterwards, we present mathematical models of the basic operations over linear fuzzy space, which are used in the definitions of basic fuzzy plane geometrical objects as well as in the definitions of the objects themselves. Finally, we present the mathematical models of basic spatial relations. The proofs of all theorems are given in papers [21, 26].

2.1 The Basics

In this subsection we present the fundamental concepts of the linear fuzzy space: fuzzy point, linear fuzzy space, and linear fuzzy space ordering (relation).

Definition 1 *Fuzzy point* $P \in \mathbb{R}^2$, denoted by $\tilde{P}$ is defined by its membership function $\mu_{\tilde{P}} \in F^2$, where the set F^2 contains all membership functions $u : \mathbb{R}^2 \to [0, 1]$ satisfying following conditions:

1. $(\forall u \in F^2)(\exists_1 P \in \mathbb{R}^2)u(P) = 1$,
2. $(\forall X_1, X_2 \in \mathbb{R}^2)(\lambda \in [0, 1])u(\lambda X_1 + (1 - \lambda)X_2) \geqslant \min(u(X_1), u(X_2))$,
3. function u is upper semi continuous,
4. $[u]^\alpha = \left\{X | X \in \mathbb{R}^2, u(X) \geqslant \alpha\right\}$ α—cut of function u is convex.

Here, a point from $\mathbb{R}^2$ with membership function $\mu_{\tilde{P}}(P) = 1$, is denoted by P (P is the core of the fuzzy point $\tilde{P}$), the membership function of point $\tilde{P}$ is denoted by $\mu_{\tilde{P}}$, while $[P]^\alpha$ stands for the α-cut (a set from $\mathbb{R}^2$) of the fuzzy point.

Definition 2 *Linear fuzzy space* is the set $H^2 \subset F^2$ of all functions which, in addition to the properties given in Definition 1, are:

1. Symmetrical with respect to the core $S \in \mathbb{R}^2$
   $(\mu(S) = 1)$,
   $\mu(V) = \mu(M) \wedge \mu(M) \neq 0 \implies d(S, V) = d(S, M)$,
   where $d(S, M)$ is the distance in $\mathbb{R}^2$.
2. Inverse-linearly decreasing regarding points' distance from the core according to:
   If $r \neq 0$
   $$\mu_{\tilde{S}}(V) = \max\left(0, 1 - \frac{d(S,V)}{(r_S)}\right),$$
   if $r = 0$
   $$\mu_{\tilde{S}}(V) = \begin{cases} 1, & \text{if } S = V \\ 0, & \text{if } S \neq V \end{cases}$$
   where $d(S, V)$ is the distance between point V and the core S, $(V, S \in R^n)$ and $r \in \mathbb{R}$ is a constant.

Elements of that space are represented as ordered pairs $\tilde{S} = (S, r_S)$ where $S \in \mathbb{R}^2$ is the core of $\tilde{S}$, and $r_S \in \mathbb{R}$ is the distance from the core for which the function value becomes 0; hereinafter, the parameter r_S will be called a *fuzzy support radius*.

Definition 3 Let the linear fuzzy space H be defined on $\mathbb{R}$. Fuzzy relations $\leq^{RF}$ and $\leq^{LF}$ for the set H are defined by the membership functions

$$\mu(\tilde{A} \leq^{RF} \tilde{B}) = \begin{cases} 0, & \text{if } A > B \\ \frac{B-A}{r_A - r_B}, & \text{if } A \leqslant B \wedge A + r_A > B + r_B \\ 1, & if A \leqslant B \wedge A + r_A \leqslant B + r_B \end{cases}$$

$$\mu(\tilde{A} \preceq^{LF} \tilde{B}) = \begin{cases} 0, & \text{if } A > B \\ \frac{B-A}{r_B - r_A}, & \text{if } A \leqslant B \wedge A - r_A > B - r_B \\ 1, & \text{if } A \leqslant B \wedge A - r_A \leqslant B - r_B \end{cases}$$

respectively, where $\tilde{A} = (A, r_A)$ and $\tilde{B} = (B, r_B)$ are points from H, A is the core of $\tilde{A}$ and r_A is a parameter determining the membership function of point $\tilde{A}$.

2.2 Basic Fuzzy Plane Geometry Objects

What follows now are the definitions of basic fuzzy plane geometry objects, together with basic operations over the linear fuzzy space H^2 defined on $\mathbb{R}^2$, and their properties which are used in definitions of these objects.

Definition 4 Let $\tilde{A}, \tilde{B} \in H^2$. Then, the operator $+ : H^2 \times H^2 \to H^2$ called *fuzzy points addition*, is given by:

$$\tilde{A} + \tilde{B} = (A + B, r_A + r_B),$$

where $A + B$ is vector addition, and $r_A + r_B$ is scalar addition.

Definition 5 Let H^2 be a linear fuzzy space. Then, function $f : H^2 \times H^2 \times [0, 1] \to H^2$ called a linear combination of the fuzzy points $\tilde{A}, \tilde{B} \in H^2$, is given by:

$$f(\tilde{A}, \tilde{B}, u) = \tilde{A} + u \cdot (\tilde{B} - \tilde{A}),$$

where $u \in [0, 1]$ and the operator $\cdot$ is scalar multiplication of the fuzzy point.

Definition 6 Let $\tilde{A}, \tilde{B} \in H^2$ and $\tilde{A} \neq \tilde{B}$. Then point $T_{AB} \in \mathbb{R}^2$ is called an internal homothetic center if the following holds

$$T_{AB} = A + \frac{a_r}{a_r + b_r}(B - A),$$

where $\tilde{A} = (A, a_r)$ and $\tilde{B} = (B, b_r)$.

Definition 7 Let H^2 be a linear fuzzy space and function f a linear combination of fuzzy points $\tilde{A}$ and $\tilde{B}$. Then fuzzy set $\widetilde{AB}$ is called a *fuzzy line* if the following holds

$$\widetilde{AB} =_{u \in [0,1]} f(\tilde{A}, \tilde{B}, u).$$

Theorem 1 *Let H^2 be a linear fuzzy space, a fuzzy line $\widetilde{AB}$ defined by fuzzy points $\tilde{A}$ and $\tilde{B} \in H^2$. Then the following holds $\widetilde{AB} = \widetilde{BA}$.*

Definition 8 Let $\widetilde{AB}$ be a fuzzy line defined on the linear fuzzy space H^2 and $X \in \mathbb{R}^2$. Then fuzzy point $\widetilde{X}' \subset \widetilde{AB}$ is called a *fuzzy image of point X on fuzzy line* $\widetilde{AB}$, and the real number $u \in [0, 1]$ is called an *eigenvalue of fuzzy image X* on the fuzzy line $\widetilde{AB}$ if the following holds:

1. $\widetilde{X}' = \tilde{A} + u(\tilde{B} - \tilde{A})$,
2. $d\left(X, [\widetilde{X}']^1\right) = \min\{d(X, Y) | \forall Y \in [\widetilde{AB}]^1\}$,
3. $u = \min\left(1, \max\left(0, \frac{(x_1-a_1)(b_1-a_1)+(x_2-a_2)(b_2-a_2)}{(b_1-a_1)^2+(b_2-a_2)^2}\right)\right)$,
   where $X = (x_1, x_2)$, $\tilde{A} = ((a_1, a_2), a_r)$ i $\tilde{B} = ((b_1, b_2), b_r)$.

If the eigenvalue of fuzzy image X is equal to 0, then the fuzzy image is the starting fuzzy point, if the eigenvalue is equal to 1, then it is the final point, otherwise it is an inner point of the fuzzy line.

Theorem 2 *Let $\widetilde{AB} \in L^2$ be a fuzzy line, and $\widetilde{X}' \in H^2$ a fuzzy image of the point $X \in R^2$ on $\widetilde{AB}$ and $u \in [0, 1]$ the eigenvalue of the fuzzy image X on $\widetilde{AB}$. Then point X belongs to fuzzy set $\widetilde{AB}$ according to the following*

$$\mu_{\widetilde{AB}}(X) = \begin{cases} \mu_{\tilde{A}}(X), & \text{if } u_m = 0 \\ \mu_{\tilde{X}_T}(X), & \text{if } 0 < u_m < 1 \\ \mu_{\tilde{B}}(X), & \text{if } u_m = 1, \end{cases}$$

where the fuzzy point $\tilde{X}_T = \tilde{A} + u_m(\tilde{B} - \tilde{A})$ and $u_m = u + \frac{(b_r-a_r)}{x_r'}\frac{d(X,X')^2}{d(A,B)^2}$.

Definition 9 Let $\tilde{A}, \tilde{B}, \tilde{C} \in H^2$ be fuzzy points with noncollinear cores ($\tilde{A} \neq \tilde{B} \neq \tilde{C}$) and function f a linear combination of two fuzzy points. Then the fuzzy set $\widetilde{ABC}$ is called a *fuzzy triangle* if the following holds

$$\widetilde{ABC} = \bigcup_{u=0}^{1} f(\tilde{A}, \bigcup_{v=0}^{1} f(\tilde{B}, \tilde{C}, v), u)$$

Its membership function is denoted by $\mu_{\widetilde{ABC}}(X)$ and determined according to the following formula:

$$\mu_{\widetilde{ABC}}(X) = \max_{u\in[0,1], v\in[0,1]} \{\mu_{\tilde{Y}}(X) | \tilde{Y} = f(\tilde{A}, f(\tilde{B}, \tilde{C}, v), u)\}.$$

α-cut of fuzzy triangle $\widetilde{ABC}$ is denoted by $\left[\widetilde{ABC}\right]^\alpha$.

Definition 10 Let $\widetilde{ABC}$ be a fuzzy triangle defined on the linear fuzzy space H^2.

(i) The Fuzzy point $\tilde{X} \subset \widetilde{ABC}$ is called an *edge point of fuzzy triangle* $\widetilde{ABC}$ if for all $\alpha \in [0, 1]$ a point $Y \in \left[\tilde{X}\right]^\alpha$ exists, such that all its neighborhoods contain at least one point from $\left[\widetilde{ABC}\right]^\alpha$ and at least one point outside of $\left[\widetilde{ABC}\right]^\alpha$.

(ii) The Fuzzy point $\tilde{X} \subset \widetilde{ABC}$ is called an *inner point of fuzzy triangle* $\widetilde{ABC}$ if it is not an edge point.

(iii) The union of all edge points of fuzzy triangle $\widetilde{ABC}$ is called a fuzzy edge of fuzzy triangle $\widetilde{ABC}$, denoted by $\partial \widetilde{ABC}$.

Theorem 3 *Let $\widetilde{ABC}$ be a fuzzy triangle defined on H^2. Then, for every fuzzy point $\tilde{X} \in \partial\widetilde{ABC}/\{\tilde{A}, \tilde{B}, \tilde{C}\}$ and $\alpha \in [0, 1]$ the single point $T \in \left[\tilde{X}\right]^\alpha$ exists such that all its neighborhoods contain at least one point from $\left[\widetilde{ABC}\right]^\alpha$ and at least one point outside of $\left[\widetilde{ABC}\right]^\alpha$.*

Theorem 4 *Let $\widetilde{ABC}$ be a fuzzy triangle defined on the linear fuzzy space H^2. Then for all $X \in \mathbb{R}^2$ the following holds*

$$\mu_{\widetilde{ABC}}(X) = \mu_{\widetilde{CAB}}(X) = \mu_{\widetilde{BCA}}(X).$$

A direct consequence of this proposition is that a fuzzy triangle can be represented by three fuzzy points, i.e., the set $\{\tilde{A}, \tilde{B}, \tilde{C}\}$.

Definition 11 Let H be a fuzzy space defined on $\mathbb{R}$, a fuzzy relation $\preceq^{RF}$ be a fuzzy ordering in the linear fuzzy space H, $C \in \mathbb{R}^2$ and $\tilde{R} \in H$. Then the union of all fuzzy points $\tilde{A} \in H^2$ such that

$$\mu(\tilde{d}(C, \tilde{A}) \preceq^{RF} \tilde{R} = 1.$$

is called a *fuzzy circle* with center C and radius $\tilde{R}$.

The fuzzy circle is represented by the ordered pair $(C, \tilde{R})$.

Theorem 5 *Let $(C, \tilde{R})$ be a fuzzy circle defined on the linear fuzzy space H^2. Then the value of the fuzzy circle membership function at point $X \in R^2$ is determined according to the following formula*

$$\mu_{(C,\tilde{R})}(X) = \max\left(0, \min\left(1, 1 - \frac{d(X, C) - R}{r_r}\right)\right),$$

where $\tilde{R} = (R, r_r)$.

Definition 12 Let H^2 be a linear fuzzy space and $\tilde{A} = \{\widetilde{A_1}, \dots, \widetilde{A_n}\}$ an ordered set of fuzzy points $\widetilde{A_i} \in H^2$. Then a *linear fuzzy path* $s(\tilde{A})$ is defined as

$$s(\tilde{A}) = \bigcup_{i=1}^{n-1} \widetilde{A_i A_{i+1}}$$

If $X \in \mathbb{R}^2$, then a membership function $\mu_{s(\tilde{A})}$ of the linear fuzzy path $s(\tilde{A})$ is

$$\mu_{s(\tilde{A})}(X) = \max_{i=1}^{n-1} \mu_{\widetilde{A_i A_{i+1}}}(X).$$

Definition 13 Let H^2 be a linear fuzzy space and $\tilde{A} = \{\widetilde{A_1}, \ldots, \widetilde{A_n}\}$ a set of fuzzy points $\widetilde{A_i} \in H^2$. Then a *closed linear fuzzy path* $c(\tilde{A})$ is defined as

$$c(\tilde{A}) = s(\tilde{A}) \cup \widetilde{A_n A_1}.$$

If $X \in \mathbb{R}^2$, then the membership function $\mu_{c(\tilde{A})}$ of a *closed linear fuzzy path* $c(\tilde{A})$ is

$$\mu_{c(\tilde{A})}(X) = \max\{\mu_{s(\tilde{A})}(X), \mu_{\widetilde{A_n A_1}}(X)\}.$$

Definition 14 Let H^2 be a linear fuzzy space and $\tilde{A} = \{\widetilde{A_1}, \ldots, \widetilde{A_n}\}$ an ordered set of fuzzy points $\widetilde{A_i} \in H^2$. Then a *linear fuzzy polygon* $p(\tilde{A})$ is defined as

$$\mu_{p(\tilde{A})}(X) = \begin{cases} 1, & \text{if } X \text{ is in interior of polygon } (\tilde{A})^0 \\ \mu_{c(\tilde{A})}(X), & \text{otherwise,} \end{cases}$$

where $c(\tilde{A})^0$ is a core of a fuzzy set $c(\tilde{A})$.

2.3 Basic Spatial Relations Between Fuzzy Plane Geometry Objects

Spatial relations (predicates) are functions that are used to establish mutual relations between fuzzy geometric objects. The basic spatial relations are *coincide*, *between* and *collinear*. In this section we present their definitions and basic properties.

Definition 15 Let λ be the Lebesgue measure on the set $[0, 1]$ and H^2 is a linear fuzzy space. The fuzzy relation $coin : H^2 \times H^2 \to [0, 1]$ is *fuzzy coincidence* represented by the following membership function:

$$\mu_{coin}(\tilde{A}, \tilde{B}) = \lambda\left(\left\{\alpha | \left[\tilde{A}\right]^\alpha \cap \left[\tilde{B}\right]^\alpha \neq \emptyset\right\}\right).$$

Since the lowest α is always 0, then the membership function of the *fuzzy coincidence* is given by:

$$\mu_{coin}(\tilde{A}, \tilde{B}) = \max\left\{\alpha | \left[\tilde{A}\right]^\alpha \cap \left[\tilde{B}\right]^\alpha \neq \emptyset\right\}.$$

Theorem 6 provides the method for its calculation.

Theorem 6 *Let the fuzzy relation coin be a fuzzy coincidence. Then the membership function of the fuzzy relation fuzzy coincidence is determined according to the following formula*

$$\mu_{coin}(\tilde{A},\tilde{B}) = \begin{cases} 0, & if\ |a_r| + |b_r| = 0 \wedge d(A,B) \neq 0 \\ \max\left(0, 1 - \frac{d(A,B)}{|a_r|+|b_r|}\right), & if\ |a_r| + |b_r| \neq 0 \\ 1, & if\ |a_r| + |b_r| = 0 \wedge d(A,B) = 0 \end{cases}$$

Definition 16 Let λ be a Lebesgue measure on the set $[0, 1]$, H^2 a linear fuzzy space, and L^2 a set of all fuzzy lines defined on H^2. Then the fuzzy relation $contain : H^2 \times L^2 \to [0, 1]$ is *fuzzy contain* represented by the following membership function

$$\mu_{contain}(\tilde{A}, \widetilde{BC}) = \lambda\left(\left\{\alpha | \left[\tilde{A}\right]^{\alpha} \cap \left[\widetilde{BC}\right]^{\alpha} \neq \emptyset\right\}\right).$$

Theorem 7 *Let $\tilde{A}, \tilde{B}, \tilde{C} \in H^2$ be fuzzy points defined on H^2 linear fuzzy space, $u \in [0, 1]$ and $\widetilde{A}'$ be fuzzy image of point A on fuzzy line $\widetilde{BC}$. Points T_{AB} and T_{AC} are internal homothetic center for fuzzy points $\tilde{A}$ and $\tilde{B}$ and $\tilde{A}$ and $\tilde{C}$ respectively. Then the membership function of the fuzzy relation fuzzy contain is determined according to the following formula*

$$\mu_{contain}(\tilde{A}, \widetilde{BC}) = \begin{cases} \mu_{coin}(\tilde{A}, \widetilde{A}'), & if\ u \in \{0, 1\} \\ \mu_{\tilde{A}}(A^*), & if\ u \in]0, 1[, \end{cases}$$

where point A^* is a projection of the core of $\tilde{A}$ on the line passing through the points T_{AB} and T_{AC}.

Definition 17 Let $\tilde{A}, \tilde{B}, \tilde{C} \in H^2$ be fuzzy points defined on the linear fuzzy space H^2, and let λ be a Lebesgue measure on the set $[0, 1]$. The fuzzy relation $coli$: $H^2 \times H^2 \times H^2 \to [0, 1]$ is *fuzzy collinearity* between the three fuzzy points and it is represented by the following membership function:

$$\mu_{coli}(\tilde{A}, \tilde{B}, \tilde{C})$$

$$= \lambda\left(\left\{\alpha | \exists u \in \mathbb{R}, \exists X \in \left[\tilde{A}\right]^{\alpha}, \exists Y \in \left[\tilde{B}\right]^{\alpha}, \exists Z \in \left[\tilde{C}\right]^{\alpha}, A = B + u(C - B)\right\}\right).$$

Theorem 8 *Let $\tilde{A}, \tilde{B}, \tilde{C} \in H^2$, fuzzy relation $contain$ be fuzzy contain. Then the membership function of the fuzzy relation fuzzy collinearity is determined according to the following formula*

$$\mu_{coli}(\tilde{A}, \tilde{B}, \tilde{C}) = \max\left(\mu_{contain}(\tilde{A}, \widetilde{BC}), \mu_{contain}(\tilde{B}, \widetilde{AC}), \mu_{contain}(\tilde{C}, \widetilde{AB})\right).$$

2.4 Measurement in the Linear Fuzzy Space

Measurement of the space, especially the distance between plane geometry objects is defined as a generalization of the concept of physical distance. In this section we present the basic distance functions between fuzzy plane geometry objects and their main properties defined in [26] and based on our work presented in [8, 21, 24].

Definition 18 Let H^2 be a linear fuzzy space and $\tilde{d} : H^2 \times H^2 \to H^+$, $L, R : [0,1] \times [0,1] \to [0,1]$ be symmetric, associative, and non-decreasing for both arguments, and $L(0,0) = 0$, $R(1,1) = 1$. The ordered quadruple $(H^2, \tilde{d}, L, R)$ is called fuzzy metric space and the function $\tilde{d}$ is a *fuzzy metric*, if and only if the following conditions hold:

1. $\tilde{d}(\tilde{X}, \tilde{Y}) = \tilde{0} \Leftrightarrow \left[\tilde{X}\right]^1 = \left[\tilde{Y}\right]^1$.
2. $\tilde{d}(\tilde{X}, \tilde{Y}) = \tilde{d}(\tilde{Y}, \tilde{X})$ for each $\tilde{X}, \tilde{Y} \in H^2$.
3. $\forall \tilde{X}, \tilde{Y} \in H^2$:
    1. $\tilde{d}(\tilde{X}, \tilde{Y})(s+t) \geqslant L(d(x,z)(s), d(z,y)(t))$ if $s \leqslant \lambda_1(x,z) \wedge t \leqslant \lambda_1(z,y) \wedge s+t \leqslant \lambda_1(x,y)$,
    2. $\tilde{d}(\tilde{X}, \tilde{Y})(s+t) \leqslant R(d(x,z)(s), d(z,y)(t))$ if $s \geqslant \lambda_1(x,z) \wedge t \geqslant \lambda_1(z,y) \wedge s+t \geqslant \lambda_1(x,y)$,

where the α-cut of fuzzy number $\tilde{d}(x,y)$ is given by $\left[\tilde{d}(\tilde{X}, \tilde{Y})\right]^\alpha = [\lambda_\alpha(x,y), \rho_\alpha(x,y)]\,(x, y \in \mathbb{R}^+, 0 < \alpha \leqslant 1)$. The fuzzy zero, $\tilde{0}$ is a *non-negative* fuzzy number with $[\tilde{0}]^1 = 0$.

Example 1 The following distance functions are fuzzy metrics:

1. $\tilde{d}(\tilde{X}, \tilde{Y}) =_{DF} (d(X,Y), (r_X + r_Y))$
2. $\tilde{d}(\tilde{X}, \tilde{Y}) =_{DF} (d(X,Y), \max(r_X, r_Y))$
3. $\tilde{d}(\tilde{X}, \tilde{Y}) =_{DF} (d(X,Y), |r_X - r_Y|)$

A distance 3. also satisfies a set of rules which define the classic metrics.

Definition 19 Let H^2 be a linear fuzzy space, L^2 a set of all fuzzy lines defined on H^2, $\tilde{d}$ a fuzzy distance between fuzzy points, and μ_L a membership function of the fuzzy relation *minimal* (a fuzzy set describing the minimal element in a set of distances of fuzzy points). The function $dist : H^2 \times L^2 \to H^+$ is called the *distance between a fuzzy point and a fuzzy line* if the following holds:

$$dist(\tilde{T}, \widetilde{AB}) = \tilde{d}(\tilde{T}, \tilde{X}),$$

where $\tilde{X} \in \widetilde{AB}$ such that $\mu_L\left(\tilde{d}\left(\tilde{T}, \tilde{X}\right)\right) = hgt\left(\left\{\tilde{d}(\tilde{T}, \tilde{Y}) | \forall \tilde{Y} \in \widetilde{AB}\right\}\right)$.

Definition 20 Let H^2 be a linear fuzzy space, T^2 a set of all fuzzy triangles defined on H^2, $\tilde{d}$ the fuzzy distance between fuzzy points and μ_L the membership function of the fuzzy relation *minimal*. The function $dist : H^2 \times T^2 \to H^+$ is called the distance between a fuzzy point and a fuzzy triangle if the following holds:

$$dist(\tilde{T}, \widetilde{ABC}) = \tilde{d}(\tilde{T}, \tilde{X}),$$

where $\tilde{X} \in \widetilde{ABC}$ such that $\mu_L\left(\tilde{d}(\tilde{T}, \tilde{X})\right) = hgt\left(\left\{\tilde{d}(\tilde{T}, \tilde{Y}) | \forall \tilde{Y} \in \widetilde{ABC}\right\}\right)$.

Definition 21 Let H^2 be a linear fuzzy space, L^2 a set of all fuzzy lines on H^2, $\tilde{d}$ the fuzzy distance between fuzzy points, and μ_L the membership function of the fuzzy relation *minimal*. The function $dist : L^2 \times L^2 \rightarrow H^+$ is called the distance between two fuzzy lines if the following holds:

$$dist(\widetilde{AB}, \widetilde{CD}) = \tilde{d}(\tilde{X}, \tilde{Y}),$$

where $\tilde{X} \in \widetilde{AB}$ and $\tilde{Y} \in \widetilde{CD}$ such that

$$\mu_L\left(\tilde{d}(\tilde{X}, \tilde{Y})\right) = hgt\left(\left\{\tilde{d}(\tilde{Q}, \tilde{W}) | \forall\tilde{Q} \in \widetilde{AB} \wedge \forall\tilde{W} \in \widetilde{CD}\right\}\right)$$

Definition 22 Let H^2 be a linear fuzzy space, L^2 a set of all fuzzy lines on H^2, T^2 a set of all fuzzy triangles, $\tilde{d}$ the fuzzy distance between fuzzy points and μ_L the membership function of the fuzzy relation *minimal*. The function $dist : L^2 \times T^2 \rightarrow H^+$ is called the distance between a fuzzy line and a fuzzy triangle if the following holds:

$$dist(\widetilde{AB}, \widetilde{CDE}) = \tilde{d}(\tilde{X}, \tilde{Y}),$$

where $\tilde{X} \in \widetilde{AB}$ and $\tilde{Y} \in \widetilde{CDE}$ satisfy the condition

$$\mu_L\left(\tilde{d}(\tilde{X}, \tilde{Y})\right) = hgt\left(\left\{\tilde{d}(\tilde{Q}, \tilde{W}) | \forall\tilde{Q} \in \widetilde{AB} \wedge \forall\tilde{W} \in \widetilde{CDE}\right\}\right).$$

Definition 23 Let H^2 be a linear fuzzy space, T^2 a set of all fuzzy triangles on H^2, $\tilde{d}$ the fuzzy distance between fuzzy points and μ_L the membership function of the fuzzy relation *minimal*. The function $dist : T^2 \times T^2 \rightarrow H^+$ is called the distance between two fuzzy triangles if the following holds:

$$dist(\widetilde{ABC}, \widetilde{DEF}) = \tilde{d}(\tilde{X}, \tilde{Y}),$$

where fuzzy points $\tilde{X} \in \widetilde{ABC}$ and $\tilde{Y} \in \widetilde{DEF}$ such that

$$\mu_L\left(\tilde{d}(\tilde{X}, \tilde{Y})\right) = hgt\left(\left\{\tilde{d}(\tilde{Q}, \tilde{W}) | \forall\tilde{Q} \in \widetilde{ABC} \wedge \forall\tilde{W} \in \widetilde{DEF}\right\}\right).$$

3 Applications of the Linear Fuzzy Space in Spatial and Temporal Modelling

In this section we analyze the linear fuzzy space with regards to its applications in spatial and temporal modelling. The first subsection presents the utilization of the linear fuzzy space for spatial modelling, while the second subsection considers its applications in temporal modelling.

3.1 Spatial Modelling—Model of 2D Fuzzy Spatial Relations

In this subsection we present a model of fuzzy spatial relations, which can be used to describe imprecise spatial data from 2D images that enables the semantic interpretation of images. For our model of 2D fuzzy spatial relations, we used the ontology proposed in [9]. The ontology contains the following basic concepts:

- *SpatialRelation*—fuzzy spatial relation.
- *SpatialObject*—fuzzy spatial object (for example, a fuzzy polygon).
- *ReferenceSystem*—reference system in which spatial relation is described (one spatial relation can be described in many ways, depending on the view perspective).
- *TargetObject*—instance of the *SpatialObject* concept. Represents the object for which spatial relation is determined with respect to some *ReferenceObject* which is also an instance of the *SpatialObject* concept.
- *DirectionalRelation*—concept that extends *SpatialRelation* concept. Allows modelling of unary spatial relation that represents position of a target object on the image.
- *BinarySpatialRelation*—concept that extends *SpatialRelation* concept. Allows modelling of binary spatial relations between the reference and target objects.
- *BinaryDirectionalRelation*—concept that extends *BinarySpatialRelation*. Allows the position of the target object to be modelled with respect to the reference object, e.g. "left of", "above", etc.
- *DistanceRelation*—concept that extends *BinarySpatialRelation*. Allows the distance of the target object from the reference object to be modelled, e.g., "(very) close to", "at a distance of ...", etc.

Using these basic concepts of fuzzy spatial relations and the concepts of linear fuzzy space (a fuzzy point, distance, a fuzzy polygon) we have defined the following models for elementary spatial relations [10].

Definition 24 Let P be a two-dimensional matrix which represents a digital image. We call fuzzy set F a *fuzzy spatial relation* if its membership function μ_F maps every element $P[i, j]$ of the matrix P in the interval [0, 1] :

$$\mu_F(P\,[i, j]) : P \rightarrow [0, 1].$$

Definition 25 Let P be a two-dimensional matrix that represents a digital image whose width and height are w and h, respectively. Then, for every pixel $P[i, j]$ determined by coordinates i and j, the membership function value of the fuzzy spatial relation AtLeft ("at left side of the image") is calculated as:

$$\mu_{AtLeft,d,g}(P\,[i, j]) = \begin{cases} 1, & \text{if } i \leqslant d \\ 1 - (i - d)/g, & \text{if } d < i \leqslant d + g \\ 0, & \text{if } i \geqslant d + g, \end{cases}$$

where d is a constant which defines the core, and g is a constant that defines the fuzziness of this fuzzy set.

Definition 26 Let P be a two-dimensional matrix that represents a digital image, whose width and height are w and h, respectively. Then, for every pixel $P[i, j]$ determined by coordinates i and j, the membership function value of the *AtBottom* ("at the bottom of the image") fuzzy spatial relation is calculated as:

$$\mu_{AtBottom,d,g}(P\,[i,\,j]) = \begin{cases} 1, & \text{if } j < d - g \\ \frac{(j-(d-g))}{g}, & \text{if } d - g \leqslant j < d \\ 0, & \text{if } j \geqslant d \end{cases}$$

where d is a constant which defines the core and g is a constant that defines the fuzziness of this fuzzy set.

Fuzzy spatial relations *AtRight* and *AtTop* can be defined in an analogous manner.

Definition 27 Let P be a two-dimensional matrix that represents a digital image whose width and height are w and h, respectively. Then for every pixel $P[i, j]$ determined by coordinates i and j, the membership function value of the fuzzy spatial relation *AtCenter* ("at the center of the image") is calculated as:

$$\mu_{AtCenter,d,g}(P\,[i,\,j]) = \begin{cases} 1, & if i^2 + j^2 \leqslant d^2 \\ 1 - \frac{(l-d)}{g}, & if d^2 < i^2 + j^2 \leqslant (d+g)^2 \\ 0, & if i^2 + j^2 \geqslant (d+g)^2 \end{cases}$$

where l is the distance of the pixel $P[i, j]$ from the center of the image, d is a constant which defines the core of this fuzzy set, and g is a constant that defines the fuzziness of this fuzzy set.

Definition 28 Let P be a two-dimensional matrix that represents a digital image whose width and height are w and h respectively. Let $\tilde{p}$ be a linear fuzzy polygon in a linear fuzzy space. For polygon $\tilde{p}$ that represents a reference object, we define two points: *maxTop* (polygon point with the smallest j coordinate), *maxBottom* (polygon point with largest j coordinate). If every pixel $P[i, j]$ is determined by coordinates i and j, then the membership function value of the fuzzy spatial relation *RightOf* ("right of") is calculated as:

$$\mu_{RightOf,g}(\tilde{p},\, P\,[i,\,j]) = \begin{cases} 1, & j_{maxTop} \leqslant j \leqslant j_{maxBottom}, \\ & i > right(j, \tilde{p}) \\ 1 - \frac{g * cos^{-1}(\theta)}{\pi}, & j < j_{maxTop} or j > j_{maxBottom}, \\ & i > right(j, \tilde{p}) \\ 0, & \text{otherwise,} \end{cases}$$

where function $right(i, \tilde{p})$ returns the maximal value of the i coordinate from the points of polygon $\tilde{p}$ whose j coordinates are equal to the j coordinate of pixel $P[i, j]$. The value of θ represents the angle which $P[i, j]$ forms with the point *maxTop* (if $j < jmaxTop$), or the point *maxBottom* (if $j > jmaxBottom$), and the constant g represents fuzziness of the fuzzy set.

Definition 29 Let P be a two-dimensional matrix that represents a digital image whose width and height are w and h respectively. Let $\tilde{p}$ be a linear fuzzy polygon in a linear fuzzy space. For polygon $\tilde{p}$ that represents a reference object, we define two points: *maxLeft* (polygon point with the smallest i coordinate), *maxRight* (polygon point with largest i coordinate). If every pixel $P[i, j]$ is determined by coordinates i and j, then the membership function value of the fuzzy spatial relation *Above* ("above") is calculated as:

$$\mu_{Above,g}(\tilde{p}, P[i, j]) = \begin{cases} 1, & i_{maxLeft} \leqslant i \leqslant j_{maxRight}, \\ & j < above(i, \tilde{p}) \\ 1 - \frac{g * cos^{-1}(\theta)}{\pi}, & i < i_{maxLeft} or i > i_{maxRight}, \\ & j > above(i, \tilde{p}) \\ 0, & \text{otherwise}, \end{cases}$$

where the function $above(i, \tilde{p})$ returns the maximal value of j coordinate from the points of polygon $\tilde{p}$ whose i coordinates are equal to the i coordinate of pixel $P[i, j]$. The Value θ represents the angle which $P[i, j]$ forms with the point *maxLeft* (if $i < imaxLeft$), or the point *maxRgiht* (if $i > imaxRight$), and the constant g represents fuzziness of this fuzzy set.

Fuzzy spatial relations *LeftOf* and *Below* are defined analogously.

Definition 30 Let P be a two-dimensional matrix that represents a digital image, whose width and height are w and h, respectively. Let $\tilde{p}$ be a linear fuzzy polygon in a linear fuzzy space. Then, for every pixel $P[i, j]$ determined by coordinates i and j, l is the distance between $P[i, j]$ and a reference object $\tilde{p}$, and the membership function value of the fuzzy spatial relation *(Very)CloseTo* ("(very) close to") is calculated as:

$$\mu_{CloseTo,d,g}(\tilde{p}, P[i, j]) = \begin{cases} 1, & \text{if } l \leqslant d \\ 1 - \frac{l-d}{g}, & \text{if } d < l \leqslant d + g \\ 0, & \text{if } l > d + g, \end{cases}$$

where d is a constant that defines the core of fuzzy set, and g is a constant that defines the fuzziness of this fuzzy set. The values of d and g for the relation *VeryCloseTo* should be smaller than those for the relation *CloseTo*.

Definition 31 Let P be a two-dimensional matrix that represents a digital image whose width and height are w and h respectively. Let $\tilde{p}$ be a linear fuzzy polygon in a linear fuzzy space. Then, for every pixel $P[i, j]$ determined by coordinates i and

j, l is the distance between $P[i, j]$ and the reference object $\tilde{p}$, and a membership function value of the fuzzy spatial relation *(Very)FarFrom* ("(very) far from") is calculated as:

$$\mu_{FarFrom,d,g}(\tilde{p}, P[i, j]) = \begin{cases} 1, & \text{if } l > d \\ \frac{l-(d-g)}{g}, & \text{if } d - g < l \leqslant d \\ 0, & \text{if } l < d - g, \end{cases}$$

where d is a constant that defines the core of fuzzy set, and g is a constant that defines fuzziness of this fuzzy set. The value of d for the relation *VeryFarFrom* should be bigger than for the relation *FarFrom*.

Definition 32 Let P be a two-dimensional matrix that represents a digital image, whose width and height are w and h, respectively. Let $\tilde{p}$ be a linear fuzzy polygon in a linear fuzzy space. Then, for every pixel $P[i, j]$ determined by coordinates i and j, l is the distance between $P[i, j]$ and the reference object $\tilde{p}$, and the membership function value of the fuzzy spatial relation *AtDistanceOf* ("at distance of") is calculated as:

$$\mu_{AtDistanceOf,d,g}(\tilde{p}, P[i, j], r) = \begin{cases} 1, & \text{if } r - d < l \leqslant r + d \\ 1 - \frac{l-(r+d)}{g}, & \text{if } r + d < l \leqslant r + d + g \\ \frac{l-(r-d-g)}{g}, & \text{if } r - d - g < l \leqslant r - d \\ 0, & \text{otherwise}, \end{cases}$$

where r is a parameter of the relation and represents the desired distance from the reference object. The constant d defines the core of the fuzzy set (allowed deviation from the desired distance), the constant g defines the fuzziness of the fuzzy set.

The previously defined elementary fuzzy spatial relations can be combined using the operators *AND*, *OR*, *NOT*, and *SUBTRACT*, to obtain more complex fuzzy spatial relations The definitions of these operators follow.

Definition 33 Let F and G be two elementary fuzzy spatial relations. The AND (conjunction) is a binary operator that composes the relations F and G in such a way that the value of the membership function for each pixel $P[i, j]$ is calculated as the minimum membership function values of F and G for the pixel $P[i, j]$:

$$AND(\mu_F(P[i, j]), \mu_G(P[i, j])) = \min(\mu_F(P[i, j]), \mu_G(P[i, j])) .$$

Definition 34 Let F and G be two elementary fuzzy spatial relations. The OR (disjunction) is a binary operator that composes the relations F and G in such a way that the value of the membership function for each pixel $P[i, j]$ is calculated as the maximum membership function values of F and G for the pixel $P[i, j]$:

$$OR(\mu_F(P[i, j]), \mu_G(P[i, j])) = \max(\mu_F(P[i, j]), \mu_G(P[i, j])).$$

Definition 35 Let F be an elementary fuzzy spatial relation. Then NOT (complement) is a unary operator that behaves in such a way that the value of the membership function for each pixel $P[i, j]$ is calculated as complement of the membership function value of F for the pixel $P[i, j]$:

$$NOT(\mu_F(P[i, j])) = 1 - \mu_F(P[i, j]).$$

Definition 36 Let F and G be two elementary fuzzy spatial relations. Then $SUBTRACT$ (subtraction) is a binary operator that composes the relations F and G in such a way that the value of the membership function for each pixel $P[i, j]$ is calculated as the minimum membership function value of F and a complementary membership function value of G for the pixel $P[i, j]$:

$$SUBTRACT\,(\mu_F(P[i, j]), \mu_G(P[i, j])) = \min(\mu_F(P[i, j]), 1 - \mu_G(P[i, j])).$$

3.2 *Temporal Modelling—Linear Fuzzy Space-Based Time Series*

Most of the real-world tasks that utilize time series rely on multivariate time series models. The common model is as follows [37].

Definition 37 Let $Z_t = \left[Z_{1,t}, Z_{2,t}, \ldots, Z_{m,t}\right]'$, be an m—dimensional jointly stationary real-valued vector process such that $E(Z_{i,t}) = \mu_i$ is a constant for each $i = 1, 2, \ldots, m$ and the cross-covariance between $Z_{i,t}$ and $Z_{j,s}$ for all $i = 1, 2, \ldots, m$ and $j = 1, 2, \ldots, m$ are functions only on the time difference $(s - t)$.

On the other hand, the original definition of the univariate first fuzzy time series model is [28]:

Definition 38 Let $Y(t)(t = \ldots, 0, 1, 2, \ldots)$, a subset of R^1 be the universe of discourse on which fuzzy sets $f_i(t)(i = 1, 2, \ldots)$ are defined and $F(t)$ is the collection of $f_i(t)(i = 1, 2, \ldots)$. Then $F(t)$ is called a fuzzy time series on $Y(t)(t = \ldots, 0, 1, 2, \ldots)$.

Our time series model is a combination of the previous two where we basically apply the same common multivariate model which is modified to support imprecise measurement values. In our model, we simply replace a crisp point with a linear fuzzy space point. In addition, we use fuzzy points to represent linguistic variables, which enables us to represent and manipulate linguistic data in our model. Accordingly, our time series model is as follows.

Definition 39 Let $Y(t)(t = \ldots, 0, 1, 2, \ldots)$, a subset of R^1 be the universe of discourse. Let $H^l\,(l = 1, 2)$ be a linear fuzzy space. Furthermore, let $f_i(t)(i = 1, 2, \ldots)$ be fuzzy sets defined as points on a linear fuzzy space over the given universe of discourse, and $\widetilde{F}_j(t)(j = 1, 2, \ldots m)$ be collections of these fuzzy points. Then, $\widetilde{F}_t = \left[\tilde{F}_{1,t}, \tilde{F}_{2,t}, \ldots, \tilde{F}_{m,t}\right]'$ is called a *linear fuzzy space based fuzzy time series* on $Y(t)(t = \ldots, 0, 1, 2, \ldots)$.

This definition enables all features of linear fuzzy space to be utilized. For example, a process vector can be of a mixed type (some components can be crisp, some can be fuzzy linguistic or non-linguistic) because in the linear fuzzy space only crisp values are treated as a special case. All others are fuzzy points or sets of fuzzy points. Also, spatial relations defined on the linear fuzzy space hold, providing for efficient representations and processing of measurement errors.

In our model, machine learning techniques can be used to create complex, non-linear relations.

4 Examples of Linear Fuzzy Space Applications

In this section we present two examples intended to show how linear fuzzy space theory can be used to solve some practical tasks. The first one shows an application in medical image interpretation while the second one shows an application in forecasting air pollution.

4.1 2D Image Segmentation and Feature Extraction Based on Linear Fuzzy Space

In this example we utilize linear fuzzy space to conduct region segmentation and object extraction on 2D images containing various types of imprecision and uncertainty. The goal is to extract one or more regions based on specific texture characteristics and the mutual spatial relations of these regions. The procedure consists of three main parts: a region segmentation, the modelling of compact or linked regions and finally the use of fuzzy spatial relations to extract the desired objects and regions. The linear fuzzy space is applied here mainly for the sake of modelling the topological relations between the regions, by utilizing the model described in Sect. 3.1. For the region segmentation, which utilizes machine learning techniques to classify image pixels, we used the fuzzy linear space for class modelling. The feature vector of the basic spatial element—a pixel—is created by extracting a matrix of neighboring pixels, a 7×7, 5×5 or 3×3 matrix transformation using wavelet or 2D frequent transformation, filtering the obtained coefficients with a zig zag algorithm, and finally normalization. The training set, which contains ordered pairs of feature vectors and class membership values, is divided into two subsets. The first one is a training subset containing 80% of the examples, and the second one is a validation subset containing 20% of the examples. For the concrete application we used

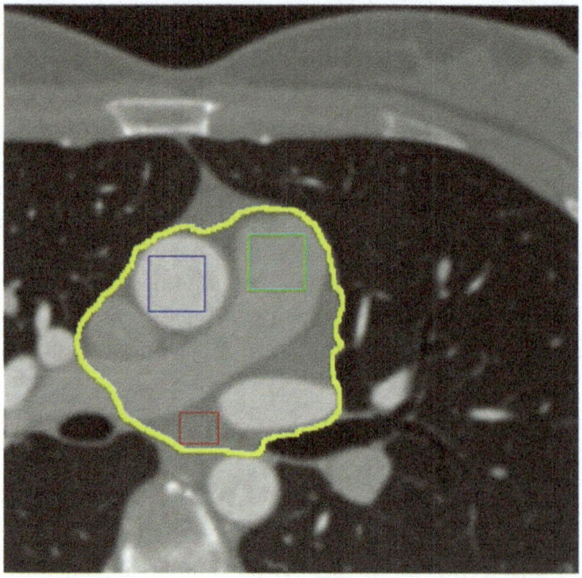

Fig. 1 An annotated DICOM image

DICOM images with manually annotated regions, as shown on Fig. 1, where blue, green, and red rectangles are regions that safely belong to a certain class, while the yellow line denotes the region in which detection occurs.

In this example, the fuzzy classifier is a fully connected artificial neural network with nine inputs, one hidden layer with 20 neurons, and a 20% dropout. The activation function for all neurons is the same—Rectified Linear Unit (ReLU). Training was performed through 1000 epochs with an iteration size of 500 examples using the Adam adaptive stochastic gradient optimization method and a mean square error fitness function [14]. Figure 2 presents the change in recognition accuracy with respect to the training epoch.

For the modelling of compact regions, we applied the algorithm for labelling image components [32] to determine which pixels belong to a fuzzy class (those with a corresponding membership function > 0). Once the compact regions are determined, it is possible to apply the spatial relations defined in the linear fuzzy space that are described in Sect. 3.1 to filter the regions of interest (those that satisfy certain topological relations). This enables semantic interpretation based on querying images for topological relations among the regions.

Figures 3, 4, 5 and 6 exhibit examples of the results. Each figure consists of four sections (from left to right). The first is an original DICOM image, the second is the image with extracted regions, the third is a spatial relation and the last one contains the query result (for example, blue texture at the top). Figure 3 shows the result which corresponds to the query: region (texture=blue AND position=bottom).

Figure 4 shows the result obtained by the query: region (texture=red AND position=right).

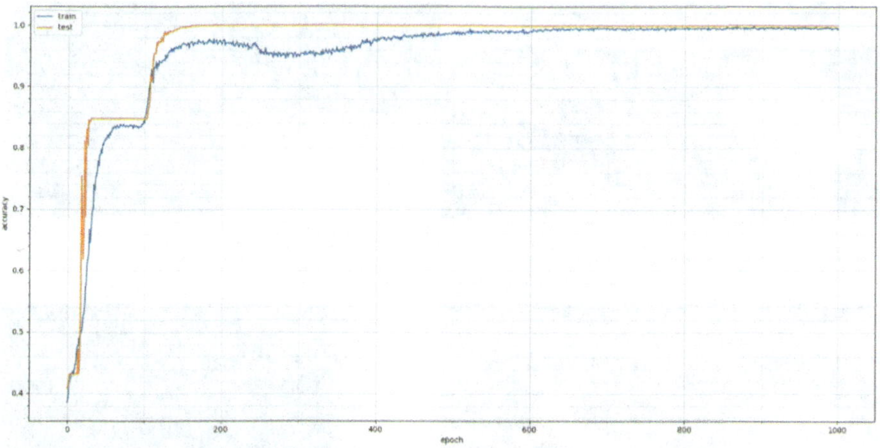

Fig. 2 Accuracy in the training process

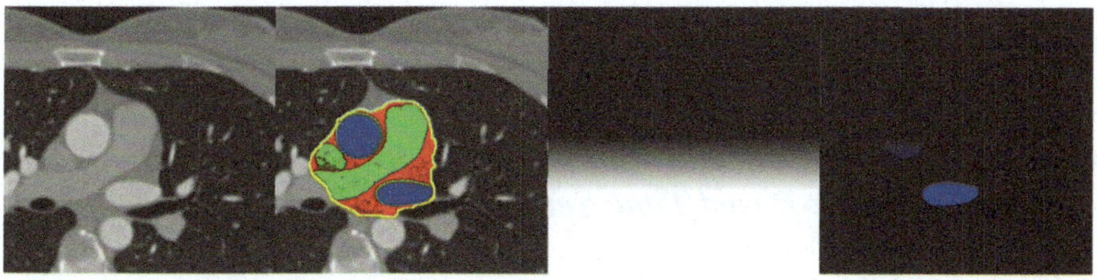

Fig. 3 Blue regions at bottom

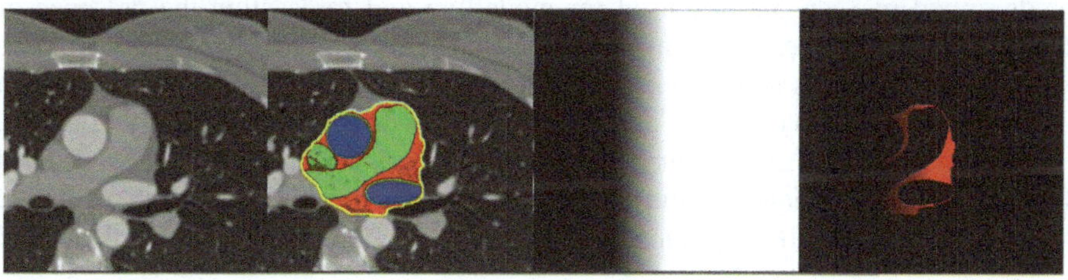

Fig. 4 Red regions on the right

Figure 5 presents the result obtained by the query: region (texture=blue AND position=bottom), while Fig. 6 presents the result obtained by the query: region (texture=blue AND position=left).

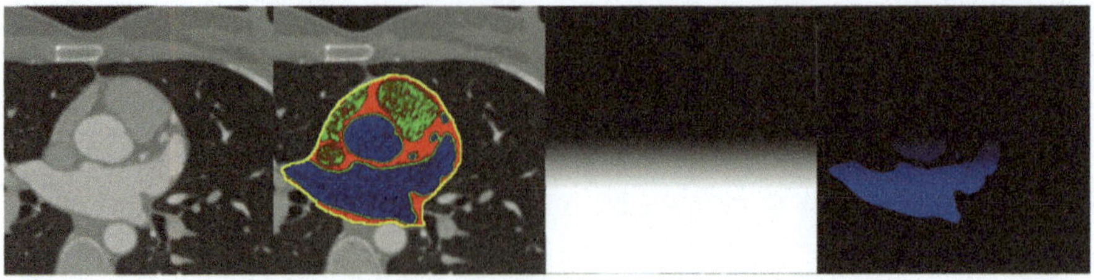

Fig. 5 Blue regions at the bottom

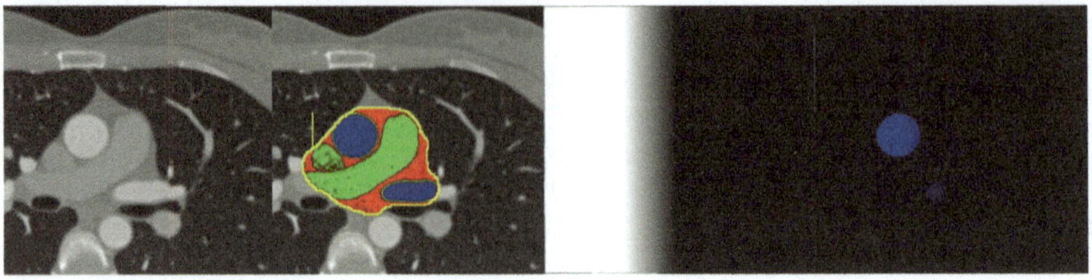

Fig. 6 Blue regions on the left

4.2 *Forecasting Concentrations of Particles Using Linear Fuzzy Space-Based Time Series*

In this example we demonstrate how the linear fuzzy space is used for time series-based forecasting. Fuzzy time series defined by means of the fuzzy linear space as described in Sect. 3.2. are used for modelling and forecasting the presence of PM10 particles in the air. In this example, the time series is a sequence of real-valued measurements of four meteorological parameters GDAS (The Global Data Assimilation System) (Table 1), and five parameters describing air quality (Table 2).

Concentrations of criteria air pollutants (SO2, NO, NO2, NOx and PM10) used for the analyses presented in this chapter were obtained from the Institute of Public Health Belgrade (IPHB) regular monitoring network. The data was complemented

Table 1 GDAS meteorological parameters

ID	Description	Unit
prss	Pressure at surface	hPa
tpp6	Accumulated precipitation (6 h accumulation)	m
rh2m	Relative Humidity at 2m AGL	%
t02m	Temperature	degC
tcld	Total cloud cover (3 or 6h average)	%

Table 2 IPHB air pollution parameters

ID	Description	Unit
SO2	Sulfor dioxide	$\mu g/m^3$
NO	Nitrogen oxide	$\mu g/m^3$
NO2	Nitrogen dioxide	$\mu g/m^3$
NOx	Other nitrogen oxides	$\mu g/m^3$
PM10	Suspended particles smaller than 2.5 μm	$\mu g/m^3$

Table 3 Linguistic variable CAQI index

Linguistic value	Left point	Right point
Very low	0+b, 2b	25-b, 2b
Low	25+b, 2b	50, 2b
Medium	50+b, 2b	90, 2b
High	90+b, 2b	180, 2b
Very high	180+b, 2b	500, 2b

with Global Data Interpolation System (GDAS1) meteorological surface parameters interpolated for each measurement site of the IPHB network.

The sampling rate for the meteorological data is 3 h, while the sampling rate for acquiring air pollution parameters is 1 h. Our time series data set consists of these two sets of data covering the period 1/1/2018 to 31/12/2018. The first step in our analysis was to normalize the two data sets to the same sample rate (one hour), which was done by utilizing a cubic interpolation spline.

Fuzzy classification of the air quality is determined based on The Common Air Quality Index (CAQI) [1]. In our examples we adopted only PM10 concentration for the CAQI.

Following the fuzzy linear space-based time series model, defined in Sect. 4.1, we set up the following time series model for our example.

Let $\tilde{G}_t = \left[\tilde{G}_{1,t}, \tilde{G}_{2,t}, \tilde{G}_{3,t}, \tilde{G}_{4,t}\right]$ and $\tilde{I}_t = \left[\tilde{I}_{1,t}, \tilde{I}_{2,t}, \tilde{I}_{3,t}, \tilde{I}_{4,t}, \tilde{I}_{5,t}\right]$ be GDAS meteorological parameters and air pollution parameters respectively, measured at the time point t. Here, all values are real-valued.

Furthermore, let the CAQI index be a linguistic variable described by fuzzy intervals, with starting (left) and ending (right) points represented by the points from H^1, i.e., ordered pairs (v, r).

In our example we apply the CAQI index linguistic variable represented by Table 3. Here, parameter b governs fuzzy sets i.e., for b=0 or very small, fuzzy sets become crisp ones.

Then, the model of our example time series can be expressed as follows.

$F_t = \left[F_{1,t} = G_{1,t}, F_{2,t} = G_{2,t}, \ldots, F_{4,t} = G_{4,t}, F_{5,t} = I_{1,t}, \ldots, F_{9,t} = I_{5,t}\right]$

$F_{i,t} = (x, 0) \in H^1$ and $x \in R^1$, $(i = 1, \ldots, 8)$

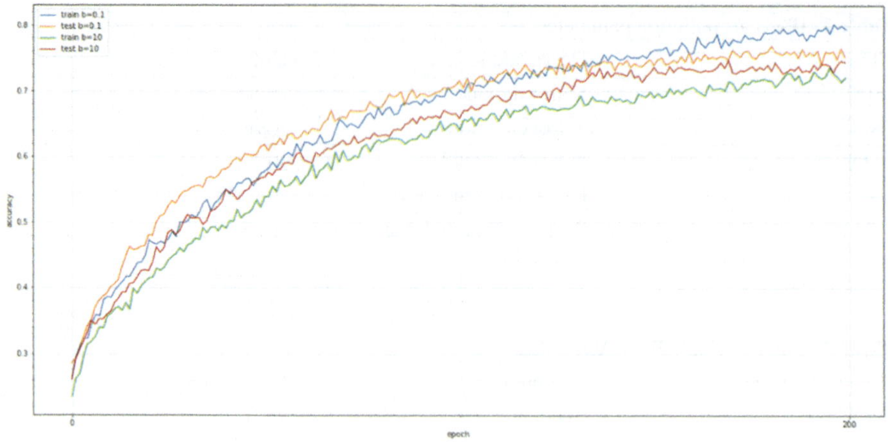

Fig. 7 Accuracy in the training process for unbalanced data set

$F_{9,t} = f_{9,j}(t), (j = 1, \ldots, 5)$
For each unit time element, a feature vector FV is created:

$$FV(tm) = [F(tm - w)..F(tm)], tm \in \{w, w + s, \ldots, w + n * s\},$$

where w is a number of consecutive/previous measurements, and s is the step.

Feature vector elements represent small time series of the length w that in our example equal 24. The feature vector sample rate denoted here as tm, is a parameter which controls the frequency of sampling selection. It can't be smaller than the sampling frequency for the starting time series. In this example we opted for $s = 2$.

In what remains of our chapter, we present the results of two experiments in which the value of the CAQI index is represented by the linguistic variable defined in Table 3. Here, the input feature corresponds to a 24 × 13 real-valued matrix, i.e., an array of 312 real numbers. The data set obtained in that way is divided into two subsets, the first one aimed at training and the second one at validation, with a ratio of 80:20%.

In the experiments, we used a fully connected neural network as a fuzzy classifier, with 240 input nodes, 5 output nodes, and two hidden layers, each containing 20 nodes. In each layer, a RELU (rectified linear unit) activation function was used with a 20% dropout for avoiding overfitting. The Adam adaptive stochastic gradient optimization method and a mean square error fitness function were used for training, which lasted for 200 epochs with an iteration size of 500 examples.

Figure 7 represents the recognition accuracy change with respect to the training epoch for the unbalanced data set. It shows no significant difference in trend.

For the classifier evaluation in this example, we used two types of confusion matrix. The first one is "classic":

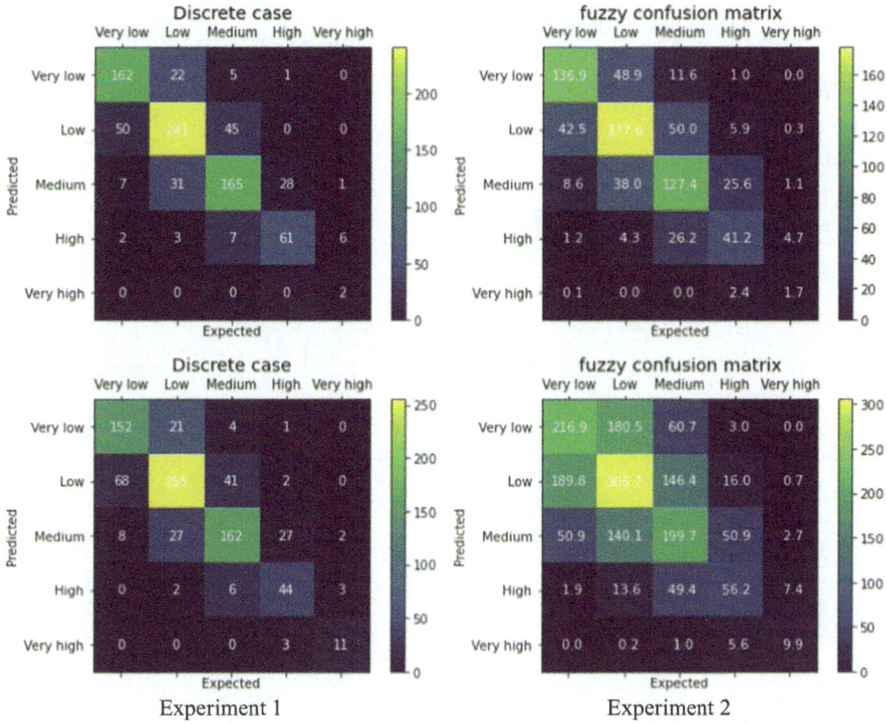

Fig. 8 The confusion matrices for unbalanced data set

$$M_{i,j} = \sum_{k=0}^{n} (P_k = i) \text{ and } (T_k = j).$$

The second one is a fuzzy modification inspired by the idea of a local fuzzy confusion matrix [40]. We define our fuzzy confusion matrix as:

$$M_{i,j} = \sum_{k=0}^{n} \min(\mu_{k,i}, \mu_{k,j}),$$

where the logical operator *and* is mapped to its fuzzy counterpart (in our case *min*), and the operator = is set membership function.

The confusion matrices for this experiment (Fig. 8) indicate that the model performs slightly better in recognizing very high pollution for the unbalanced dataset in Experiment 2 (with borders between the classes being less steep/more fuzzy).

Inspection of the dataset shows a lack of data balance, especially considering the class 'Very high' (only 2%). Therefore, new, balanced datasets (Table 4) were created by randomly repeating and removing examples.

Table 4 Training samples distribution per fuzzy sets

	Original (%)	Balanced (%)
Very low	27	22
Low	35	19
Medium	26	20
High	10	21
Very high	2	19

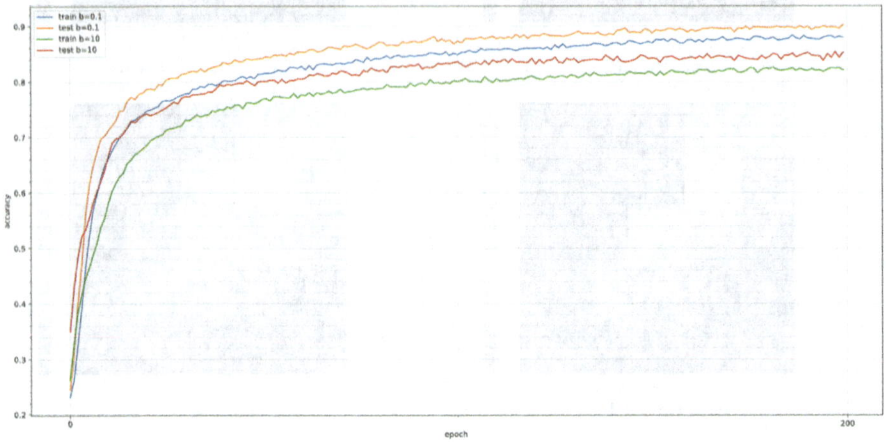

Fig. 9 The classification accuracy for training using the balanced data set

Figure 9 shows the classification accuracy for the training performed on the balanced data set, while Fig. 10 shows confusion matrices for the same data set.

Some useful conclusions for improving the model can be derived from these preliminary experiments. Firstly, there is an indication that the model's accuracy depends significantly on dataset balance. Secondly, the model performs better with fuzzy sets describing linguistic variables being closer to a "crisp" one. The results indicate that the introduction of a new function for accuracy evaluation that does not favor crisp class borders, as well as more sophisticated use of the concepts of fuzzy linear space for data sets balancing, should be considered in the future.

5 Conclusion

The basic properties of dynamic systems include imprecisions and uncertainties in both spatial and temporal data which require adequate representations and further manipulations. Fuzzy theory is one of the most frequently used solutions.

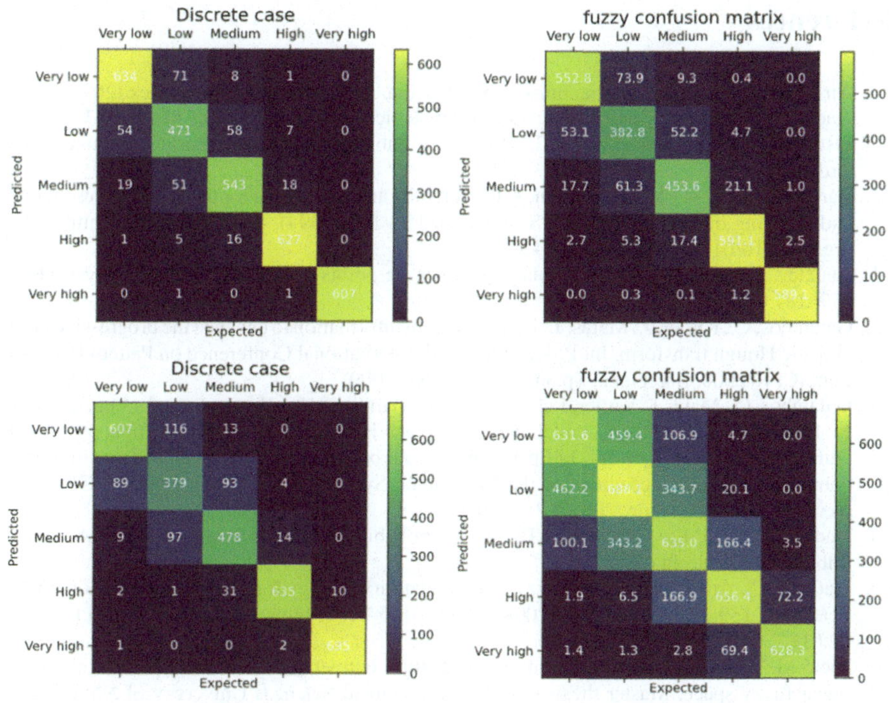

Experiment 1 Experiment 2

Fig. 10 The confusion matrices for the balanced data set

The linear fuzzy space is an original fuzzy-based approach that provides an effective and compact model of imprecise data, including spatial objects and their relations. In this chapter we have presented the main theoretical results of the linear fuzzy space and illustrated its applicability for the modelling of spatial and temporal systems through two representative examples.

The first example, which pertains to medical image analysis, confirms that the linear fuzzy space has the full capacity to model complex spatial relations. It is useful for semantic image analysis and spatial databases. The second example is an attempt to apply the linear fuzzy space in temporal modelling through fuzzy time series. Initial results indicate that this approach is promising if properly combined with machine learning techniques.

Future research aims to further develop linear fuzzy space theory and its applications in various tasks. These will include fuzzy differential equations, prioritization of training sets, fuzzy rules-based regularization, the fitness function of fuzzy classifiers, and fuzzy decision trees based on the linear fuzzy space.

Acknowledgements The authors acknowledge funding provided by the Science Fund of the Republic of Serbia #GRANT No. 6524105, AI—ATLAS.

References

1. Air Quality Index.: Wikipedia. https://en.wikipedia.org/wiki/Air_quality_index (2020)
2. Andree, B.P.J.: Theory and Application of Dynamic Spatial Time Series Models. (Tinbergen Institute, Research Series; vol. 762). Rozenberg Publishers and the Tinbergen Institute, Amsterdam (2020)
3. Barro, S., Marín, R., Mira, J., Patón, A.R.: A model and a language for the fuzzy representation and handling of time. Fuzzy Sets Syst. **61**(2), 153–175 (1994). ISSN 0165-0114. https://doi.org/10.1016/0165-0114(94)90231-3
4. Buckley, J.J., Eslami, E.: Fuzzy plane geometry I: points and lines. Fuzzy Sets Syst. **86**(2), 179–187 (1997)
5. Galambos, C., Kittler, J., Matas, J.: Using gradient information to enhance the progressive probabilistic Hough transform. In: Proceedings 15th International Conference on Pattern Recognition. ICPR-2000, Barcelona, Spain, pp. 560–563 (2000)
6. Galambos, C., Matas J., Kittler, J.: Progressive probabilistic Hough transform for line detection. In: IEEE Computer Society Conference Computer Vision and Pattern Recognition, p. 1 (1999)
7. Guibas, L., Salesin, D., Stolfi, J.: Epsilon geometry: building robust algorithms from imprecise computations. In: Proceedings of the 5th Annual Symposium on Computational Geometry, New York, NY, USA, pp. 208–217 (1989)
8. Hadžić, O., Pap, E.: Fixed Point Theory in Probabilistic Metric Spaces. Kluwer Academic Publishers, Dordecht (2001)
9. Hudelot, C., Atif, J., Bloch, I.: Fuzzy spatial relation ontology for image interpretation. Fuzzy Sets Syst. **159**(15), 1929–1951 (2008). ISSN 0165-0114. https://doi.org/10.1016/j.fss.2008.02.011
10. Jocić M.: Digital image segmentation and feature extraction based on geometric shapes in linear fuzzy space. Master thesis, Faculty of Technical Sciences University of Novi Sad (in Serbian) (2012)
11. Jocić, M., Dimitrijević, D., Pantović, M., Madić, D., Konjović, Z.: Linear fuzzy space based scoliosis screening. In: Proceedings of the ICIST 2014, Kopaonik, Serbia, pp. 180–187 (2014)
12. Jocić, M., Obradović, Đ, Konjović, Z., Pap, E.: 2D fuzzy spatial relations and their applications to DICOM medical images. In: IEEE 11th International Symposium on Intelligent Systems and Informatics (SISY), pp. 39–44 (2013)
13. Jocić, M., Pap, E., Szakál, A., Obradović, Đ., Konjović, Z.: Managing big data using fuzzy sets by directed graph node similarity. Acta Polytech. Hung. **14**(2), 183–200 (2017)
14. Kingma, D.P., Ba, J.: Adam. A Method for Stochastic Optimization. Computer Science, Mathematic (2015)
15. Kiryati, N., Eldar, Y., Bruckstein, A.M.: A probabilistic Hough transform. Pattern Recognit. **24**(4), 303–316 (1991)
16. Löffler, M.: Data imprecision in computational geometry. Ph.D. thesis, Utrecht University, Utrecht (2009)
17. Löffler, M., Van Kreveld, M.: Geometry with imprecise lines. In: 24th European Workshop on Computational Geometry (2008)
18. Löffler, M., Van Kreveld, M.: Largest and smallest convex hulls for imprecise points. Algorithmica **56**(2), 235–269 (2008)
19. Marín, R., Barro S., Bosch, A., Mira, J.: Modeling the representation of time from a fuzzy perspective. Cybern. Syst. **25**(2), 217–231 (1994). https://doi.org/10.1080/01969729408902325
20. Matas, J., Galambos, C., Kittler J.: Robust detection of lines using the progressive probabilistic hough transform. Comput. Vis. Image Understand. **78**(1), 119–137 (2000)
21. Obradović, Đ.: Model of the computer simulation system for managing geospace under uncertainty conditions. Ph.D. dissertation, Faculty of Technical Sciences University of Novi Sad (in Serbian) (2011)
22. Obradović, Đ., Konjović, Z., Pap, E.: Extending PostGIS by imprecise point objects. In: IEEE 8th international symposium on intelligent systems and informatics, Subotica, Serbia, pp. 23–28 (2010)

23. Obradović, Đ., Konjović, Z., Pap, E., Jocić, M.: Linear fuzzy space polygon based image segmentation and feature extraction. In: Proceedings of the IEEE 10th Jubilee International Symposium on Intelligent Systems and Informatics (SISY), pp. 543–548 (2012)
24. Obradović, Đ., Konjovic, Z., Pap, E., Ralević, N.M.: The maximal distance between imprecise point objects. Fuzzy Sets Syst **170**(1), 76–94 (2011)
25. Obradović, Đ., Konjović, Z., Pap, E., Rudas, I.J.: Linear Fuzzy Space Based Road Lane Model and Detection. Knowl.-Based Syst. **38**, 37–47 (2012)
26. Obradović, Đ., Konjović, Z., Pap, E., Rudas, I.: Fuzzy geometry in linear fuzzy space, intelligent systems: models and applications.iN: Revised and Selected Papers from the 9th IEEE International Symposium on Intelligent Systems and Informatics SISY 2011. Springer, Berlin, Heidelberg, PP. 137–153 (2013)
27. Obradović, Đ., Konjović, Z., Segedinac, M.: Extensible software simulation system for imprecise geospatial process. In: Konjović, Z. (Eds.) ICIST 2011 Proceedings, pp. 71–77 (2011)
28. Panigrahi, S., Behera, H.S.: Fuzzy time series forecasting: a survey. In: Behera, H., Nayak, J., Naik B., Pelusi, D. (eds.) Computational Intelligence in Data Mining. Advances in Intelligent Systems and Computing, vol. 990. Springer, Singapore (2020). https://doi.org/10.1007/978-981-13-8676-3_54
29. Rosenfeld, A.: The fuzzy geometry of image subsets. Pattern Recogn. Lett. **2**(5), 311–317 (1984)
30. Sahin, A., Dodurka, M.F., Kumbasar, T., Yesil, E., Siradag, S.: Review study on fuzzy time series and their applications in the last fifteen years. In: International Fuzzy Systems Symposium (FUZZYSS'15), Istanbul, pp. 166–171 (2015)
31. Sahin, A., Kumbasar, T., Yesil, E., Dodurka, M.F., Karasakal, O., Siradag, S.: An enhanced fuzzy linguistic term generation and representation for time series forecasting. In: IEEE International Conference on Fuzzy Systems (FUZZ-IEEE), Istanbul, pp. 1–8 (2015). https://doi.org/10.1109/FUZZ-IEEE.2015.7337904
32. Samet, H., Tamminen, M.: Efficient component labeling of images of arbitrary dimension represented by linear bintrees. In: IEEE Transactions on Pattern Analysis and Machine Intelligence, vol. 10, issue 4, p. 579 (1988). https://doi.org/10.1109/34.3918
33. Schneider, M.: Spatial Data Types for Database Systems: Finite Resolution Geometry for Geographic Information Systems, 1st edn. Springer, Berlin (1997)
34. Schneider, M.: Fuzzy topological predicates, their properties, and their integration into query languages. In: Proceedings of the 9th ACM International Symposium on Advances in Geographic Information Systems, pp. 9–14 (2001)
35. Schneider, M.: Design and implementation of finite resolution crisp and fuzzy spatial objects. Data Knowl. Eng. **44**, 81–108 (2003)
36. Schockaert, S., Cock, M.D.: Temporal reasoning about fuzzy intervals. Artif. Intell. **172**, 1158–1193 (2008)
37. Singh, P.: A brief review of modelling approaches based on fuzzy time series. Int. J. Mach. Learn. Cyber. **8**, 397–420 (2017)
38. Song, Q., Chissom, B.S.: Fuzzy time series and its model. Fuzzy Sets Syst. **54**(3), 269–277 (1993). https://doi.org/10.1016/0165-0114(93)90372-O
39. Stephens, R.S.: Probabilistic approach to the Hough transform. Image Vis. Comput. **9**(1), 66–71 (1991)
40. Trajdos, P., Kurzynski, M.: A dynamic model of classifier competence based on the local fuzzy confusion matrix and the random reference classifier. Int. J. Appl. Math. Comput. Sci. **26**(1), 175–189 (2016). https://doi.org/10.1515/amcs-2016-0012
41. Verstraete, J., Tré, G., Caluwe, R., Hallez, A.: Field based methods for the modeling of fuzzy spatial data. In: Fuzzy Modeling with Spatial Information for Geographic Problems, pp. 41–69 (2005)
42. Wei, W.W.S.: Multivariate time series analysis and applications. Wiley Series in Probability and Statistic. Wiley, Hoboken (2019). ISBN:9781119502951. https://doi.org/10.1002/9781119502951
43. Yang, K., Sam, Ge S., He, H.: Robust line detection using two-orthogonal direction image scanning. Comput. Vis. Image Understand. **115**(8), 1207–1222 (2011)

Image Fuzzy Segmentation Using Aggregated Distance Functions and Pixel Descriptors

Endre Pap, Ljubo Nedović, and Nebojša Ralević

Abstract This paper is a review of recent research on image fuzzy segmentation using the fuzzy c-means clustering algorithm based on a distance function constructed by applying the aggregation function on the sequence of the initial distance functions and pixel descriptors. In image segmentation algorithms, distance functions compare pixels and represent a decision criterion for the classification of pixels into image segments. Determination of the segmentation criterion is based on the information fusion process, where the application of the appropriated aggregation function enables the adjustment of the segmentation criteria according to the intuitively expected decision. Initial distance functions represents the basic criteria which are relevant for segmentation, and applied aggregation function represent a model for this basic criteria fusion into one final decision criteria. With regards to a new distance function construction by applying aggregation functions, in this article we present relevant properties of the following aggregation functions: minimum, maximum, weighted arithmetic mean, generalized means, product of powers, weighted arithmetic mean of powers and OWA aggregation functions. Beside the pixel color or color components, other pixel descriptors are important for image segmentation and other image processing tasks. For experimental verification of the methodology used in image segmentation, the fuzzy c-means clustering algorithm is used.

Keywords Aggregation function · Distance function · OWA aggregation function · Generalized mean · Weighted arithmetic mean of powers · Product of powers · Image · Fuzzy segmentation · Pixel descriptor

---

E. Pap (✉)
Department for Postgraduate Studies, Singidunum University, 32 Danijelova St., 11000 Belgrade, Serbia
e-mail: epap@singidunum.ac.rs

L. Nedović · N. Ralević
Faculty of Technical Sciences, Department of Fundamentals Sciences, University of Novi Sad, Trg Dositeja Obradovića 6, 21000 Novi Sad, Serbia
e-mail: nljubo@uns.ac.rs

N. Ralević
e-mail: nralevic@uns.ac.rs

E. Pap (ed.), *Artificial Intelligence: Theory and Applications*,
Studies in Computational Intelligence 973,
https://doi.org/10.1007/978-3-030-72711-6_14

1 Introduction

Image segmentation is a process of classifying pixels into clusters called picture segments. The goal of segmentation is partitioning the image into visual regions, where each region contains pixels with similar characteristics, and each region represents a meaningful whole. Each segment can be, but does not necessarily need to be, a homogeneous and connected object in the image. One segment can also be a group of homogeneous and connected regions with similar characteristics, for example, with similar color. Other than color, the obtained segments also have additional interesting features such as shape, texture, size etc. In various complex image processing tasks, a segmented image may be the final result to be used by an expert, or an intermediate step where the segmented image is used for further processing such as shape recognition, object edge recognition, object classification etc.

There are many data clustering algorithms. For example, k-means, ISODATA, QMODEL, FCM, fuzzy QMODEL algorithm etc., see [1, 2, 20]. For image segmentation, the most frequently used algorithms are probably the k-means algorithm and the fuzzy c-means algorithm (FCM). In our previous research [15, 28, 30, 33–35], we used the FCM algorithm developed by J. C. Dunn in 1973, and improved by J. C. Bezdek in 1981, see [3, 4, 11]. FCM algorithm is a fuzzified k-means algorithm.

In image segmentation and other clustering algorithms, distance functions and metrics plays a key role in determining whether a pixel belongs to a certain segment or whether an object belongs to a certain cluster. There are many types of distance function and they may have many additional properties which have an effect the their application, see [5, 12, 16, 34]. Using distance functions in an image segmentation algorithm we compare two pixels or one pixel with the segment, and based on the obtained value of the distance function, the pixel is classified into appropriated segment.

For comparing pixels and image objects using the selected distance function, beside pixel color, other pixel or image object characteristics significantly contribute to image segmentation and other image processing algorithms. These characteristics are called pixel descriptors, or image and image objects descriptors, see [14, 27, 29, 36]. In [15, 30, 33], pixel descriptors are used for image segmentation.

In our previous work [15, 28, 30, 31, 33–35] we constructed a new distance function for image segmentation by using an aggregation function, see [7, 8, 10, 18, 21–23, 26]. Applying the aggregation function to the sequence of distance functions, a new distance function is obtained, see [34]. The properties of the obtained distance function depend both on the properties of the initial distance functions and the properties of the applied aggregation function, and depending on the properties of the distance function applied in image segmentation, very good results are achieved in [15, 30, 33, 34]. In this review article we present our results on the construction of new distance functions and their applications in image segmentation.

In Sect. 2, the notion of distance functions and aggregation functions is presented. In Sect. 3 we present a method for the construction of a new distance function by applying an aggregation function to the sequence od initial distance functions, with basic results about the properties of the constructed distance function. In Sect. 4 we present pixel descriptors introduced and used for image segmentation in [15, 30, 33], and a short description of the FCM algorithm. In Sect. 5 we present the experimental results in image segmentation published in [15, 30, 33, 34], which were obtained using distance functions constructed as shown in Sect. 3. In Sect. 6 we formulate conclusions about the applications of the constructed distance functions and some additional remarks about the methodology presented in Sect. 3.

2 Distance Functions and Aggregation Functions

2.1 Distance Functions

Distance functions have a diverse range of applications in the fields of science and engineering, see [16, 34].

Definition 1 (*Distance function*) For arbitrary $X \neq \emptyset$, function $d : X^2 \to [0, \infty[$ is a *distance function* on the set X if d has the following properties:

(d1) $\forall x, y \in X, \quad d(x, y) = d(y, x),$ (*symmetry*)
(d2) $\forall x \in X, \quad d(x, x) = 0.$ (*reflexivity*)

With certain additional properties, distance functions are suitable for specific applications, see [3, 5, 6, 13, 16, 19, 34, 37]. In the following definition we list additional possible properties of distance functions which are significant for their application in image segmentation.

Definition 2 Distance function $d : X^2 \to [0, \infty[$ may have following additional and important properties.

(d3) Identity of indiscernibles:
$\forall x, y \in X, \quad d(x, y) = 0 \Rightarrow x = y.$
(d4) Triangle inequality:
$\forall x, y, z \in X, \quad d(x, z) \leqslant d(x, y) + d(y, z).$
(d5) C-triangle inequality: for some $C \in [1, \infty[$ it holds:
$\forall x, y, z \in X, \quad d(x, y) \leqslant C\,(d(x, z) + d(z, y)).$
(d6) Boundedness:
$\exists a > 0, \ d : X^2 \to [0, a].$

Remark 1 A distance function $d : X^2 \to [0, \infty[$ is a metric on X if for d, the identity of indiscernibles (d3) and triangle inequality (d4) hold. With some of the listed or with other additional properties, distance functions are known as *dissimilarity functions*, *pseudo-metric*, etc., see [16].

Remark 2 In image segmentation and other image processing tasks we deal with the bounded distance function d on X, where underlying space X is a set of pixels of an image and a pixel is represented as n-tuple of color components and other pixel features (descriptors). Then, it is convenient to use distance functions with values in [0, 1] because of the comparability of experimental results obtained with using different bounded distance functions. In [34] it is proved proved that for bounded distance function $\overline{d} : X^2 \to [0, a]$ with $a > 0$, function d defined on X by $d(x, y) = \frac{1}{a}\overline{d}(x, y)$, $x, y \in X$ is again a distance function with values in [0, 1], and that d inherits from $\overline{d}$ properties (d3), (d4) and (d5) listed in Definition 2, also as some other properties of function $\overline{d}$. Distance functions $d : X^2 \to [0, 1]$ will hereinafter be referred to as *normalized distance functions*.

2.2 Aggregation Functions

Aggregation functions are functions which combine several values into a single representative one in very weakly limited way, and extended aggregation functions are functions with a variable number of arguments in which every restriction on a function with n inputs is an aggregation function, see [9, 21, 23, 26]. There are many types of aggregation functions, see [8, 10, 22, 24, 39], with wide applications in engineering and the natural and social sciences, see [17, 18, 32, 37–40]. Choosing an appropriated aggregation function and applying it on a sequence of information, we model a process of information fusion which corresponds to our intuitive expectations.

Usually, an aggregation function is defined as a function with arguments and values in [0, 1], $[0, \infty[$, $[0, \infty]$, or any bounded or unbounded interval $[a, b]$ of real numbers, even in Boolean algebras and posets. Due to applications stated in this article, the interval [0, 1] will be used, see [9, 21].

Definition 3 (*Aggregation function*) For fixed $n \in \mathbb{N}$, an n-ary *aggregation function* is a function $A : [0, 1]^n \to [0, 1]$ with the following properties.

- (a1) Boundary conditions $\mathsf{A}(0, \ldots, 0) = 0$ and $\mathsf{A}(1, \ldots, 1) = 1$ hold.
- (a2) A function A is monotonically non-decreasing in each component, i.e., $\forall i \in \{1, \ldots, n\}, \ a_i \leqslant b_i \quad \Rightarrow \quad \mathsf{A}(a_1, \ldots, a_n) \leqslant \mathsf{A}(b_1, \ldots, b_n)$ hold for arbitrary n-tuples $(a_1, \ldots, a_n) \in [0, 1]^n$, $(b_1, \ldots, b_n) \in [0, 1]^n$.

In case $n = 1$, by definition is $\mathsf{A}(x) = x$ for all $x \in [0, 1]$.

Definition 4 (*Extended aggregation function*) An *extended aggregation function* is a function $\mathsf{A} : \bigcup_{n=1}^{\infty}[0, 1]^n \to [0, 1]$ such that every restriction $\mathsf{A}_{[n]} : [0, 1]^n \to [0, 1]$, $n \in \mathbb{N}$ of this function is an n-ary aggregation function.

An n-ary aggregation function $\mathsf{A}_{[n]}$ may have additional properties. In the following definition we list the properties which are relevant for the application presented in this paper.

Definition 5 n-ary aggregation function $\mathsf{A}_{[n]} : [0,1]^n \to [0,1]$ may have some of the following properties.

(a3) Function $\mathsf{A}_{[n]}$ is continuous.
(a4) Function $\mathsf{A}_{[n]}$ is subadditive, i.e., for all $(a_1, \ldots, a_n), (b_1, \ldots, b_n) \in [0,1]^n$ that satisfy $(a_1 + b_1, \ldots, a_n + b_n) \in [0,1]^n$ hold
$\mathsf{A}_{[n]}(a_1 + b_1, \ldots, a_n + b_n) \leqslant \mathsf{A}_{[n]}(a_1, \ldots, a_n) + \mathsf{A}_{[n]}(b_1, \ldots, b_n)$.
(a5) Function $\mathsf{A}_{[n]}$ is positively subhomogeneous, i.e., for each $t \geqslant 0$ and all $(a_1, \ldots, a_n) \in [0,1]^n$ that satisfy $(ta_1, \ldots, ta_n) \in [0,1]^n$ hold
$\mathsf{A}_{[n]}(ta_1, \ldots, ta_n) \leqslant t\mathsf{A}_{[n]}(a_1, \ldots, a_n)$.
(a6) $\mathsf{A}_{[n]}(a_1, \ldots, a_n) = 0 \;\Rightarrow\; \forall i \in \{1, \ldots, n\},\; a_i = 0$.
(a7) $\mathsf{A}_{[n]}(a_1, \ldots, a_n) = 0 \;\Rightarrow\; \exists i \in \{1, \ldots, n\},\; a_i = 0$.

Remark 3 Restriction $\mathsf{A}_{[1]}(x)$, $x \in [0,1]$ of extended aggregation function A obviously satisfies all properties from Definition 5. Because of that, in the examinations of the properties of extended aggregation functions we will consider restrictions $\mathsf{A}_{[n]}(x)$ for $n \geqslant 2$.

3 Construction of New Distance Functions

In this section we present the methodology for the construction of a new distance function by applying an aggregation function to the sequence of initial distance functions, published in [34]. In image segmentation, initial distance functions represent particular criteria for the comparison of pixels, and a newly constructed distance function represents the final criterion and a measure of pixel similarity, obtained by aggregation-based fusion of particular pixels' similarity measures. In addition, we present the properties of distance functions constructed by some of the aggregation functions investigated in [15, 30, 33, 34]. Many other aggregation functions are investigated from the before mentioned perspective in [15, 28, 34].

Let d_i, $i \in \{1, \ldots, n\}$, be distance functions on the set X, and let $\mathsf{A}_{[n]}$ be an arbitrary n-ary aggregation function. Let us observe the function $d_{[n]} : X^2 \to [0,1]$ defined by

$$d_{[n]}(x, y) = \mathsf{A}_{[n]}\left(d_1(x, y), \ldots, d_n(x, y)\right) \tag{1}$$

for all $x, y \in X$.

The following theorem establishes some properties of function $d_{[n]}$ depending on the properties of initial distance functions d_i, $i \in \{1, \ldots, n\}$, and the properties of the applied aggregation function $\mathsf{A}_{[n]}$, see [34].

Theorem 1 *Let $d_i : X^2 \to [0,1]$, $i \in \{1, \ldots, n\}$, be a sequence of arbitrary distance functions, and let $\mathsf{A}_{[n]}$ be an arbitrary n-ary aggregation function. Then for function $d_{[n]} : X^2 \to [0,1]$ defined with* (1) *the following hold.*

(ad1) *Function $d_{[n]}$ is a distance function.*

Table 1 Properties and applicability of minimum and maximum

	(a3)	(a4)	(a5)	(a6)	(a7)	(ad1)	(ad2)	(ad3)	(ad4)	(ad5)	(ad6)
$\min_{[n]}$	yes	no	yes	no	yes	yes	yes	no	no	no	yes
$\max_{[n]}$	yes	yes	yes	yes	yes	yes	yes	yes	yes	yes	yes

(ad2) *If for each of the functions d_i, $i \in \{1, \dots, n\}$, the "identity of indiscernibles"* (d3) *holds and aggregation function $\mathsf{A}_{[n]}$ has the property* (a7), *then the function $d_{[n]}$ also has the property "identity of indiscernibles"* (d3).

(ad3) *If for at least one of the considered distance functions d_i, $i \in \{1, \dots, n\}$, "identity of indiscernibles"* (d3) *holds and if the aggregation function $\mathsf{A}_{[n]}$ has the property* (a6), *then the function $d_{[n]}$ also has the property "identity of indiscernibles"* (d3).

(ad4) *Let each of the functions d_i, $i \in \{1, \dots, n\}$, be metric. Let n-ary aggregation function $\mathsf{A}_{[n]} : [0, 1]^n \to [0, 1]$ have the subadditivity property* (a4) *and the property* (a7). *Then the distance function $d_{[n]}$ is also metric.*

(ad5) *Let each of the distance functions d_i, $i \in \{1, \dots, n\}$, satisfy the C-triangle inequality* (d5). *Let $\mathsf{A}_{[n]} : [0, 1]^n \to [0, 1]$ be a subadditive and positively sub-homogeneous n-ary aggregation function, i.e., have the properties* (a4) *and* (a5). *Then distance function $d_{[n]}$ also has the C-triangle inequality property* (d5).

(ad6) *Let $\mathsf{A}_{[n]}$ be a continuous n-ary aggregation function. If the space X^2 is equipped with a topological structure and each of the distance functions d_i, $i \in \{1, \dots, n\}$, is continuous on X^2, then the distance function $d_{[n]}$ is also continuous on X^2.*

Remark 4 Statements from (ad1) to (ad5) of Theorem 1 are proved in [34], and the statement (ad6) is obvious.

In our previous works, the properties of some known and constructed aggregation functions were examined. Hereinafter we present a number of them which are applied for construction of distance function which are used for image segmentation.

3.1 Minimum and Maximum

The properties from Definition 5 of well known n-ary aggregation functions minimum $\min_{[n]} : [0, 1]^n \to [0, 1]$ and maximum $\max_{[n]} : [0, 1]^n \to [0, 1]$ are examined in [31, 34]. They are summarized in Table 1 together with applicability of $\min_{[n]}$ and $\max_{[n]}$ in Theorem 1.

Table 2 Properties and applicability of means

	(a3)	(a4)	(a5)	(a6)	(a7)	(ad1)	(ad2)	(ad3)	(ad4)	(ad5)	(ad6)
$\mathsf{MA}_{[n]}$	yes	yes	yes	yes	yes	yes	yes	yes	yes	yes	yes
$\mathsf{MH}_{[n]}$	yes	no	yes	no	yes	yes	yes	no	no	no	yes
$\mathsf{MG}_{[n]}$	yes	no	yes	no	yes	yes	yes	no	no	no	yes
$\mathsf{GM}^{\alpha}_{[n]}$, $\alpha > 1$	yes	yes	yes	yes	yes	yes	yes	yes	yes	yes	yes
$\mathsf{GM}^{\alpha}_{[n]}$, $\alpha < 1$	yes	no	yes	yes	yes	yes	yes	yes	no	no	yes

3.2 *Generalized Means*

Let us consider a very significant type of n-ary aggregation function called *generalized mean* or *generalized average*, with some of its special cases, see [8, 10, 21, 22]. n-ary generalized mean $\mathsf{GM}^{\alpha}_{[n]} : [0, 1]^n \to [0, 1]$ is for $\alpha \neq 0$ defined with

$$\mathsf{GM}^{\alpha}_{[n]}(a_1, \ldots, a_n) = \left(\frac{a_1^{\alpha} + \cdots + a_n^{\alpha}}{n}\right)^{\frac{1}{\alpha}}, \quad \text{for } \alpha > 0$$

and

$$\mathsf{GM}^{\alpha}_{[n]}(a_1, \ldots, a_n) = \begin{cases} \left(\frac{a_1^{\alpha} + \cdots + a_n^{\alpha}}{n}\right)^{\frac{1}{\alpha}} & , \ \forall i,\ a_i > 0 \\ 0 & , \ \exists i,\ a_i = 0 \end{cases}, \quad \text{for } \alpha < 0.$$

For some special cases of α we obtain the well known functions as arithmetic mean, geometric mean and harmonic mean.

[MA] For $\alpha = 1$ function $\mathsf{GM}^{\alpha}_{[n]}$ comes down to n-ary *arithmetic mean*

$$\mathsf{MA}_{[n]}(a_1, \ldots, a_n) = \frac{1}{n}(a_1 + \cdots + a_n).$$

[MH] For $\alpha = -1$ function $\mathsf{GM}^{\alpha}_{[n]}$ comes down to n-ary *harmonic mean*

$$\mathsf{MH}_{[n]}(a_1, \ldots, a_n) = \begin{cases} \frac{n}{\frac{1}{a_1} + \cdots + \frac{1}{a_n}} & , \ \forall i,\ a_i \neq 0 \\ 0 & , \ \exists i,\ a_i = 0 \end{cases}.$$

[MG] For $\alpha \to 0$ function $\mathsf{GM}^{\alpha}_{[n]}$ converges to *geometric mean*

$$\mathsf{MG}_{[n]}(a_1, \ldots, a_n) = (a_1 \cdots a_n)^{\frac{1}{n}}.$$

Properties from Definition 5 of means are examined in [31, 34]. They are summarized in Table 2 together with their applicability in Theorem 1.

Table 3 Properties and applicability of OWA

	(a3)	(a4)	(a5)	(a6)	(a7)	(ad1)	(ad2)	(ad3)	(ad4)	(ad5)	(ad6)
$\mathsf{OWA}_{[n]}^{\omega_{[n]}}$	yes	no	yes	no[2]	yes	yes	yes	no[2]	no	no	yes

YES if $\omega_{n,i} > 0$ for all $i \in \{1, \ldots, n\}$

3.3 *Ordered Weighted Average*

Let us consider n-ary aggregation function called *ordered weighted average* introduced by Ronald Yager, see [39], which has an important role in computer science.

Definition 6 For the fixed $n \in \mathbb{N}$ and sequence of real numbers $\omega_{[n]} = \big(\omega_{n,1}, \ldots, \omega_{n,n}\big)$ such that $\omega_{n,i} \geqslant 0, i \in \{1, \ldots, n\}$, and $\sum_{i=1}^{n} \omega_{n,i} = 1$, ordered weighted average aggregation function $\mathsf{OWA}_{[n]}^{\omega_{[n]}} : [0, 1]^n \to [0, 1]$ is defined by

$$\mathsf{OWA}_{[n]}^{\omega_{[n]}}(a_1, \ldots, a_n) = \omega_{n,1}a_1^* + \cdots + \omega_{n,n}a_n^*,$$

where $(a_1^*, \ldots, a_n^*)$ is a non-decreasing[1] permutation of input $(a_1, \ldots, a_n) \in [0, 1]^n$. Properties from Definition 5 of $\mathsf{OWA}_{[n]}^{\omega_{[n]}}$ function and applicability in Theorem 1 are examined in [30], and they are summarized in Table 3. In [30] it is experimentally confirmed that the application of $\mathsf{OWA}_{[n]}^{\omega_{[n]}}$ in the construction of distance functions produces good results in image segmentation.

3.4 *Weighted Arithmetic Mean of Powers*

In this subsection we present the properties and applicability of the aggregation function called *weighted arithmetic mean of powers* which is introduced and applied for image segmentation in [15].

Definition 7 For coefficients $\omega_{[n]} = \big(\omega_{n,1}, \ldots, \omega_{n,n}\big)$ and $\lambda_{[n]} = \big(\lambda_{n,1}, \ldots, \lambda_{n,n}\big)$ such that

(i) $\omega_{n,i} \geqslant 0, i \in \{1, \ldots, n\}$ and $\sum_{i=1}^{n} \omega_{n,i} = 1$,

(ii) $\lambda_{n,i} > 0, i \in \{1, \ldots, n\}$,

weighted arithmetic mean of powers $\mathsf{WAMP}^{\omega_{[n]},\lambda_{[n]}} : \bigcup_{n=1}^{\infty}[0, 1]^n \to [0, 1]$ is an n-ary aggregation function defined by

[1]Some authors regard it as non-increasing.

$$\mathsf{WAMP}_{[n]}^{\omega_{[n]},\lambda_{[n]}}(a_1,\ldots,a_n)=\sum_{i=1}^{n}\omega_{n,i}a_i^{\lambda_{n,i}}$$

for all $(a_1,\ldots,a_n)\in[0,1]^n$.

Remark 5 In applications of the $\mathsf{WAMP}_{[n]}^{\omega_{[n]},\lambda_{[n]}}$ function, weight coefficients $\omega_{n,i}$ control the participation of the i-th argument, and the degree coefficients $\lambda_{n,i}$ define the growth tempo of the influence of the i-th component.

The properties from Definition 5 of function $\mathsf{WAMP}_{[n]}^{\omega_{[n]},\lambda_{[n]}}$ and its applicability in Theorem 1 are examined in [15]. Aggregation function $\mathsf{WAMP}_{[n]}^{\omega_{[n]},\lambda_{[n]}}$ generally has properties (a3) and (a7), doesn't have property (a5), and the properties (a4) and (a6) satisfy for some types of coefficients $\omega_{[n]}$ and $\lambda_{[n]}$.

(a4) Aggregation function $\mathsf{WAMP}_{[n]}^{\omega_{[n]},\lambda_{[n]}}$ has the property (a4) (subadditivity) for the coefficients $\lambda_{[n]}$ satisfying $\forall i\in\{1,\ldots,n\},\ \lambda_{n,i}\leqslant 1$.

(a6) Aggregation function $\mathsf{WAMP}_{[n]}^{\omega_{[n]},\lambda_{[n]}}$ has the property (a6) for the coefficients $\omega_{[n]}$ satisfying $\forall i\in\{1,\ldots,n\},\ \omega_{n,i}>0$.

Consequently, in parts of Theorem 1 the function $\mathsf{WAMP}_{[n]}^{\omega_{[n]},\lambda_{[n]}}$ is generally applicable in (ad1), (ad2) and (ad6), not applicable in (ad5), and in (ad3) and (ad4) is applicable for some types of coefficients $\omega_{[n]}$ and $\lambda_{[n]}$.

(ad3) Aggregation function $\mathsf{WAMP}_{[n]}^{\omega_{[n]},\lambda_{[n]}}$ is applicable in part (ad3) of Theorem 1 for the coefficients $\omega_{[n]}$ satisfying $\forall i\in\{1,\ldots,n\},\ \omega_{n,i}>0$.

(ad4) Aggregation function $\mathsf{WAMP}_{[n]}^{\omega_{[n]},\lambda_{[n]}}$ is applicable in part (ad4) of Theorem 1 for the coefficients $\lambda_{[n]}$ satisfying $\forall i\in\{1,\ldots,n\},\ \lambda_{n,i}\leqslant 1$.

3.5 *Product of Powers*

One more n-ary aggregation function suitable for application in image segmentation, called *powers product* is introduced and examined in [15]. This function is applied in image segmentation in [33], where excellent segmentation results are obtained.

Definition 8 For the arbitrary sequence $\lambda_{[n]}=\left(\lambda_{n,1},\ldots,\lambda_{n,n}\right)$ of positive coefficients $\lambda_{n,i}>0,\ i\in\{1,\ldots,n\}$, function $\prod_{[n]}^{\lambda_{[n]}}:[0,1]^n\to[0,1]$ that is defined by

$$\prod\nolimits_{[n]}^{\lambda_{[n]}}(a_1,\ldots,a_n)=\prod_{i=1}^{n}a_i^{\lambda_{n,i}}$$

Table 4 Properties and applicability of powers product functions

	(a3)	(a4)	(a5)	(a6)	(a7)	(ad1)	(ad2)	(ad3)	(ad4)	(ad5)	(ad6)
$\prod_{[n]}^{\lambda_{[n]}}$	yes	no	no	no	yes	yes	yes	no	no	no	yes
$\prod_{[n]}^{\lvert\lambda_{[n]}\rvert}$	yes	no	yes	no	yes	yes	yes	no	no	no	yes
$\prod_{[n]}$	yes	no	no	no	yes	yes	yes	no	no	no	yes

for $(a_1, \ldots, a_n) \in [0, 1]^n$ is a n-ary aggregation function called *powers product*. Specialy, for the sequence $\lambda_{[n]}$ of positive coefficients with the property $\sum_{i=1}^{n} \lambda_{n,i} = 1$, the function $\prod_{[n]}^{\lambda_{[n]}}$ is called *1-powers product* and it is denoted by $\prod_{[n]}^{\lvert\lambda_{[n]}\rvert}$.

Remark 6 For the sequence of coefficients $\lambda_{[n]}$ that satisfy $\lambda_{n,i} = 1, \forall i \in \{1, \ldots, n\}$, function $\prod_{[n]}^{\lambda_{[n]}}$ reduces on n-ary product aggregation function $\prod$.

Properties from Definition 5 of powers product aggregation functions $\prod_{[n]}^{\lambda_{[n]}}$, $\prod_{[n]}^{\lvert\lambda_{[n]}\rvert}$ and $\prod_{[n]}$ and their applicability in Theorem 1 are examined in [15] and summarized in Table 4.

4 Image Segmentation

In our previous papers, the constructed distance functions are applied in image segmentation, but various other applications are possible.

4.1 Pixel Descriptors

Beside pixel color or color components, other pixel descriptors also significantly contribute in image segmentation, as well as in other image processing tasks such as image denoising, image objects texture recognition, etc., see [27, 29, 36].

4.1.1 Pixel Saturation

Pixel saturation is a simple pixel descriptor. It is not defined in a unique way in literature. In [15], for color image segmentation is defined in the following way. Let the image be produced using the RGB color technique, i.e., let $r_{i,j}$, $g_{i,j}$ and $b_{i,j}$ represent the red, green and blue color components of pixel $p_{i,j}$ respectively. Saturation $\mathsf{S}(p_{i,j})$ of the pixel $p_{i,j}$ is defined as

$$\mathsf{S}(p_{i,j}) = \frac{1}{255\sqrt{3}}\sqrt{r_{i,j}^2 + g_{i,j}^2 + b_{i,j}^2} \in [0, 1].$$

4.1.2 The Average Color of Neighbouring Pixels

In [30], a one pixel 8-nearest neighbors oriented descriptor is introduced and applied in black and white image segmentation. For any color component c, denoted by $c_{i,j} \in \{0, \ldots, 255\}$ color c level of the observed pixel $p_{i,j}$ in i-th row and j-th column in image matrix representation $P = \left[p_{i,j}\right]_{n \times m}$. The mentioned pixel $p_{i,j}$ descriptor is called *neighbors average c color* and is denoted by $\mathsf{NAC}^{(c)}(p_{i,j})$. It represents the average c color level of 8-nearest neighbors for $p_{i,j}$ in the interior of the image, or an average c color level of 3-nearest neighbors for $p_{i,j}$ in the corner of the image, or an average c color level of 5-nearest neighbors for $p_{i,j}$ on the edge of the image. Observe that $\mathsf{NAC}^{(c)}(p_{i,j}) \in [0, 255]$. For example, for pixel $p_{i,j}$ in interior of image, where $1 < i < n$ and $1 < j < m$, is

$$\mathsf{NAC}^{(c)}\left(p_{i,j}\right) = \frac{1}{8}\left(\sum_{k=j-1}^{j+1} c_{i-1,k} + c_{i,j-1} + c_{i,j+1} + \sum_{k=j-1}^{j+1} c_{i+1,k}\right),$$

while for example

$$\mathsf{NAC}^{(c)}(p_{1,1}) = \frac{1}{3}\left(c_{1,2} + c_{2,1} + c_{2,2}\right),$$

and

$$\mathsf{NAC}^{(c)}(p_{1,2}) = \frac{1}{5}\left(c_{1,1} + c_{1,3} + c_{2,1} + c_{2,2} + c_{2,3}\right),$$

for the pixel $p_{1,1}$ in the upper-left corner of the image, i.e. for the pixel $p_{1,2}$ on the upper edge of the image.

For multi-color component images with color components $c_1, \ldots, c_k$, it is useful to join particular descriptors $\mathsf{NAC}^{(c_i)}$, $i \in \{1, \ldots, k\}$, in one single descriptor. Aggregation functions are a natural choice of model for the aforementioned joining. In [15], for application in RGB image segmentation, as joining aggregation function are simple arithmetic mean of $\mathsf{NAC}^{(\mathsf{R})}$, $\mathsf{NAC}^{(\mathsf{G})}$ and $\mathsf{NAC}^{(\mathsf{B})}$ used, i.e., as single NAC descriptor is used

$$\mathsf{NAC}(p_{i,j}) = \frac{1}{3}\left(\mathsf{NAC}^{(\mathsf{R})}(p_{i,j}) + \mathsf{NAC}^{(\mathsf{G})}(p_{i,j}) + \mathsf{NAC}^{(\mathsf{B})}(p_{i,j})\right).$$

The application of NAC in image segmentation is presented in [15, 30], where it is experimentally proved that the use of NAC contributes to good segmentation results.

4.1.3 α-Similarity Indicator Counter

In [15], another pixel 8-nearest neighbors oriented descriptor is introduced and applied in color image segmentation with good experimental results.

For the observed pixel $p_{i,j}$ and color component c, and for the arbitrary threshold level $\alpha \in \{0, \ldots, 255\}$, we first define a function called *α-similarity indicator* denoted by $I^{(c)}_{p_{i,j};\alpha}$, which values 0 and 1 indicates "α-similarity" between the observed pixel $p_{i,j}$ and each of its nearest neighbor pixels. Let $c_{i,j}$ be the c-color level of pixel $p_{i,j}$, and let $c_{i,j,m}$, $m \in \{1, \ldots, 8\}$, be the same c-color level of the 8 nearest neighbor pixels for the pixel $p_{i,j}$ (if they exist). Function $I^{(c)}_{p_{i,j};\alpha} : \{n_{i,j,1}, \ldots, n_{i,j,8}\} \to \{0, 1\}$ is defined by

$$I^{(c)}_{p_{i,j};\alpha}(m) = \begin{cases} 1 \,, & \left|c_{i,j} - c_{i,j,m}\right| \leq \alpha \\ 0 \,, & \left|c_{i,j} - c_{i,j,m}\right| > \alpha \end{cases} \,, \quad m \in \{1, \ldots, 8\}.$$

For the pixel $p_{i,j}$ which is not in the interior of the image and because of that does not have some of the m-th neighbors, we define $I^{(c)}_{p_{i,j};\alpha}(m) = 0$.

Remark 7 For the pixel $p_{i,j}$ and the m-th neighbor pixel, $I^{(c)}_{p_{i,j};\alpha}(m)$ takes a value of 1 when the difference in c color code is less than or equal to the threshold value α and represents a signal of α-similarity in c color level between the pixel $p_{i,j}$ and its m-th neighbor pixel.

For the color component c, using $I^{(c)}_{p_{i,j};\alpha}(m)$ indicators, we define *α-indicator counter* descriptor $\mathsf{IC}^{(c)}_{\alpha}(p_{i,j})$ of pixel $p_{i,j}$ as a number of 8 nearest neighbor pixels that are α-similar with $p_{i,j}$ in the c color component. For the given threshold value $\alpha \in \{0, \ldots, 255\}$, α-indicator counter $\mathsf{IC}^{(c)}_{\alpha}(p_{i,j})$ for the c color component is defined by

$$\mathsf{IC}^{(c)}_{\alpha}(p_{i,j}) = \sum_{m=1}^{8} I^{(c)}_{p_{i,j};\alpha}(m).$$

In the same way as for neighbors average color from Sect. 4.1.2, for multi-color images with color components $c_1, \ldots, c_k$, it is useful to join particular descriptors $\mathsf{IC}^{(c_i)}_{\alpha}$, $i \in \{1, \ldots, k\}$, in one single with some appropriate aggregation function. In [15], the arithmetic mean is used again for the application in RGB image segmentation, i.e., as single IC_{α} descriptor is used

$$\mathsf{IC}_{\alpha}(p_{i,j}) = \frac{1}{3}\left(\mathsf{IC}^{(\mathsf{R})}_{\alpha}(p_{i,j}) + \mathsf{IC}^{(\mathsf{G})}_{\alpha}(p_{i,j}) + \mathsf{IC}^{(\mathsf{B})}_{\alpha}(p_{i,j})\right).$$

The pixel descriptor IC_{α} is applied in [15, 33] with good results in image segmentation, and we believe that application of IC_{α} descriptor can produce good results in other image processing tasks such as image denoising, image objects edge detection etc.

4.2 FCM Algorithm

In our previous work, distance functions constructed using the methodology described in Sect. 3 were applied to image segmentation using the FCM clustering algorithm. The goal of image segmentation is to divide image pixels into regions which are meaningful and compact in some sense. In image segmentation (and data clustering generally), distance functions represent a criterion by which pixels are classified into segments, see [16, 34].

The fuzzy c-means clustering algorithm (FCM) is a fuzzified version of k-means clustering algorithm, see [3, 4, 34]. Besides the image, the input parameters of the FCM algorithm are the following.

- User predefined number c of clusters that are going to be obtained by segmentation.
- Weight coefficient m, which affects the quality and speed of segmentation, see [3].
- Distance function $d : P^2 \to [0, 1]$, where P is the set of pixels, and each pixel is represented as n-tuple of its color component and other descriptors which are significant from the aspect of segmentation criteria. For pixels $p_1, p_2 \in P$, value $d(p_1, p_2) \in [0, 1]$ represents a measure of the difference or dissimilarity between them.

Beside the segmented image, the following output parameters of segmentation are the main indicators of the quality of the obtained segmentation and the speed of the FCM algorithm.

- PI—performance index, the main measure of the compactness of the clustered data. The lower value of this coefficient means a more compact clusters. Beside the PI, there are other numerical quality indicators for the obtained segmentation.
- IT—the number of performed iterations in the FCM algorithm, as a measure of the speed of producing the segmented image.

5 Experimental Results

In this section we present selected results of image segmentation using the FCM clustering algorithm with distance functions constructed with the methodology presented in Sect. 3. In all of the presented experiments, the number of clusters $c = 4$ and the weight coefficient $m = 2.0$ are selected for the FCM algorithm.

5.1 Image Segmentation Using OWA Constructed Distance Function

In [30] a gray-scale X-ray image segmentation using the distance function constructed by applying an OWA aggregation function is presented (see Sect. 3.3).

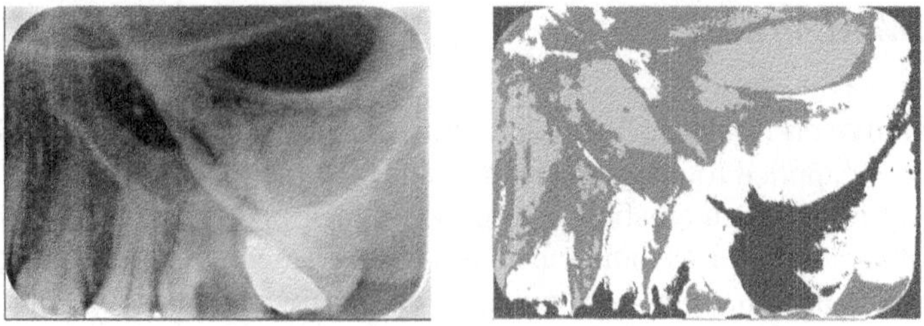

(a) Original X-ray image. (b) $d_{\omega_{0.8}}$ segmented image.

Fig. 1 Segmentations by using OWA aggregated distance functions

As the initial distance function, a normalized L_1 metric $d_1 : \{0, \ldots, 255\} \to [0, 1]$, $d_1(x, y) = \frac{1}{255} \cdot |x - y|$ is used, and applied on gray-level code $c_{i,j} \in \{1, \ldots, 255\}$ of the pixel $p_{i,j}$ and on $\mathsf{NAC}(p_{i,j}) \in [0, 255]$ descriptor of pixel $p_{i,j}$ (see Sect. 4.1.2). On the set $X = \{1, \ldots, 255\} \times [0, 255]$ of pixels $p_{i,j} = (c_{i,j}, \mathsf{NAC}(p_{i,j}))$, for some values of $\lambda \in \,]0, 1[$ and $\omega_{[2]}^{\lambda} = (\lambda, 1 - \lambda)$, using aggregation function $\mathsf{OWA}_{[2]}^{\omega_{[2]}^{\lambda}}$, a new distance function $d_\lambda : X^2 \to [0, 1]$ is constructed as

$$\begin{aligned} d_\lambda(p_{i,j}, p_{k,n}) &= \mathsf{OWA}_{[2]}^{\omega_{[2]}^{\lambda}} \left(L_1(c_{i,j}, c_{k,n}), L_1(\mathsf{NAC}(p_{i,j}), \mathsf{NAC}(p_{k,n}))\right) \\ &= \frac{1}{255} \left(\lambda |c_{i,j} - c_{k,n}| + (1 - \lambda) |\mathsf{NAC}(p_{i,j}) - \mathsf{NAC}(p_{k,n})|\right). \end{aligned}$$

Distance function d_λ is applied in the FCM algorithm for image segmentation. For example, for $\lambda = 0.8$, a performance index value of $\mathsf{PI} = 902.5$ is obtained, which is better compared with the performance index value $\mathsf{PI} = 1007.5$ obtained using the L_1 metric on the set of gray-level codes of pixels. The original image and the image segmented using $d_{0.8}$ distance function are shown in Fig. 1.

5.2 *Image Segmentation Using* **WAMP** *Constructed Distance Function*

In [15] is presented a color image segmentation using a distance function constructed applying WAMP aggregation function (see Sect. 3.4). The image is produced using the RGB color technique.

Other than the red $r_{i,j}$, green $g_{i,j}$ and blue $b_{i,j}$ color levels of the pixel $p_{i,j}$, the saturation $\mathsf{S}(p_{i,j})$ (see Sect. 4.1.1) and α a-similarity indicator counter $\mathsf{IC}_\alpha(p_{i,j})$ (see Sect. 4.1.3) are used as pixel $p_{i,j}$ features for segmentation. Therefore, the pixel $p_{i,j}$ is observed as 5-tuple $p_{i,j} = (r_{i,j}, g_{i,j}, b_{i,j}, \mathsf{S}(p_{i,j}), \mathsf{IC}_\alpha(p_{i,j}))$ in 5-dimensional space $X = \{0, \ldots, 255\}^3 \times [0, 1]^2$ of image pixels.

As the each of the initial distance function for each color component and pixel descriptor is normalized L_1 metric used, i.e.,

(a) Original image. (b) d_E segmentation. (c) $d_\alpha^{\omega_{[5]},\lambda_{[5]}}$ segmentation.

Fig. 2 Segmentations by using WAMP aggregated distance functions

$$d_1\left(p_{i,j}, p_{k,n}\right) = \frac{1}{255}\left|c_{i,j} - c_{k,n}\right|$$

for each of the color component levels $c \in \{r, g, b\}$, and

$$d_2\left(p_{i,j}, p_{k,n}\right) = \left|\mathsf{D}(p_{i,j}) - \mathsf{D}(p_{k,n})\right|$$

for each of the utilized pixel descriptors $\mathsf{D} \in \{\mathsf{S}, \mathsf{IC}_\alpha\}$ and fixed $\alpha \in \{0, \ldots, 255\}$. Further, for sequences of non-negative coefficients

$$\omega_{[5]} = (\omega_{5,1}, \omega_{5,2}, \omega_{5,3}, \omega_{5,4}, \omega_{5,5})$$

and positive coefficients

$$\lambda_{[5]} = (\lambda_{5,1}, \lambda_{5,2}, \lambda_{5,3}, \lambda_{5,4}, \lambda_{5,5}),$$

applying the WAMP aggregation function from Sect. 3.4, a new distance function $d_\alpha^{\omega_{[5]},\lambda_{[5]}} : X^2 \to [0, 1]$ is constructed by

$$\begin{aligned} d_\alpha^{\omega_{[5]},\lambda_{[5]}}\left(p_{i,j}, p_{k,n}\right) &= \omega_{5,1} d_1^{\lambda_{5,1}}(r_{i,j}, r_{k,n}) + \omega_{5,2} d_1^{\lambda_{5,2}}(g_{i,j}, g_{k,n}) \\ &+ \omega_{5,3} d_1^{\lambda_{5,3}}(r_{i,j}, r_{k,n}) + \omega_{5,4} d_2^{\lambda_{5,4}}(\mathsf{S}(p_{i,j}), \mathsf{S}(p_{k,n})) \\ &+ \omega_{5,5} d_2^{\lambda_{5,5}}(\mathsf{IC}_\alpha(p_{i,j}), \mathsf{IC}_\alpha(p_{k,n})) \end{aligned}$$

for pixels $p_{i,j}, p_{k,n} \in X$. In [15] there are three RGB color images segmented using $d_\alpha^{\omega_{[5]},\lambda_{[5]}}$ distance function in the FCM algorithm. For appropriate selections of $\omega_{[5]}$, $\lambda_{[5]}$ and α good results are obtained in comparison with the use of the Euclidean distance function (metric) d_E applied to pixel color components without using pixel descriptors.

For illustration purposes in this article we present the segmentation of another image shown in Fig. 2a, used from public database https://www2.eecs.berkeley.edu/Research/Projects/CS/vision/bsds/, *Berkeley University of California, Department of Electrical Engineering and Computer Sciences, Computer Vision Group, segmentation dataset—training images*, see [25]. Image segmented using normalized Euclidean d_E metric is shown in Fig. 2b. In Fig. 2c an image segmented using the distance function $d_\alpha^{\omega_{[5]},\lambda_{[5]}}$ with parameters $\omega_{[5]} = (0.25, 0.25, 0.25, 0.0, 0.25)$,

Table 5 Comparative output parameters for WAMP segmentation

	PI	IT
d_E	467.6547	22
$d_\alpha^{\omega_{[5]},\lambda_{[5]}}$	7.7195	21

$\lambda_{[5]} = (2.0, 2.0, 2.0, 2.0, 2.0)$ and $\alpha = 20$ is shown. Output parameters PI and IT of the segmentation with distance function $d_\alpha^{\omega_{[5]},\lambda_{[5]}}$ and with Euclidean distance function d_E are presented in Table 5.

The results in Table 5 show that the application of distance function $d_\alpha^{\omega_{[5]},\lambda_{[5]}}$ and IC_α descriptor produce a much better PI indicator of segmentation quality compared with the result from the Euclidean distance function d_E.

5.3 *Image Segmentation Using Distance Function Constructed with Product-Type Aggregation Function*

In [33], a black and white image segmentation is presented using a distance function constructed by applying the powers product aggregation function from Sect. 3.5.

The X-ray image shown in Fig. 1a is segmented. Beside pixel $p_{i,j}$ gray level code $c_{i,j}$, descriptors $\mathsf{NAC}(p_{i,j})$ and $\mathsf{IC}_\alpha(p_{i,j})$ from Sect. 4.1 are again used as pixel characteristics. Therefore, the pixel $p_{i,j}$ is observed as 3-tuple $p_{i,j} = \big(c_{i,j}, \mathsf{NAC}(p_{i,j}), \mathsf{IC}_\alpha(p_{i,j})\big)$ from the space $X = \{0, \dots, 255\} \times [0, 1]^2$ of pixels. As initial distance functions d_1 and d_2 from Sect. 5.2 are used, and a new distance function d is constructed by applying powers product aggregation function on d_1 and d_2. With coefficients $\lambda_{[3]} = \big(\lambda_{3,1}, \lambda_{3,2}, \lambda_{3,3}\big)$, distance function $d_\alpha^{\lambda_{[3]}} : X^2 \to [0, 1]$ is constructed by $d_\alpha^{\lambda_{[3]}}(p_{i,j}, p_{k,n}) = d_1^{\lambda_{3,1}}(c_{i,j}, c_{k,n}) \cdot d_2^{\lambda_{3,2}}(\mathsf{NAC}(p_{i,j}), \mathsf{NAC}(p_{k,n})) \cdot d_2^{\lambda_{3,3}}(\mathsf{IC}_\alpha(p_{i,j}), \mathsf{IC}_\alpha(p_{k,n}))$.

Application of function $d_\alpha^{\lambda_{[3]}}$ in the segmentation of the image shown in Fig. 1a gave excellent results. For example, for coefficients $\lambda_{[3]}(A) = (2.0, 2.0, 2.0)$ and $\lambda_{[3]}(B) = (0.5, 2.0, 2.0)$ with $\alpha = 20$ selected, output parameters PI and IT of segmentation with $d_\alpha^{\lambda_{[3]}(A)}$ and $d_\alpha^{\lambda_{[3]}(B)}$ distance function and with Euclidean distance function d_E are presented in Table 6. Image 1a segmented with usage of $d_\alpha^{\lambda_{[3]}(A)}$ and $d_\alpha^{\lambda_{[3]}(B)}$ distance functions are shown in Fig. 3.

Table 6 Comparative output parameters for powers product segmentation

	PI	IT
d_E	1007.5	24
$d_\alpha^{\lambda[3](A)}$	0.000012702	35
$d_\alpha^{\lambda[3](B)}$	0.0021	17

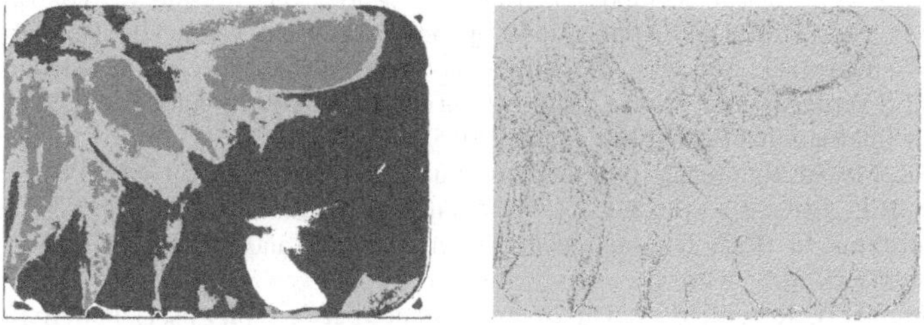

(a) $d_\alpha^{\lambda_{[3]}(A)}$ segmented image. (b) $d_\alpha^{\lambda_{[3]}(B)}$ segmented image.

Fig. 3 Segmentations by using powers product aggregated distance functions

6 Conclusions and Additional Remarks

The research presented in this article shows that the constructions of new distance functions by applying an aggregation function on a sequence of initial distance functions produce good results in image segmentation. Beside the construction of distance functions, this method can also be applied on the construction of other measure-type functions such as fuzzy metrics, fuzzy measures etc., see [32]. The presented pixel descriptors significantly contribute to image segmentation quality, and can also be applied in other image processing tasks.

Acknowledgements The first author was supported by the Science Fund of the Republic of Serbia, #Grant No. 6524105, AI-ATLAS. The second and third author has been supported by the Ministry of Education, Science and Technological Development through the project no. 451-03-68/2020-14/200156: "Innovative scientific and artistic research from the FTS (activity) domain".

References

1. Ball, G.H., Hall, D.J.: ISODATA, a novel method of data analysis and pattern classification. Stanford Research Institute, Menlo Park, CA (1965)
2. Bezdek, J.C.: Pattern recognition with fuzzy objective function algorithms. Kluwer Academic Publishers, Norwell (1981)
3. Bezdek, J.C., Ehrlich, R., Full, W.: FCM: The fuzzy c-means clustering algorithm. Comput. Geosci. **10**, 191–203 (1984)
4. Bezdek, J.C., Keller, J., Krisnapuram, R., Pal, N.: Fuzzy models and algorithms for pattern recognition and image processing. Springer, Berlin, US (1999)

5. Bloch, I.: Fuzzy geodesic distance in images. In: Martin, T.P., Ralescu, A.L. (eds.) Fuzzy Logic in Artificial Intelligence Towards Intelligent Systems, pp. 153–166. Springer, Berlin, Heidelberg (1997)
6. Bloch, I.: On fuzzy distances and their use in image processing under imprecision. Pattern Recogn. **32**, 1873–1895 (1999)
7. Bouchon-Meunier, B., Kacprzyk, J.: Aggregation and fusion of imperfect information. Physica-Verlag, Heidelberg (1998)
8. Calvo, T., Kolesárová, A., Komorníková, M., Mesiar, R.: Aggregation operators: properties, classes and construction methods. In: Calvo, T., Mayor, G., Mesiar, R. (eds.) Aggregation Operators: New Trends and Applications, pp. 3–104. Physica-Verlag HD, Heidelberg (2002)
9. Calvo, T., Mayor, G., Suñer, J.: Globally monotone extended aggregation functions. In: Magdalena, L., Verdegay, J.L., Esteva, D. (eds.) Enric Trillas: A Passion for Fuzzy Sets, pp. 49–66. Springer International Publishing, Cham (2015)
10. Calvo, T., Mesiar, R.: Generalized medians. Fuzzy Sets Syst. **124**, 59–64 (2001)
11. Cannon, R.L., Dave, J.V., Bezdek, J.C.: Efficient implementation of the fuzzy c-means clustering algorithms. In: IEEE Transactions on Pattern Analysis and Machine Intelligence PAMI-8, pp. 248–255 (1986)
12. Chaudhuri, B.B., Rosenfeld, A.: On a metric distance between fuzzy sets. Pattern Recogn. Lett. **17**, 1157–1160 (1996)
13. Ćurić, V., Lindblad, L., Sladoje, N., Sarve, H., Borgefors, G.: A new set distance and its application to shape registration. Pattern Anal. Appl. **17**, 141–152 (2014)
14. Delić, M., Lindblad, J., Sladoje, N.: αLBP—a novel member of the local binary pattern family based on α-cutting. In: Proceedings of the 9th International Symposium on Image and Signal Processing and Analysis ISPA 2015, pp. 13–18 (2015)
15. Delić, M., Nedović, Lj., Pap, E.: Extended power-based aggregation of distance functions and application in image segmentation. Inf. Sci. **494**, 155–173 (2019)
16. Deza, M.M., Deza, E.: Encyclopedia of distances. Springer, Berlin, Heidelberg (2012)
17. Dubois, D., Prade, H.: Aggregation of possibility measures. In: Kacprzyk, J., Fedrizzi, M. (eds.) Multiperson Decision-Making Models Using Fuzzy Sets and Possibility Theory, pp. 55–63. Springer, Netherlands (1990)
18. Dubois, D., Prade, H.: On the use of aggregation operations in information fusion processes. Fuzzy Sets Syst. **142**, 143–161 (2004)
19. Dubuisson, M.P., Jain, A.K.: A modified Hausdorff distance for object matching. In: Pattern recognition, 1994. Vol. 1-conference A: Computer vision & image processing IAPR, pp. 566–568. IEEE (1994)
20. Dunn, J.C.: A fuzzy relative of the ISODATA process and its use in detecting compact well-separated clusters. Cybern. Syst. **3**, 32–57 (1973)
21. Grabisch, M., Marichal, J.L., Mesiar, R., Pap, E.: Aggregation Functions. Cambridge University Press, Cambridge (2009)
22. Grabisch, M., Marichal, J.L., Mesiar, R., Pap, E.: Aggregation functions: means. Inf. Sci. **181**, 1–22 (2011)
23. Grabisch, M., Marichal, J.L., Mesiar, R., Pap, E.: Aggregation functions: construction methods, conjunctive, disjunctive and mixed classes. Inf. Sci. **181**, 23–43 (2011)
24. Klement, E.P., Mesiar, R., Pap, E.: Triangular Norms. Springer Science & Business Media, Dordrecht (2013)
25. Martin, D., Fowlkes, C., Tal, D., Malik, J.: A database of human segmented natural images and its application to evaluating segmentation algorithms and measuring ecological statistics. In: Proceedings 8th IEEE International Conference on Computer Vision. ICCV 2001, pp. 416–423 (2001)
26. Mesiar, R., Kolesárová, A., Calvo, T., Komorníková, M.: A review of aggregation functions. In: Bustince, H., Herrera, F., Montero, J. (eds.) Fuzzy Sets and Their Extensions: Representation, Aggregation and Models, pp. 3–104. Springer, Berlin, Heidelberg (2008)
27. Nanni, L., Lumini, A., Brahnam, S.: Survey on LBP based texture descriptors for image classification. Expert Syst. Appl. **39**, 3634–3641 (2012)

28. Nedović, L., Delić, M.: Image segmentation by applying median-aggregated distance functions. In: Proceedings of the 3rd Conference on Mathematics in Engineering: Theory and Applications. META 2018, pp. 1–6 (2018)
29. Nedović, L., Delić, M.: New pixel descriptors based on neighbor similarity. In: Proceedings of the 5th Conference on Mathematics in Engineering: Theory and Applications. META 2020, pp. 1–7 (2020)
30. Nedović, L., Delić, M., Ralević, N.M.: OWA-aggregated distance functions and their application in image segmentation. In: Proceedings of 16th IEEE International Symposium on Intelligent Systems and Informatics. SISY 2018, pp. 311–316 (2018)
31. Nedović, L., Dragić, Đ.: Minimum, maximum and means aggregation of distance functions. In: Proceedings of the 4th Conference on Mathematics in Engineering: Theory and Applications. META 2020, pp. 41–46 (2019)
32. Nedović, L., Pap, E.: Aggregation of sequence of fuzzy measures. Iranian J. Fuzzy Syst. **17**, 39–55 (2020)
33. Nedović, L., Pap, E., Ralević, N.M., Delić, M.: X-ray image segmentation using product-type aggregation of distance functions. In: Proceedings of of XVLI Symposium on Operational Research. SYM-OP-IS 2019, pp. 361–366 (2019)
34. Nedović, L., Ralević, N.M., Pavkov, I.: Aggregated distance functions and their application in image processing. Soft Comput. **22**, 4723–4739 (2018)
35. Pap, E., Delić, M., Nedović, L.: Information fusion by extended power-based aggregation in image segmentation. In: Proceedings of International Scientific Conference on Information Technology and Data Related Research. SINTEZA 2019, pp. 162–167 (2019)
36. Pietikäinen, M., Hadid, A., Zhao, G., Ahonen, T.: Computer Vision Using Local Binary Patterns. Springer, London (2011)
37. Ralević, N.M., Karaklić, D., Pištinjat, N.: Fuzzy metric and its applications in removing the image noise. Soft Comput. **23**, 12049–12061 (2019)
38. Rudas, I.J., Pap, E., Fodor, F.: Information aggregation in intelligent systems: an application oriented approach. Knowl.-Based Syst. **38**, 3–13 (2013)
39. Yager, M.R., Kacprzyk, J.: The Ordered Weighted Averaging Operation: Theory, Methodology and Applications. Springer, New York, U.S. (1997)
40. Zhou, L., Chen, H.: A generalization of the power aggregation operators for linguistic environment and its application in group decision making. Knowl.-Based Syst. **26**, 216–224 (2012)

A Generative Model for the Creation of Large Synthetic Image Datasets Used for Distance Estimation

Nebojša Nešić, Mladen Vidović, Ivan Radosavljević, Aleksandra Mitrović, and Đorđe Obradović

Abstract This paper presents a generative model for the creation of large synthetic image datasets. The model is implemented as a 3D scene, representing an urban environment, in Blender. The images are rendered using the Cycles rendering engine which allows for the creation of high-fidelity data. During the process of image rendering, the data was automatically labeled for the purpose of distance estimation. The process of data labeling was achieved by extracting object metadata from the 3D scene, allowing for the generative model to be reconfigured to generate datasets for other purposes such as classification, object detection, semantic segmentation, etc. The data acquired in the manner described previously was used for the end-to-end training of a convolutional neural network, designed to estimate distance from stereoscopic images. Evaluation of the neural network's performance showed that the generative model presented in this paper is viable for the generation of large image datasets for the training of predictive models, eliminating the need for time-consuming manual data acquisition and labeling.

Keywords Image · Distance estimation · 3D scene · Blender · Modular assets · Neural network

N. Nešić (✉) · M. Vidović · I. Radosavljević · A. Mitrović · Đ. Obradović
Singidunum University, Danijelova 29, Belgrade, Republic of Serbia
e-mail: nnesic@singidunum.ac.rs

M. Vidović
e-mail: mvidovic@singidunum.ac.rs

I. Radosavljević
e-mail: iradosavljevic@singidunum.ac.rs

A. Mitrović
e-mail: amitrovic@singidunum.ac.rs

Đ. Obradović
e-mail: djobradovic@singidunum.ac.rs

E. Pap (ed.), *Artificial Intelligence: Theory and Applications*,
Studies in Computational Intelligence 973,
https://doi.org/10.1007/978-3-030-72711-6_15

1 Introduction

Depth estimation on 2D images has several uses. Some of the most common ones include the reconstruction of a 3D environment, navigation and obstacle avoidance or visual odometry. As these applications use depth data as an input value, the accuracy of the depth estimation is crucial to the functioning of the applications. Based on their training type, models for depth estimation can either be supervised or unsupervised.

Supervised training processes for depth estimation models require labeled data, which is usually acquired by using a device for distance measurement. These devices include cameras and time-of-flight based tools, such as lidar, radar, sonar [2]. Radars and sonars can be used to accurately measure the distance to a single object, but they are not suited for high resolution measurements such as those required for the construction of point clouds. Optical devices such as lidars do not suffer from this problem but their accuracy can be hindered by environmental conditions such as adverse weather [8]. Cameras, as another example of optical devices, can also be used to measure the distance to a single object and to create point clouds. The accuracy of the measurements acquired by cameras is largely dependent on environmental conditions and proper device calibration. However, an additional advantage of cameras, compared to the aforementioned devices, is that they can be used to collect more general information about the environment allowing for the creation of datasets for a wider variety of uses, for example popular datasets such as KITTI [7] and COCO [11]. The process of collecting data for a dataset using any of the previously mentioned devices can be time consuming and error-prone due to the accuracy of collected data being inherently dependent on proper device calibration and the possibility of device malfunction. These drawbacks limit the availability of large datasets containing depth ground truth [12]. Unsupervised distance estimation training processes provide a solution for the limited availability of labeled datasets, since they rely on unlabeled images as training inputs. However, these methods are often less accurate than supervised depth estimation methods.

One solution for the limited availability of large amounts of labeled image data, aside from the usage of unsupervised training processes, is the creation of synthetic datasets [22]. These datasets can be based on existing data or generated without pre-existing data. A common method to generate datasets from existing data is to either heavily augment existing data [4, 9, 18] or to train a generative model that generates new data using sample data [3] or random noise as an input [1, 14]. A significant advantage of this method is that a large amount of data can be generated by using a small sample of real data. However, the quality of the generated data depends heavily on the quality of the sample data. Synthetic datasets created without pre-existing data rely on domain-specific knowledge encoded in a generative model. These generative models are usually manually created, either in the form of a rule-based system, an explicitly parameterized generative model, or by manually modelling the data itself. A drawback of these methods is their reliance on domain-specific knowledge and the time-consuming nature of the manual process for creating generative models. An example of such implementation is shown in [20]. The authors use advanced rendering techniques to create photo-realistic images. Each image is rendered from a unique procedurally generated 3D world scene, allowing them to create a diverse labeled dataset that can be used as an input for training semantic segmentation models. Another example of generating a synthetic dataset for training semantic segmentation models is presented in [15], in which a game engine, Unity, is used to create a virtual world from which the dataset is generated. Additionally this dataset contains depth maps allowing for depth estimation models to be trained.

Most of the work in the field of depth estimation from images can be divided into two categories: monocular and multi-view. Monocular depth estimation relies on acquiring depth data from a single point of view, using a single image capturing device. These methods rely on image properties such as variations in color or contrast as depth cues [13, 19]. Another monocular depth estimation approach relies on using multiple consecutive frames of a video recorded from a single capturing device [21]. Multi-view depth estimation models rely on images taken from multiple viewpoints in order to infer depth values from images. Usually, this involves generating disparity maps from two or more images, with research focusing on the process of disparity map generation. The generated disparity maps are then used, together with camera parameters to estimate the depth [6, 10, 17].

In this paper, we propose a manually created generative model in the form of a 3D scene representing an urban environment. The synthetic data generated using this model is used to train, validate and evaluate a supervised binocular depth estimation model based on a convolutional neural network. The network was trained using an end-to-end learning approach.

In Sect. 2 the generative model, as well as the workflow used for its creation is presented. Section 3 describes the architecture of the convolutional neural network used to evaluate the generated dataset. In Sect. 4, the training process of the neural network is described. Section 5 consists of the results obtained by evaluating the model and a discussion of the results. Section 6 contains a brief summary of the paper and a discussion of potential directions for future research.

2 Scene Creation

In order to generate datasets, an artificial 3D scene, representing an urban environment was created using Blender [5], an open source, free software for 3D modelling. The creation of the assets used in the scene was divided into seven phases. The first phase consisted of deciding on a consistent workflow for the creation of the assets. Due to the scale and nature of the scene, a non-destructive, modular and parametrized workflow was selected. The non-destructive nature of the workflow allowed for the assets to be edited in such a way as to preserve the original form of the asset. This was realized by creating a base shape as the starting point for the making of an asset. Future steps add elements to the asset while ensuring that the base shape remains unchanged. These steps can also be undone at any point, to revert to a previous version of the asset, or the base shape itself. The modularity ensures that the final

Table 1 Modular assets and their variations

Asset name	Asset variations
Window	62
Door	15
Stair	9
Air conditioning	4
Antenna	12
Roof decoration	11

visual characteristics of the assets can easily be altered and also allows for randomization to avoid clear repetition. The parametrization of the steps that build on the original asset lends itself to greater control over the complexity of the final output mesh. This allows for additional render time or resource usage optimization of the final scene. The resulting workflow simplified the process of generating 3D assets of varying complexity and number, eliminating the need for manual modelling of each individual asset.

The second phase consisted of identifying the assets that were going to be utilized in the final scene and creating appropriate containers for each group of assets, which allowed for easier handling of the assets themselves and the simplification of the asset placement in the scene. The grouping of assets into these containers, which are implemented as collections, enables the precise and even distribution of assets throughout the final world scene.

The third phase consisted of the manual creation of individual modular assets. In most cases, these assets represent additions to a base asset, and are not used individually. In the case of the urban environment used in this project, the assets in question and the number of variations of each asset are shown in Table 1. The assets are placed into separate collections, with each collection consisting of variations of an asset. A clear example of this would be the window assets used for the building objects. Each group of buildings has several types of windows, which differ in shape, architecture, as well as their intended positioning on the building itself, i.e., top floor windows or basement windows. After the initial classification was determined, and the base versions of the windows were created, variations of those windows were generated from the base. The variations share the same geometry as the base version; however, the geometry was rearranged to create visual differences. For example, the base version of the window is shut, and has no decorations whatsoever, while the other versions are open to varying degrees, have blinds, flower pots etc. This way, the creation of a large quantity of different objects was simplified, and the level of repetition was reduced while maintaining visual consistency between the objects themselves. Figure 1 shows some of the generated window assets.

The fourth phase consisted of the manual creation of the base shape of assets onto which the modular assets were automatically distributed to create the final output asset. The base shapes were mostly general, simple shapes, such as cuboids. Details

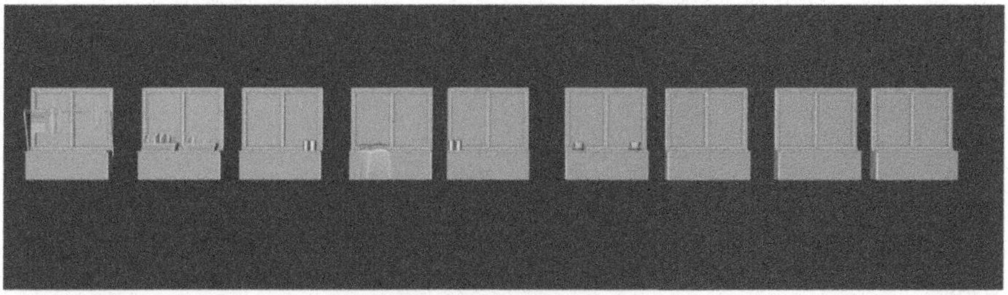

(a) Untextured window assets.

(b) Textured window assets.

Fig. 1 Example of window assets before and after applying a texture

and complexity were added by attaching modular assets, described in the previous phase, to the base shape. The simplicity of the base shapes also greatly reduced the time required to set up the groundwork for the final asset, while also minimizing the required manual input. In some cases, assets required details that did not warrant the creation of modular assets. In those cases, any amount of detail can be added to the base shape. These modifications however do not need to be predetermined. It is possible to return to the base shape at the end of creating the asset and alter it in some way, where the alterations will echo onto all of the modifications that came after. This is possible due to the non-destructive nature of the workflow.

The fifth phase consisted of using a particle system to distribute the small modular assets onto the base shapes. Certain vertices of the base assets were marked as distribution points. These vertex groups were used in the particle system as a way of limiting the areas where the system could place the modular assets. As the assets were grouped into collections, entire collections were used as distribution assets. Blender allows for multiple particle systems to be active on one object. It also allows for the usage of different groups and distribution points for each particle system. The particle system randomly selected objects from a collection to distribute onto the base asset. The selected distribution asset is called an instancer. Every asset has its respective origin point. The particle system uses the origin of the modular assets and projects one modular asset from each distribution point where the origin and the distribution point share the same location. Altering the position of the modular asset on the base shape can be done in two ways, either by moving the distribution point on the base mesh or moving the object's geometry respective to its own origin. The usage of a

Table 2 Texture variations

Texture name	Number of unique textures
Asphalt	5
Brick	16
Wood	20
Plaster	24
Glass	3
Marble	3
Metal	15
Paint	4
Plastic	6
Tiles	4
Ground	3
Foliage	7

particle system is more memory efficient than the alternative approach of copying an asset several times and manually placing it at designated points, since instances are not stored in memory, whereas copies of an object would be. This approach also simplified the process of editing assets after their distribution, as any changes made to the instancer object are automatically applied to the instances. The number of distributed objects had to be equal to or lower than the number of distribution points chosen in order to avoid overlap.

The sixth phase consisted of applying materials to the assets and finalizing the asset's appearance. Materials were created using PBR textures. PBR stands for physically based rendering, a texture workflow that enables a higher level of detail while maintaining a lower poly count for assets [16]. The texture maps used in the materials were: albedo, metalness, roughness, ambient occlusion, specular, bump and normal. The specific material properties such as color, reflectivity and surface imperfections were parametrized to enable the alteration of their influence and values. To avoid repeating patterns of textures as well as duplicates of the applied textures, a random component is added into the texture setup. The randomness is controlled by a unique object ID parameter. Each object or object instance will have a different texture applied. Table 2 shows each texture type used in the scene, as well as the number of unique texture variations of each texture type.

The seventh and final phase of creating the assets was the conversion of the base asset and several particle systems into one final asset to be placed into the world scene. A duplicate of the base asset with the active particle systems was created, to ensure that the workflow remained non-destructive. The original base asset served as a generator for different output assets of the same type. By creating the duplicate, the original asset was kept intact, whilst the particle system conversion was applied to the duplicated asset, thus creating the finalized asset. This process can be repeated indefinitely. Each iteration generates a unique asset that differs from the outputs of the previous iterations in their particle placement and selection.

The creation of every asset used in the scene followed this workflow. The differing complexity of the objects dictated the amount of particle systems and the level of modularity used. Increasing or decreasing the complexity of the assets is facilitated by the workflow itself, which in turn means the final scene output is scalable. Every type of asset had a differing number of parts that the particle system could place. For example, a building had doors, windows, lights, antennas, gutters, stairs etc. On the other hand, a street light only had a number of globes and the shape of the globe itself. However, this is mostly manually controllable, meaning that more complexity could be added to the street lights, if necessary. For example, the placement of more of their parts can be deferred to the particle system to reduce the manual input in the final setup to a minimum.

The entire process of creating a finalized building asset will be covered to better illustrate the workflow. For simplicity's sake the creation of the modular assets will not be shown as they were created using standard modeling techniques. The base mesh is created first, shown in Fig. 2. Its shape is decided by the modeler and is also created using a technique of choice. Edge loops, and in tandem vertices, are added in key locations to ensure that the distribution points exist at the desired spots on the base mesh. This is necessary for the correct placement of the distributed objects. Once the base shape is finished, one or more vertex groups are created. The vertex groups represent the points where fitting modular assets will be placed. An example of vertex groups used to place modular assets is shown in Fig. 3. There must be no overlap between the vertex groups unless the geometry of the modular assets is placed in such a way that any potential overlap is avoided. The number of particle systems added does not have to match the number of vertex groups. In this case for each particle system a vertex group is used to drive the density of the distribution. An asset or asset collection is selected for the particle system, at which point a chosen or random asset from the selected collection is immediately distributed on the base mesh. A random seed value allows for fine-tuning, and the scale of the distributed asset can also be corrected if necessary. Once the particle system is created, its settings can be applied to multiple assets, where the only difference would be the vertex group used. Before the copy of the base mesh is created, the texturing of the asset is verified. If the modular assets were textured when created, the base mesh will be the only asset lacking a texture. Once the texture is added, the asset is ready for the final step. A copy of the asset is created, and the particle systems are converted from instances to assets with geometry. After which all of the newly created objects are selected along with the base mesh and are joined into one asset. Examples of four types of buildings generated in the manner described previously are given in Fig. 4.

Once the asset library was completed, the final world scene was ready to be assembled. Firstly, a road network was created. The roads themselves were modelled using simple planes. The finalized assets were distributed along the edges of the roads using linked duplicates, which ensured low memory consumption as the assets were only written into the working memory once. This workflow is similar to using particle

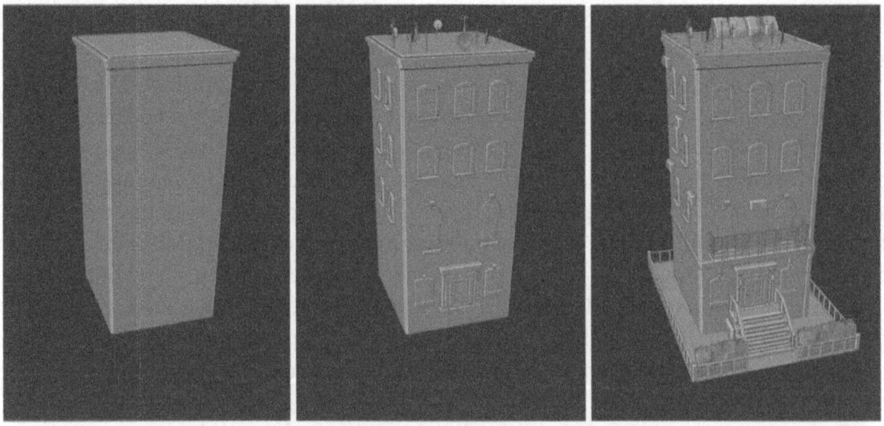

(a) Untextured base shape, base shape with modular assets, and final building asset.

(b) Textured base shape, base shape with modular assets, and final building asset.

Fig. 2 An example of a building asset at different stages of generation

systems, however the placing of the duplicate assets can be achieved directly through code rather than using vertex groups.

The lighting of the scene was done by using HDRi environment maps which allowed for a greater range of luminosity than using Blenders internal lighting system. A multitude of HDRi maps were selected to cover different times of day, weather conditions etc. The HDRi could be selected manually or randomly. A world shader was built using Blenders internal tools which served to simplify the process of changing the HDRi parameters, due to the fact that all of the relevant parameters were grouped together into the same menu. A specialized shader switch was created on every asset that could provide an artificial light source, since both day and night scenes were implemented. For example, windows had two shader modes, one where no light was present inside, and the other where the light would be turned on. The

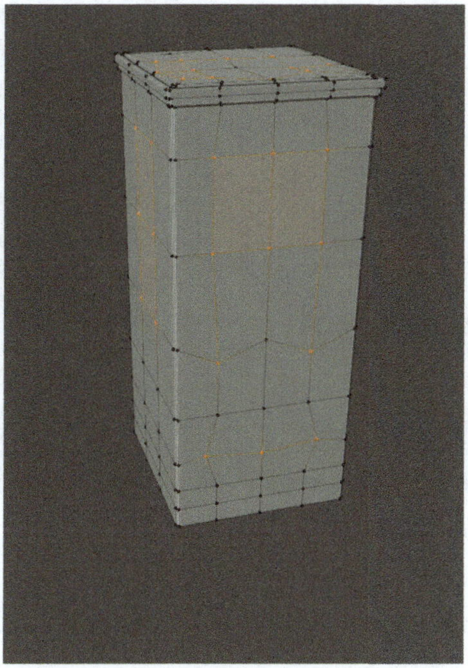

Fig. 3 Vertex groups used for modular asset placement

Fig. 4 Examples of generated building assets

Fig. 5 Example of a rendered image

shader enabled switching between these two states using either percentages or in binary fashion.

The nature of the dataset and its future use meant that rendering the scene needed to produce results that were as realistic as possible while maintaining acceptable render times. For this reason, the Cycles render engine was used. Cycles is a ray tracing engine in Blender, which uses the projection of light rays to ensure realistic lighting in the final render results. Certain Cycles parameters were altered due to the fact that they operate under diminishing returns, meaning they had a high impact on render times but low results fidelity-wise.

A stereoscopic camera setup in Blender was used to simulate a camera traversing the scene and recording images. It also measured the distance to the objects it observed, thus labeling them. The cameras are positioned into a parallel configuration. The focal length of the camera lenses is 50 mm, clip start is 0.1 m and clip end is 1000 m. The baseline is set to 6.5 cm. The traversal of the camera was achieved by positioning the camera to a random point along a fixed path. After each positioning, the camera is rotated along the X axis by 80 to 95°, and along the Z axis by up to 360°. The trajectory of the camera can be modified by altering the underlying shape of the path. The height of the path also varies, in the ranges from 1.5 m to 5.5 m. Each render pass outputs two images and metadata corresponding to the captured part of the scene. This metadata includes: an image name, the name of the target object, the distance to the target object from the camera baseline, and the coordinates of the intersection point between the ray cast from the baseline of the camera to the target object. An example of a rendered scene and scene setup is given in Figs. 5 and 6.

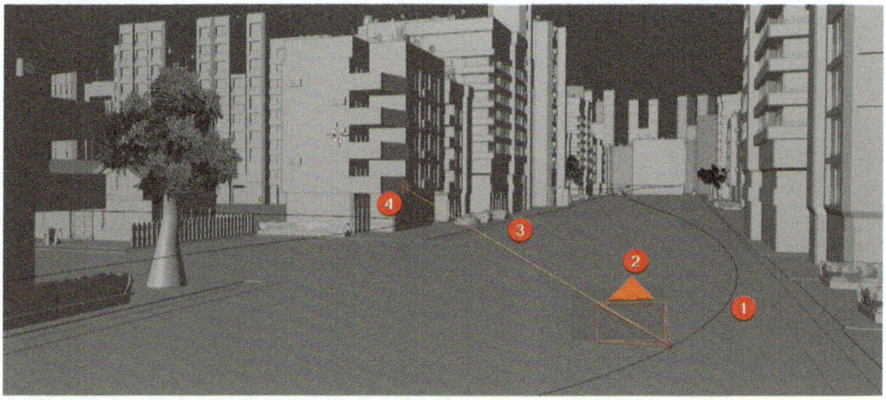

Fig. 6 Scene setup before rendering: 1. Camera trajectory, 2. Camera, 3. Ray cast from the camera, 4. Intersection point of the cast ray and the target object

3 Neural Network

The neural network utilized in our solution is a feed-forward network consisting of a convolutional part and a fully connected part. The convolutional part is used to extract features from image pairs that are then forwarded to the fully connected part of the network. Each image pair is composed of a left and a right image component of a stereoscopic view of the target object. The convolutional layers are divided into two identical branches, each handling a single input image. In order for the network not to be dependent on the size of the input images, before the first convolutional layer, the images are scaled to the fixed resolution of 960×540 pixels by passing them through an adaptive average pooling layer. The outputs of the convolutional branches are concatenated and passed through an additional convolutional layer, and an adaptive average pooling layer to further reduce the dimensions of the data passed to the fully connected layer. This layer outputs the extracted features reshaped into a $100 \times 100 \times 128$ tensor which is then flattened into a 1D tensor and is passed as an input into a fully connected part of the network. The fully connected part of the network is used to combine the features extracted from the images to calculate a distance estimation. It is composed of 6 layers with the PReLU activation function between the first five layers. The activation function was selected empirically. After the final layer, a ReLU activation function was used to generate the network output, since the expected output is a distance value, expressed in meters, larger or equal to 0, thus effectively functioning as a neural network regression model. The input size and the depth of the convolutional network was based on hardware limitations, as the model was trained and evaluated on commodity hardware. The visual representation of the neural network architecture is given in Fig. 7.

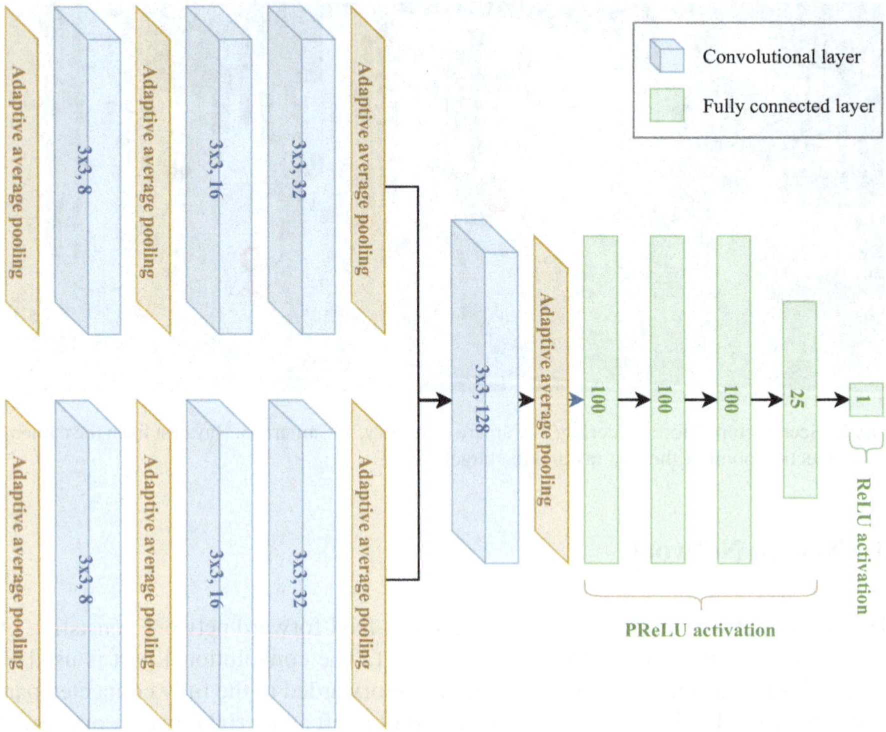

Fig. 7 Diagram of the neural network architecture

4 Training Process

The neural network was trained using the stochastic gradient descent (SGD) algorithm. The learning rate of the algorithm was set to 10^{-6}, whilst the momentum was set to 0.5. Each training sample was composed of an image pair and the distance measured from the stereo camera to the target object. The training was done in batches with each batch containing 5 training samples. The batch size and learning parameters were chosen empirically with the goal of maximizing the utilization of the graphics processing unit and minimizing IO operations.

A predictive model was trained using these presets. A dataset, generated from the 3D scene previously described in this paper, was used for training. The analysis of the generated dataset showed a disproportionate number of images with a measured distance of more than 60 m. In order to create an equal distribution of distances for training, these images were removed. The final dataset contained 10000 image pairs, divided into a training subset, containing 80% of the images and a validation subset containing 20% of the image pairs. The distribution of the number of images with respect to their distance is given in Table 3.

Table 3 The number of images per bin

Image bins (m)	Percentage of images (%)
0–10	14.69
10–20	41.48
20–30	22.14
30–40	11.92
40–50	6.36
50–60	3.40

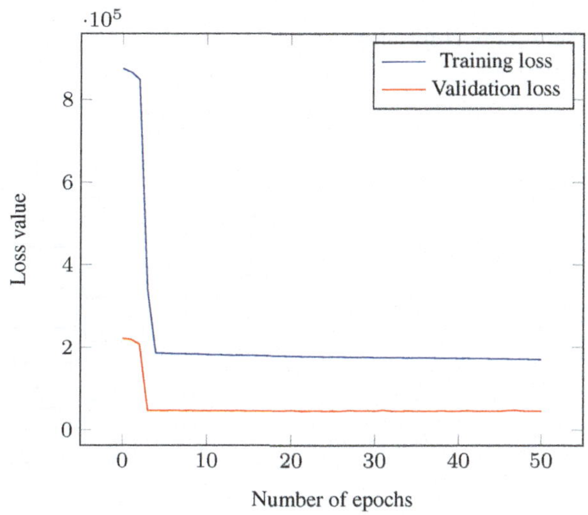

Fig. 8 Change of training and validation loss with respect to the number of epochs

During the training process, both the loss on the training dataset, and the average and median error on the validation dataset were monitored. The loss measure used during the training was the mean square error (MSE). The error measure used on the validation dataset is expressed as an absolute difference of the expected and estimated values. Once the error measured on the validation dataset stabilized, the training process was terminated. This was done in order to prevent overfitting. The change in loss and validation error with respect to the number of epochs is shown in Fig. 8.

5 Result

The trained model was tested on a separately generated test dataset consisting of 2000 image pairs. These image pairs were not part of the training or validation subset. Prior to testing, the neural network weights from the checkpoint with the

Table 4 Performance of the model with respect to each distance range

Image bins (m)	Number of images (%)	Mean error	Median error	Min error	Max error
0–10	13.96	7.95	6.86	0.79	22.91
10–20	40.68	5.78	4.64	0.02	25.72
20–30	23.69	4.70	4.13	0.01	14.63
30–40	13.07	11.76	11.63	1.36	23.69
40–50	6.59	19.85	19.50	4.23	31.96
50–60	2.01	28.05	27.90	17.85	39.88

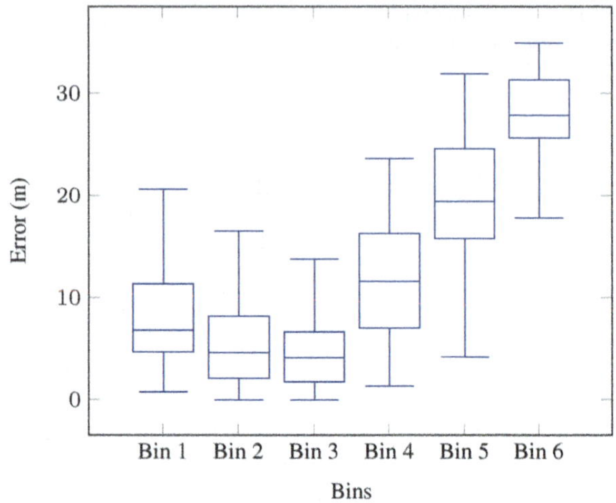

Fig. 9 Error distribution across bins

optimal performance were loaded for the purpose of testing. The results achieved on the model trained on the generated dataset are shown in Table 4, where each column represents the errors measured on the test subset, and the rows represent distance intervals in which the data was measured. The range for each interval was established empirically and is equal to 10 m. The error distribution across the intervals for the trained model is shown in Fig. 9.

The results show that the neural network performs best in the range of 10–30 m, which can be explained by the majority of images in the dataset belonging to that range. As the distance increased, the performance of the model degraded. A possible explanation for this is that the number of images in these ranges was insufficient for proper training. The limited resolution of the input images could have been insufficient to discern the features necessary for depth estimation.

Samples from the test dataset were subjected to visual inspection in order to discover other potential causes of errors in distance estimation. The image with the largest difference between the recorded error and median error in its bin is shown in Fig. 10, with an estimated distance of 37.55 m, and a real distance of 11.84 m. The

Fig. 10 An example of an image for which the distance was overestimated

image shows 3 buildings, two of which have similar textures, making it difficult to discern the edges of those two buildings. Additionally, the 3 buildings fill the entirety of the frame, due to the camera's proximity to the buildings and its orientation, eliminating any potential objects that can be used as a reference to establish the scale and depth of the features. Further visual inspection showed the existence of images similar to the previously described image. Each of these images caused the distance estimation model to either largely underestimate or overestimate the distance to the target object shown in the image.

6 Conclusion

In this paper we proposed the use of synthetic data for the end-to-end training of a distance estimation model. The synthetic dataset was generated by using a generative model based on a 3D urban environment created in Blender. The workflow for the creation of assets for the scene was described and illustrated in detail by using the example of creating a building asset for a 3D scene. The workflow is non-destructive, modular and parameterized, allowing for the creation of large quantities of varying assets while minimizing the need for manual modelling. The proposed distance estimation model was a CNN that uses stereoscopic image pairs as inputs and outputs a single value representing the estimated distance of the target object, expressed in meters. This model was trained on a dataset generated by using the previously described generative model. The model was trained using a stochastic gradient descent algorithm, on 8000 image pairs from the selected dataset. The evaluation of the model's performance was conducted using 2000 image pairs generated from the generative model. The model was most accurate on images of target objects in the range of 10–30 m. The relatively small size of the dataset, as well as smaller number of images in other ranges negatively impacted the performance of the model.

The research presented in this paper can be further expanded. The proposed generative model can be used to generate datasets that are not limited to distance estimation models. The 3D scene can be used to extract additional features for each image, such as the position, shape, size, and type of each object in the image. This, combined with the ability to create bounding boxes and bounding polygons for each object allows for the creation of datasets that can be used for classification, object detection, semantic segmentation and instance segmentation. Point clouds can also be generated, with a specified, controllable and consistent density of points, negating the drawbacks of manually collected and labeled point cloud datasets. The proposed generative model could be further enhanced by simulating adverse environmental conditions, such as precipitation and fog. The main drawback of the presented approach is the necessity for manual asset placement in the creation of the 3D scene. This issue could be resolved by implementing a process for the automatic distribution of assets based on a set of input parameters. The currently available corpus of assets could be further expanded in order to simulate settings other than urban environments.

Acknowledgements This research was supported by the Science Fund of the Republic of Serbia, #GRANT No. 65241005, AI—ATLAS.

References

1. Alzantot, M., Chakraborty, S., Srivastava, M.: Sensegen: a deep learning architecture for synthetic sensor data generation. In: 2017 IEEE International Conference on Pervasive Computing and Communications Workshops (PerCom Workshops). IEEE, pp. 188–193 (2017)
2. Areann, M.C., Bosch, T., Lescure, M., et al.: Laser ranging: a critical review of usual techniques for distance measurement [j]. Opt. Eng. **40** (2001)
3. Barua, S., M., Islam, M., Murase, K.: A novel synthetic minority oversampling technique for imbalanced data set learning. In: International Conference on Neural Information Processing. Springer, Berlin, pp. 735–744 (2011)
4. Chernyshova, Y.S., Gayer, A.V., Sheshkus, A.V.: Generation method of synthetic training data for mobile OCR system. In: 10th International Conference on Machine Vision (ICMV 2017). International Society for Optics and Photonics, SPIE, vol. 10696, pp. 640–646 (2018)
5. Community, B.O.: Blender—a 3D Modelling and Rendering Package. Blender Foundation, Stichting Blender Foundation, Amsterdam (2018)
6. Falkenhagen, L.: Depth estimation from stereoscopic image pairs assuming piecewise continuos surfaces. In: Image Processing for Broadcast and Video Production. Springer, Berlin, pp. 115–127 (1995)
7. Geiger, A., Lenz, P., Stiller, C., Urtasun, R.: Vision meets robotics: the Kitti dataset. Int. J. Robot. Res. (IJRR) **32**, 1231–1237 (2013)
8. Heinzler, R., Schindler, P., Seekircher, J., Ritter, W., Stork, W.: Weather influence and classification with automotive lidar sensors. In: 2019 IEEE Intelligent Vehicles Symposium (IV). IEEE, pp. 1527–1534 (2019)

9. Kniaz, V., Gorbatsevich, V., Mizginov, V.: Thermalnet: a deep convolutional network for synthetic thermal image generation. In: The International Archives of Photogrammetry, Remote Sensing and Spatial Information Sciences, pp. 41–45 (2017)
10. Lim, B., Woo, T., Kim, H.: Integration of vehicle detection and distance estimation using stereo vision for real-time aeb system. VEHITS, pp. 211–216 (2017)
11. Lin, T.Y., Maire, M., Belongie, S., Hays, J., Perona, P., Ramanan, D., Dollár, P., Zitnick, C.L.: Microsoft coco: common objects in context. In European conference on computer vision. Springer, Berlin, pp. 740–755 (2014)
12. Mahjourian, R., Wicke, M., Angelova, A.: Unsupervised learning of depth and ego-motion from monocular video using 3d geometric constraints. In: Proceedings of the IEEE Conference on Computer Vision and Pattern Recognition, pp. 5667–5675 (2018)
13. O' Shea, R.P., Blackburn, S.G., Ono, H.: Contrast as a depth cue. Vis. Res. **34**(12), 1595–1604 (1994)
14. Planche, B., Wu, Z., Ma, K., Sun, S., Kluckner, S., Lehmann, O., Chen T., Hutter, A., Zakharov, S., Kosch, H., et al.: Depthsynth: real-time realistic synthetic data generation from cad models for 2.5 d recognition. In: 2017 International Conference on 3D Vision (3DV). IEEE, pp. 1–10 (2017)
15. Ros, G., Sellart, L., Materzynska, J., Vazquez, D., Lopez, A.M.: The synthia dataset: a large collection of synthetic images for semantic segmentation of urban scenes. In: Proceedings of the IEEE Conference on Computer Vision and Pattern Recognition (CVPR) (2016)
16. Rosu, R.A., Behnke, S.: Easypbr: a lightweight physically-based renderer (2020)
17. Smolyanskiy, N., Kamenev, A., Birchfield, S.: The importance of stereo for accurate depth estimation: an efficient semi-supervised deep neural network approach (2020)
18. Tripathi, S., Chandra, S., Agrawal, A., Tyagi, A., Rehg, J.M., Chari, V.: Learning to generate synthetic data via compositing. In: Proceedings of the IEEE Conference on Computer Vision and Pattern Recognition, pp. 461–470 (2019)
19. Troscianko, T., Montagnon, R., Le Clerc, J., Malbert, E., Chanteau, P.L.: The role of colour as a monocular depth cue. Vis. Res. **31**(11), 1923–1929 (1991)
20. Tsirikoglou, A., Kronander, J., Wrenninge, M., Unger, J.: Procedural modeling and physically based rendering for synthetic data generation in automotive applications. arXiv: 1710.06270 (2017)
21. Wang, C., Miguel Buenaposada, J., Zhu, R., Lucey, S.: Learning depth from monocular videos using direct methods. In: Proceedings of the IEEE Conference on Computer Vision and Pattern Recognition, pp. 2022–2030 (2018)
22. Zhang, Y., Song, S., Yumer, E., Savva, M., Lee, J.Y., Jin, H., Funkhouser, T.: Physically-based rendering for indoor scene understanding using convolutional neural networks (2017)

Appraisal of Apartments in Belgrade Using Hedonic Regression: Model Specification, Predictive Performance, Suitability for Mass Appraisal, and Comparison with Machine Learning Methods

Nemanja Stanišić, Tijana Radojević, and Nenad Stanić

Abstract We present a Bayesian hedonic regression model for the appraisal of apartments located within the metropolitan area of Belgrade, Serbia. Data on 12.904 apartments were collected from a local classified website, and used to fit and validate the model. The response variable, which is the price per m^2 in euros, is assumed to be log-normally distributed. Nested random effects are used to model the hierarchical structure present in the location identifiers, and thin-plate spline functions are used to capture the nonlinear effects. Location explains around 78.62% of the total variation in advertised prices per m^2 in the training set, confirming its paramount importance. Further major factors affecting prices are: area in m^2, floor number, the total number of floors in the building, the availability of an elevator and condition. The predictive functionality of the model is demonstrated through an open access online application that accompanies this study. The model achieves Mean Average Percent Error (MAPE) of 13.06% in the validation set. Its predictive performance is compared to that of three popular machine learning (ML) methods, and its suitability for mass appraisal is examined. This is the first empirical study to present the results obtained from a hedonic regression built using data pertaining to Belgrade.

Keywords Mass appraisal · Bayesian hedonic regression · Machine learning · Mean Average Percent Error (MAPE)

N. Stanišić (✉) · T. Radojević · N. Stanić
Singidunum University, Danijelova 29, Belgrade, Republic of Serbia
e-mail: nstanisic@singidunum.ac.rs

T. Radojević
e-mail: tradojevic@singidunum.ac.rs

N. Stanić
e-mail: nstanic@singimail.rs

E. Pap (ed.), *Artificial Intelligence: Theory and Applications*,
Studies in Computational Intelligence 973,
https://doi.org/10.1007/978-3-030-72711-6_16

1 Introduction

Real estate prices are of great interest to actual and potential homeowners, real estate developers, realtors, banks, and policy makers. In the significant body of literature devoted to the appraisal of individual units [1, 2], as well as mass appraisal of real estate [3], the hedonic pricing method has emerged as one of the most commonly used methods. The general hedonic pricing method has been independently proposed by Court [4], Wallace [5], and Haas [6], and was further popularized by Zvi Griliches [7]. The core idea of the method is that the price of a good is a function of its characteristics. When applied to real estate pricing, the method's assumption is that the price of a property is a function of the property's structural characteristics, and the characteristics of its neighborhood, location and environment [8]. The function may be linear or non-linear, making the definition of hedonic methods rather wide, in that it may encompass any method, from the realms of statistics and machine learning that uses the above groups of characteristics to predict the price of a property. Nevertheless, in literature, the term hedonic is typically reserved for the subgroup of approaches that are based on regression. The specification of hedonic regression models, evaluating the suitability of different variations for real estate pricing (e.g. OLS regression, mixed-effects regression, geographically weighted regression), evaluation of their predictive accuracy across different periods and urban areas, and their comparison with state of the art machine learning methods, are all areas of active research. The result being that hedonic regression has emerged as an important research topic [8].

Despite the continuous expansion of the real estate market in Belgrade, Serbia and many large scale projects being under construction [9], there is an apparent lack of literature demonstrating the application of hedonic regression models using local empirical data. The few academic studies examining real estate prices in Belgrade employ methods other than hedonic regression, such as spatial analysis [10] or machine learning algorithms [11]. Local professionals, on the other hand, regularly use hedonic regression models, but are reluctant to disseminate key insights obtained from them. This is due to the fact that regressions are typically built using in-house developed methodologies and proprietary databases on historical transactions, and making the resulting intellectual property publicly available may result in companies losing their key competitive advantage. As a result, little is known about the factors affecting the prices of apartments in the local market. To bridge this knowledge gap, this paper presents key insights obtained from a Bayesian multilevel hedonic regression, built using a sizeable empirical data set collected from online listings of apartments located within the metropolitan area of Belgrade. Thereupon, the predictive functionality of the regression model is demonstrated through an open access online application that accompanies this study. Finally, as a part of a supplementary analysis, we compare the regression's predictive performance to that of three popular machine learning methods, and examine its suitability for mass appraisal.

2 Data and Methods

In this section, we will briefly describe the initial data set, outline the data preprocessing procedures applied to the initial data set, provide descriptive statistics for selected variables from the processed version of the data set, present the specifications of the statistical models, and specify the software used to fit the models.

2.1 Data

The initial data set comprises information on 12.904 properties located within the metropolitan area of Belgrade, Serbia, which, at the time of data collection, were advertised for sale on a local classified website. Besides the asking price in euros, the dataset includes information on the location of the apartment (as determined by the municipality, the neighborhood and the street), area in m^2, the number of rooms, the type of property (used, new or under construction), condition (luxury, renovated or regular), the floor, the number of stories in the building, the heating system (central heating, district heating, electric thermal storage, marble heaters, natural gas, Norwegian panel heaters, tiled stove or underfloor heating), the type of advertiser (agency, owner, investor or bank), the advertiser's unique identifier, the agency fee (where applicable), and true/false variables indicating the presence of the following features: an elevator, not located on the top floor, registration in the local cadaster, move-in ready condition, VAT refund, a balcony, a recessed balcony, a phone, an intercom, air conditioning, cable TV, parking, a garage, a basement, property surveillance and internet connection.

While the lists of municipalities and neighborhoods are determined and updated by the website, the information on streets are input in a textual form by the advertisers. For that reason, the values for this variable were reconciled with a comprehensive public database of the names of streets in Belgrade accessible at https://zis.beograd.gov.rs/ulicebgdout/ulicebgdout.php using the *OpenRefine* application https://openrefine.org/.

Photos and textual descriptions, for the most part, have not been considered as predictors. The only exceptions are the variable named *Duplex*, which was defined during the modeling stage based on the presence of (several possible variations of) the word "duplex" in the textual descriptions using *RegEx* search, and the variable named *Belgrade_WF*, which is a dummy indicator for the neighborhood of Belgrade Waterfront defined with the aim of capturing the price-wise idiosyncrasy of apartments located in this newly built neighborhood that couldn't be appropriately accounted for within the random effects modeling framework.

Observations with missing information on numerical variables have been deleted from the sample. Missing information on categorical variables have been coded as a separate category.

Table 1 Descriptive statistics for selected variables

Municipality	No. of properties	Mean price per m^2	Mean area in m^2	Mean number of rooms	Maximum of stories
Zvezdara	2,179	1,420.05	68.30	2.59	28
Vozdovac	1,585	1, 452.34	76.45	2.79	25
Novi Beograd	1,423	1,701.04	73.69	2.68	30
Vracar	1,403	2,234.00	88.60	3.05	22
Palilula	1,321	1,236.73	67.84	2.55	20
Savski venac	1,172	2,481.12	94.29	3.07	30
Stari grad	1,133	2,495.13	83.22	2.92	20
Zemun	906	1,200.68	68.18	2.63	22
Cukarica	825	1,243.52	78.56	2.86	19
Rakovica	814	985.41	66.42	2.59	19
Grocka	74	730.95	58.74	2.20	9
Barajevo	23	480.85	62.13	2.15	12
Lazarevac	18	770.60	72.50	2.75	7
Surcin	17	1,120.56	48.18	1.97	4
Mladenovac	6	671.32	67.00	2.17	6
Obrenovac	4	790.33	74.75	3.25	4
Sopot	1	780.00	50.00	2.00	5
Total	**12,904**	**1,653.87**	**76.17**	**2.76**	**30**

The above table presents descriptive statistics for selected variables in the complete dataset (Table 1).

To obtain an unbiased estimate of the model's predictive accuracy, the complete dataset has been split randomly into a training set comprising 10,000 apartments (around 77.50%) and a validation set comprising 2,904 apartments (around 22.50%).

2.2 Method

The first model that has been fit is the "null" model, which is specified using the "classification notation" [12] as follows:
Equation 1 *—Null model specification*

$$\text{Price\_per\_sqm}_i = u^{(2)}_{\text{municipality}(i)} + u^{(3)}_{\text{neighbourhood}(i)} + u^{(4)}_{\text{street}(i)} + u^{(5)}_{\text{advertiser}(i)} + e_i \quad (1)$$

The second model is the "saturated" model, which is specified as follows:
Equation 2 *—Saturated model specification*

$$
\begin{aligned}
Price\_per\_sqm_i &= u^{(2)}_{municipality(i)} + u^{(3)}_{neighbourhood(i)} + u^{(4)}_{street(i)} + u^{(5)}_{advertiser(i)} \\
&+ s\left(Area\_in\_sqm_i\right) + s(Num\_of\_rooms_i) + s\left(Floor\_num_i, Stories_i\right) \\
&+ s\left(Floor\_num_i, Elevator_i\right) + Elevator_i \times \beta_1 + Not\_at\_top\_floor_i \times \beta_2 \\
&+ Registered\_in\_cadaster_i \times \beta_3 + Move\_in\_ready_i \times \beta_4 + VAT\_refund_i \times \beta_5 \\
&+ Belgrade\_WF_i \times \beta_6 + Duplex_i \times \beta_7 + Belgrade\_WF \times Duplex_i \times \beta_8 \\
&+ Balcony_i \times \beta_9 + Recessed\_balcony_i \times \beta_{10} + Phone_i \times \beta_{11} + Intercom_i \times \beta_{12} \\
&+ Air\_conditioner_i \times \beta_{13} + Surveillance_i \times \beta_{14} + Internet_i \times \beta_{15} \\
&+ Cable\_TV_i \times \beta_{16} + Garage_i \times \beta_{17} + Parking_i \times \beta_{18} + Basement\_storage_i \times \beta_{19} \\
&+ Cndtn\_lux_i \times \beta_{20} + Cndtn\_renovated_i \times \beta_{21} + Cndtn\_missing\_info_i \times \beta_{22} \\
&+ Type\_new_i \times \beta_{23} + Type\_under\_cnstrctn_i \times \beta_{24} + Type\_missing\_info_i \times \beta_{25} \\
&+ Heat\_dstrct_i \times \beta_{26} + Heat\_ets_i \times \beta_{27} + Heat\_mrbl_i \times \beta_{28} \\
&+ Heat\_missing\_info_i \times \beta_{29} + Heat\_gas_i \times \beta_{30} + Heat\_nrwgn_i \times \beta_{31} \\
&+ Heat\_tiled_i \times \beta_{32} + Heat\_undrflr_i \times \beta_{33} + Advertiser\_bank_i \times \beta_{34} \\
&+ Advertiser\_investor_i \times \beta_{35} + Advertiser\_owner_i \times \beta_{36} \\
&+ Agency\_fee\_1\%_i \times \beta_{37} + Agency\_fee\_1.5\%_i \times \beta_{38} + Agency\_fee\_2.0\%_i \times \beta_{39} \\
&+ Agency\_fee\_2.5\%_i \times \beta_{40} + Agency\_fee\_missing\_info_i \times \beta_{41} + e_i,
\end{aligned} \tag{2}
$$

where the subscript i denotes individual properties, $Price\_per\_sqm_i$ is the log-normally distributed response variable, u is a random intercept, and $s()$ is a thin-plate smoothing spline function. The model specification evolved from a combination of the authors' domain knowledge, discussions with the most prominent local real-estate specialists, the relevant results available in literature, and statistical diagnostics. The R software environment for statistical computing [13] and its associated software package for Bayesian Multilevel Models Using Stan named brms [14] were used for model fitting.

3 Results

This section will present the most interesting aspects of the models' outputs, compare the predictive performance of the saturated regression model to that of three widely used machine learning methods, and examine its suitability for mass appraisal.

3.1 Regression Output

Our results confirm the findings of prior studies that location is the principal driver of property prices. Around 78.62% of the total observed variation in the prices (the value of R-squared, as defined for Bayesian fits [15], for the null model is 0.7862) can be explained by three levels of location indicators, i.e. municipalities, neighborhoods and streets (with negligible assistance from the advertiser's identifiers that are also included in the null model). As random effects in the null model are actually combined effects of location and other relevant factors that correlate with them (e.g. pricey locations are associated with newly built properties, better construction quality, more luxurious furnishings, increased probability of the property being registered in the cadaster etc.), the intrinsic importance of location is expected to be somewhat lower.

Locational coefficients are interpretable, as their exponentiated values (as presented in Fig. 1) represent relative $price - per - m^2$ indices (averagely priced units within each level have index values of 1) that can be multiplied across levels to obtain the cumulative price index. For instance, at the high end, the priciest municipality is *Stari Grad*, where apartment prices (per m^2) are 114% higher than in averagely priced municipalities. Within this municipality, *Andricev venac* is the most expensive neighborhood with prices being 27% higher than the municipality's average. Within this neighborhood, *Kralja Milana* seems to be the priciest street, with a margin of 28% above the neighborhood's average. Therefore, the cumulative price index for an averagely priced apartment located in *Kralja Milana* street is around 3.49 ($2.14 \times 1.27 \times 1.28$), which is consistent with the prime downtown location (see the corresponding street view presented in Fig. 2). On the low end, the most affordable location for an apartment, having a cumulative price index of 0.34 is *Dvadesetog Oktobra* street, which is located within the inexpensive *Barajevo* neighborhood, which, in turn, is situated within the suburban municipality of *Barajevo*. Again, the coefficient is consistent with the peripheral character of the location (see the corresponding street view presented in Fig. 3).

Municipality	Effect
Stari grad	2.14
Vracar	1.84
Savski venac	1.72
Novi Beograd	1.50
Zvezdara	1.20
Vozdovac	1.18
Zemun	1.14
Palilula	1.13
Cukarica	0.96
Rakovica	0.89
Surcin	0.89
Obrenovac	0.77
Sopot	0.74
Lazarevac	0.72
Mladenovac	0.66
Grocka	0.61
Barajevo	0.53

Neighbourhood	Effect
Andricev venac	1.27
Obilicev venac	1.24
Knez Mihailova	1.23
Trg Nikole Pasica	1.19
Kralja Milana	1.15
Kralja Petra	1.15
Studentski trg	1.14
Terazije	1.08
Toplicin venac	1.06
Kosancicev venac	1.04
Francuska	1.04
Savamala	1.03
Politika	1.02
Skupstina	1.02
Centar	1.01
Takovska	1.01
Kalemegdan	1.00
Kopitareva gradina	1.00
Trg Republike	0.98
Atelje 212	0.98
Dom omladine	0.98
Skadarlija	0.96
Cara Dusana	0.93
Dorcol	0.91
Zeleni venac	0.89
25. maj	0.88
Bajlonijeva pijaca	0.85
London	0.84
Gundulicev venac	0.83
Dunavski kej	0.81

Street	Effect
Kralja Milana	1.28
Majke Jevrosime	1.01
Andricev Venac	0.92
Kraljice Natalije	0.88

Neighbourhood	Effect
Arnajevo	1.60
Vranic	1.08
Guncati	0.82
Ravni gaj	0.81
Barajevo (mesto)	0.75

Street	Effect
Borivoja Tasica	1.10
Svetosavska	1.07
Carice Milice	1.03
Cara Lazara	0.96
Gaj	0.93
Dvadesetog Oktobra	0.85

Fig. 1 Illustration of nested random effects pertaining to location (with the two most extreme instances)

Fig. 2 Street view of Kralja Milana street (retrieved on 6.5.2020. from Google Maps)

Fig. 3 Street view of Dvadesetog oktobra street (retrieved on 6.5.2020. from Google Maps)

Further factors critically affecting apartment prices are the total area in m^2 and the total number of rooms. These two non-linear effects are shown in Figs. 4 and 5.

A larger total area, consistently with the law of diminishing marginal utility, leads to a lower price per sqm (Fig. 4). Near the high end of the price range, however, there is an evidence of an increase in the price, which can be attributed to such apartments having the characteristics of luxury goods. A larger number of rooms, up to a point, exerts a positive effect on the price (Fig. 5). Again, the law of diminishing marginal utility causes the price for an averagely sized apartment (around $76\,m^2$) to plateau at around 4 rooms, and even decline thereafter.

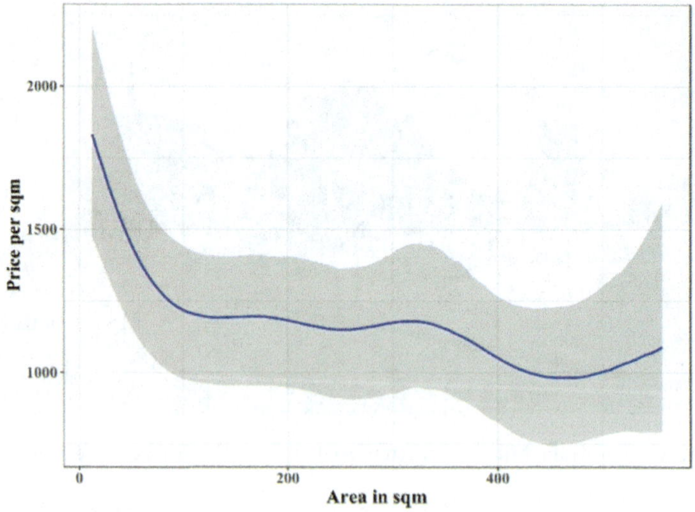

Fig. 4 Estimated price of property in euros per m^2 as a function of total area in m^2

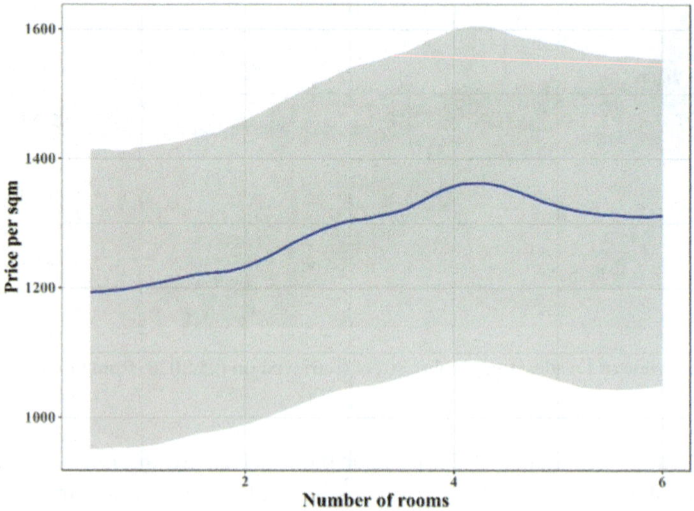

Fig. 5 Estimated price of property in euros per m^2 as a function of total number of rooms

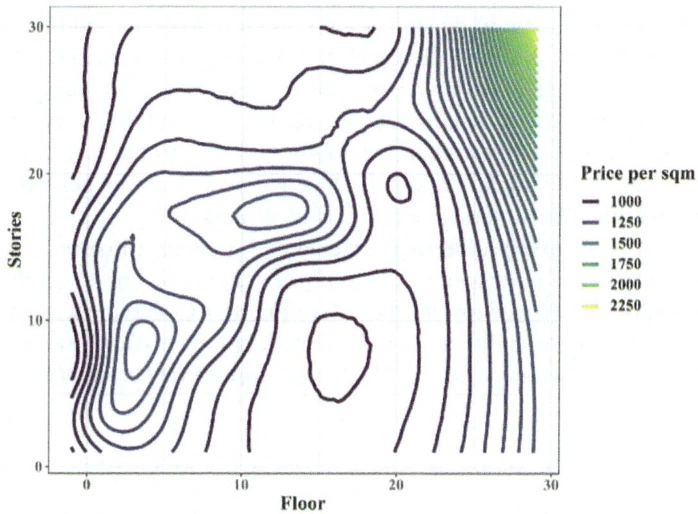

Fig. 6 Estimated price of property in euros per m^2 as a function of floor number and total number of stories in the building

The floor on which the unit is situated, the total number of stories in the building, and elevator availability constitute another set of important factors. Figures 6 and 7 present the interactive nonlinear interplay between the floor number and the total number of stories, and the floor number and elevator availability.

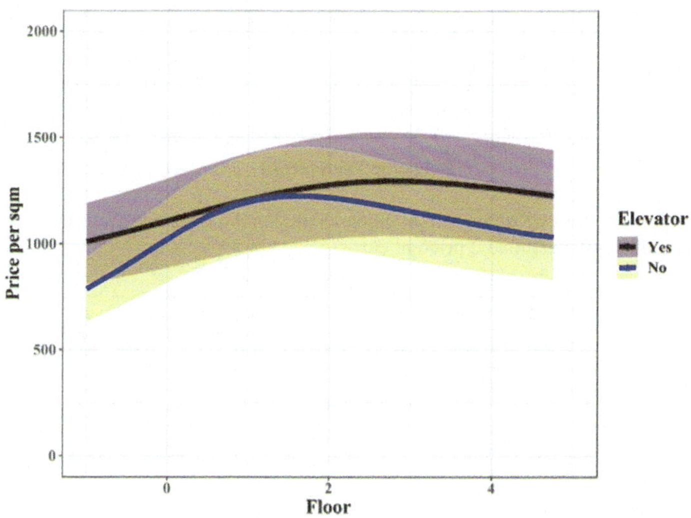

Fig. 7 Estimated price of property in euros per m^2 as a function of floor number and elevator availability

Floors 1 through 5 seem to be sold at a premium in low- and mid-rise buildings (Fig. 6). For high-rise buildings, the optimal range expands to include floors up to the 15th, whereas in residential towers top floors seem to have the highest value.

Finally, apartments located in the new upscale neighborhood *Belgrade Waterfront* are offered at a considerably higher price range relative to the rest of the market. Unlike elsewhere in Belgrade, duplex apartments located in this neighborhood are priced at a premium. The interactive effect is illustrated in Fig. 8.

Other features positively affecting prices are: (1) being a luxury apartment, as opposed to a regular or renovated one (see Fig. 9), (2) being a newly built or under-construction unit, as opposed to a used one (see Fig. 10), (3) having a favorable heating system, such as high-quality underfloor heating or low-cost and low-maintenance district or natural gas heating (see Fig. 11), (4) having a private garage (5% premium),

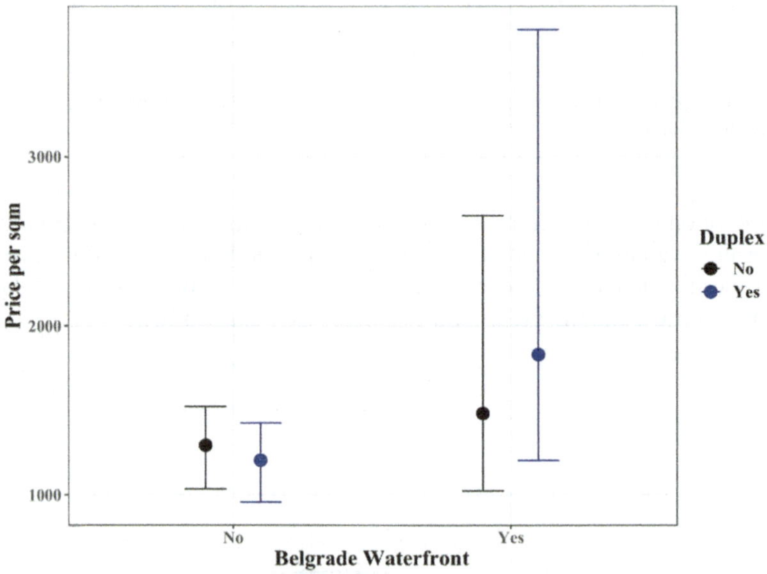

Fig. 8 The interactive effects of apartment being located in the Belgrade Waterfront neighborhood and being a duplex on its estimated price in euros

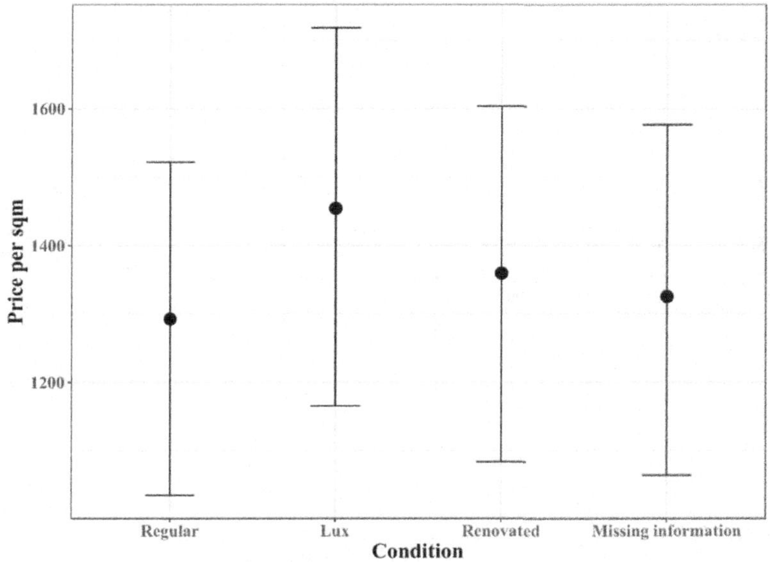

Fig. 9 Estimated price of property in euros per m^2 as a function of the apartment condition

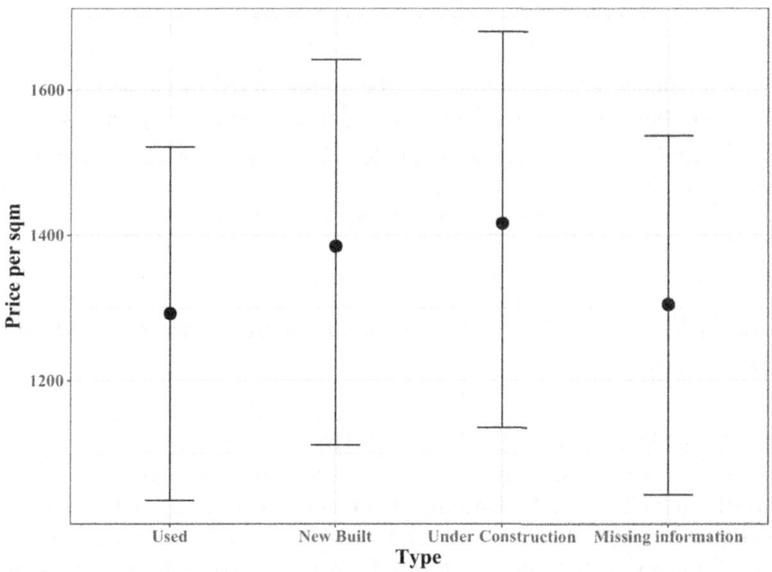

Fig. 10 Estimated price of property in euros per m^2 as a function of the type of building

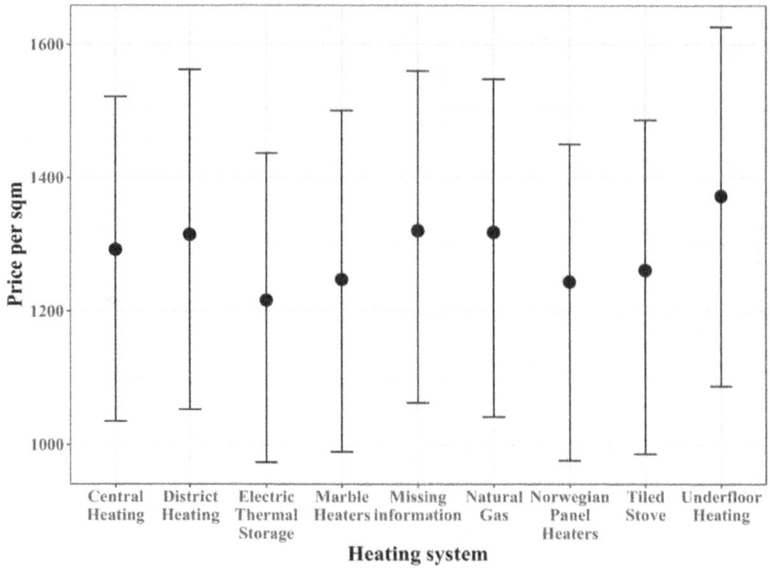

Fig. 11 Estimated price of property in euros per m^2 as a function of the heating system installed in the building

(5) having a balcony (2% premium), (6) being registered in the cadaster (2% premium), (7) not being on the top floor (2% premium), and having (8) a surveillance system (1% premium), (9) air conditioning (1% premium) and (10) intercom (1% premium) in place.

The complete output of the model is presented in Appendix A.

3.2 Model's Predictive Performance and Suitability for Mass Appraisal

The model described in this paper can be used for estimating the prices of individual out-of-sample apartments, primarily those located within the municipalities listed in the first column of Table 1. To demonstrate the predictive functionality of the model, we developed a web application that can be accessed via the following link: https://nstanisic.shinyapps.io/REValuer/. The predictive accuracy achieved in the validation set, as measured by the four metrics most commonly used in the related literature [16], is as follows: Mean Average Percent Error (MAPE) is 13.06%, Root Mean Squared Error (RMSE) is 313.62 euros per m^2, Mean Absolute Error (MAE) is 211.62 euros per m^2, and R-squared (R^2) is 80.65%.

The normal distribution and independence of errors make the model suitable for mass appraisal (e.g. for valuing mortgage loan portfolios). Specifically, as the size of the mortgage loan portfolio increases, the positive and negative errors of the individual appraisals are gradually cancelled out, decreasing the aggregate % error for larger portfolios. Figure 12 shows 100 bootstrap resamples for the cumulative ratio of estimated over actual value as a function of the size of the portfolio under consideration, while Fig. 13 displays the corresponding 95% confidence interval.

Based on the results of the analysis, in 95 out of 100 mass appraisals, the estimated aggregate value would deviate from the actual aggregate value by less than 2% for a portfolio comprising 1000 apartments, and less than 1% for a portfolio comprising 3000 apartments.

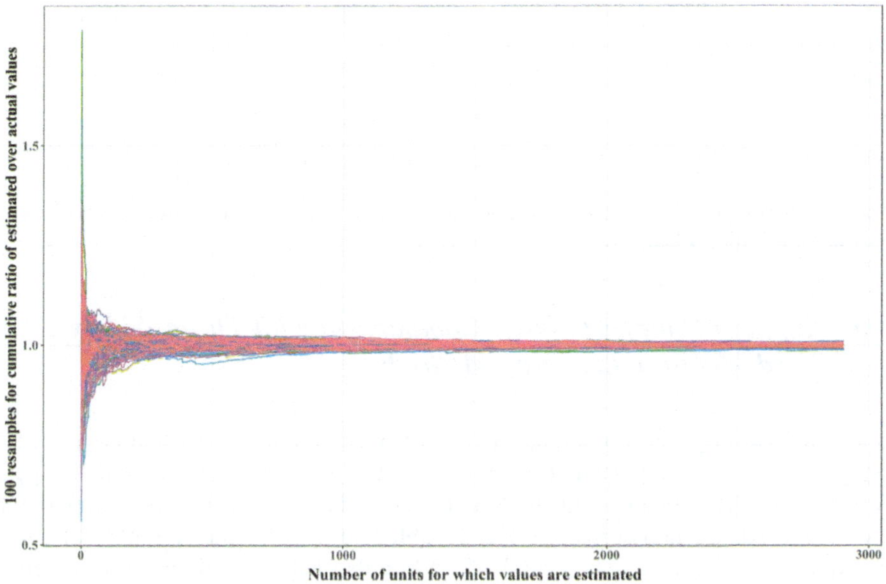

Fig. 12 100 resamples for cumulative ratio of estimated over actual values as a function of the number of units valued

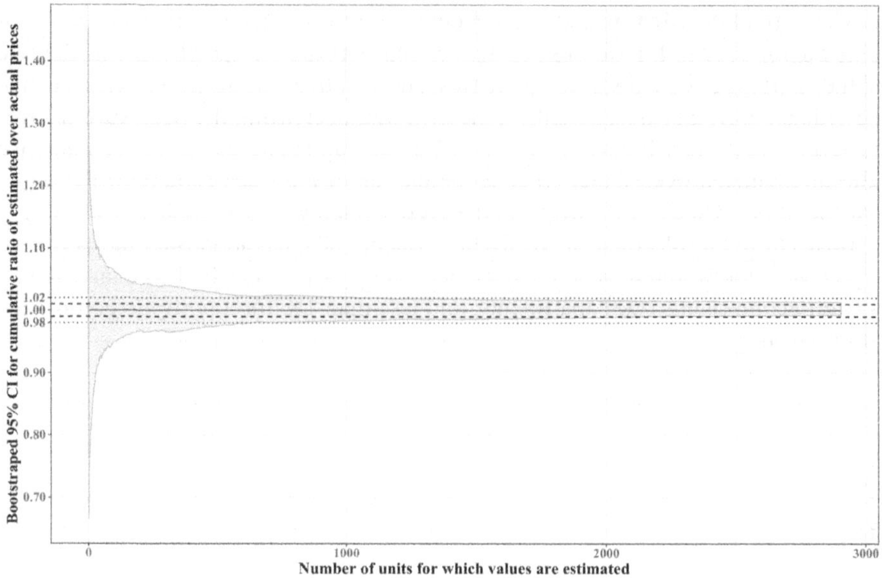

Fig. 13 Bootstrapped 95% CI for cumulative ratio of estimated over actual values as a function of the number of units valued

3.3 Supplementary Analysis: Comparison with Three Widely Used Machine Learning Methods

Machine Learning based methods have recently gained popularity and are increasingly used for mass appraisal. The existing literature on the relative predictive performance of the competing methods is mixed, offering slightly more support for the use of ML methods (please refer to Table 5 in [16] for the summary of comparative studies). Nevertheless, studies' finding that ML algorithms perform better than regression, such as [17–19], typically use ordinary least square regression (aka MRA or OLS) as the representative of regression methods, which is a suboptimal choice. Studies considering more advanced regression-based methods, such as additive regression [20] or geographically weighted regression [21], generally conclude that regression-based methods achieve comparatively similar or better accuracy than ML-based methods. To compare the model's predictive performance to that of well-known machine learning methods, we provided the same training set to the following three methods: Gradient Boosting Machine, Random Forest and XGBoost. All predictor variables are preprocessed using standardization (centering and scaling). To facilitate a fair comparison of predictive performance, all models are tuned in a consistent way, with 5 levels being considered for each tuning parameter, using 3-fold cross-validation repeated 3 times, and Root Mean Squared Error (RMSE) as the performance metric. The complete preprocessing, model fitting and tuning, and the comparative analysis of predictive performance are carried out using R's 'caret' package [22].

The comparative results are as follows:
Multilevel regression—MAPE = 13.06%, RMSE = 313.62 euros per m^2, MAE = 211.62 euros per m^2, $R^2 = 80.65\%$;
Gradient Boosting Machine—MAPE = 15.35%, RMSE = 332.82, MAE = 236.77, $R^2 = 78.15\%$;
Random forest—MAPE = 12.57%, RMSE = 304.80, MAE = 199.42, $R^2 = 81.78\%$;
XGBoost—MAPE = 13.16%, RMSE = 310.93, MAE = 207.13, $R^2 = 81.02\%$.

Another relevant performance criterion is that given by Freddie Mac, stating that: "on the test data, at least half of the predicted sale prices should be within 10% of the actual prices" [23]. The regression model meets this criterion, with 51.76% of the predicted prices being within 10% of the actual prices. The corresponding percentages for the machine learning methods are 44.39% for Gradient Boosting Machine, 56.20% for Random forest, and 52.58% for XGBoost.

The only machine learning method that performs better than the regression model across all the performance metrics is Random Forest. To illustrate the practical consequence of the observed difference in performance between the regression model and the Random Forest model, in Fig. 14, we present the overlay of the two bootstrapped 95% confidence intervals for the cumulative ratio of estimated over actual values as a function of the number of units valued. The CI for the regression model is shown in grey, whereas the corresponding CI for Random Forest model in shown in red.

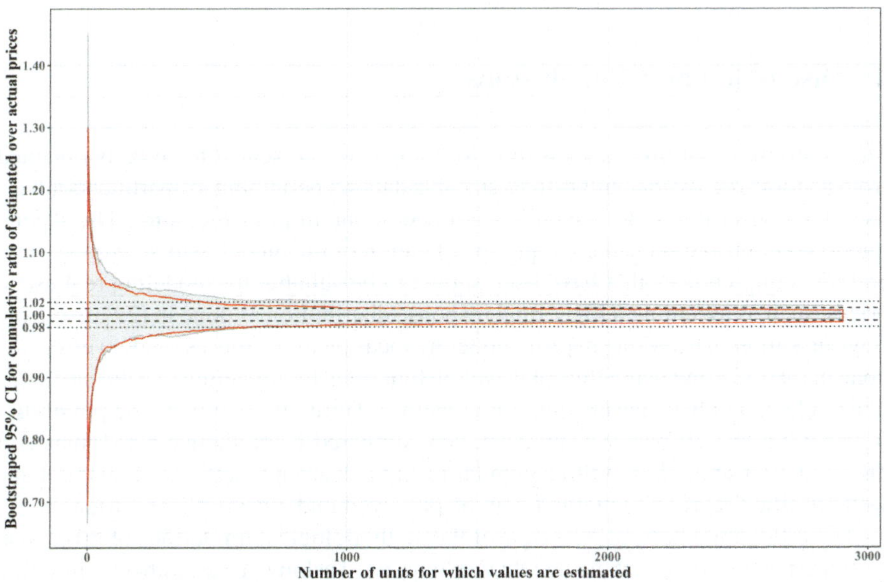

Fig. 14 Comparison of the bootstrapped 95% CI for the cumulative ratio of estimated over actual values as a function of the number of units valued (95% CI for the regression model is shown in grey and for Random Forest model in red)

If we base our decision solely on the predictive performance metrics, Random Forest should be the apparatus of choice for the mass appraisal of any portfolio of apartments similar to that analyzed in this study. However, as explained in [2], in addition to outright performance, the choice of the predictive method should also take into consideration the methodology's transparency, explainability and defensibility. We would argue that the regression method has a significant edge in this aspect, as it provides more detailed insights into the effects that each individual characteristic has on the price of a unit than any of the ML methods do, allowing professionals in the real estate industry to continually learn about preferences in the local market, spot significant trends, and check the consistency of the predictive apparatus they use. Furthermore, as explained in [24]: "regression models are based on theory and assumptions, and benefit from human intervention and subject knowledge for model specification". Specifically, models using data that have a spatial component to them can immensely benefit from being appropriately informed about the hierarchical structure existing in the location parameters [25–28]. Also, when dealing with limited data, subject knowledge can assist the regression method in estimating the parameters by suggesting a realistic model specification. Furthermore, in the event a Bayesian approach to modeling is used, subject knowledge can suggest credible values for parameters by defining appropriate priors. Thus, the overall conclusion of our study reaffirms the view presented in [21], which states that when "all-round performance and defensibility" is considered, the regression-based approach appears to be the better choice.

4 Discussion and Conclusions

The contribution of this paper is threefold. First, we presented the effects obtained from hedonic regression built using a sizeable data set pertaining to apartments advertised for sale in Belgrade, which has not been done in prior literature. The second important contribution is the unique specification of the model, which employs several modeling choices that have been shown to be suitable for modeling real estate prices, such as assuming a lognormal response distribution as done in [29], using the Bayesian approach as done in [30], modeling location indicators as random effects as done in [30–33], and using thin-plate spline functions for modeling non-linearities as done in [34], but have never been used together. Third, we examined the possibility of using the model for mass appraisal, and compared its predictive performance in the validation set to that of three popular machine learning methods, showing that a well-specified regression method may be preferred under certain circumstances.

The following is a brief overview of the methodological limitations of this study.

Experts may argue that the fact that we have used asking prices, rather than selling prices, is a major limitation of the model presented herein, because the asking prices can be set unrealistically high or low. Nevertheless, due to the apparent lack of reliable data on selling prices,[1] professional property valuers in Serbia make extensive use of asking prices as market inputs for both comparison and the income approach, making considered professional judgment on the reliability of such sources of information. Furthermore, the following two circumstances greatly alleviate the concerns: Firstly, around 96% of listings are posted by professional agencies, which have sufficient expertise, information on the local market and motivation to assists the owners in setting the asking price of the apartments at a realistic level. Secondly, both the type of entity that has posted the listing (the agency/the owner/ the investor) and their identity (the company name for agencies and investors, and a unique id for individual owners) is accounted for in the model. Modeling the effects of the type of advertiser as a fixed effect and the identity of the advertisers as a random effect allows the regression model to generate price estimates that imply a moderate pricing strategy. Still, we must stress that the estimates provided by the models are asking prices[2] as opposed to market prices,[3] meaning that appropriate adjustments need to be made before giving market price estimates. The local banks, for instance, insist that valuers apply a 5% discount on asking prices for apartments to carry out mortgage valuations, which may be a good guideline.

Another methodological limitation is that information on the age of the property, which is a potentially important predictor variable, was not available on the website.

Finally, as predictors, we mainly used structured data on the apartments, as available on the website. Photos and textual descriptions of the properties available on the website were not considered (with a single exception described in the section Data) as predictors in the model. Future research should combine both structured and unstructured data to deliver improved predictive performance.

Acknowledgements This research was supported by the Science Fund of the Republic of Serbia, #GRANT No. 65241005, AI—ATLAS.

[1]In Serbia there is an available organized database of transactional property prices, maintained by The Republic Geodetic Authority (https://katastar.rgz.gov.rs/RegistarCenaNepokretnosti/). However, the structure of data does not allow for an adequate assessment of the predictive accuracy of the models presented in this paper; except for the address, the date of transaction, the size and the cadastral parcel, any other information about the quality and the state of the property is missing. The database is available both for public and professional access, but there is a very small difference in the quality of information obtained in both. However, professional access is costly for any sort of research which demands a large amount of data about transactional prices.

[2]Asking price is "a price at which a seller commits to taking his good off the market and trading immediately" as defined in [35].

[3]Market price is "the estimated amount for which the property should exchange on the date of valuation between a willing buyer and a willing seller in an arm's-length transaction after proper marketing wherein the parties had each acted knowledgeably, prudently and without being under compulsion", as defined in [36].

Appendix

Appendix A: Estimated Regression Coefficients

Population-level effect	Estimate	Estimate Error	l-95 % CI	u-95 % CI	Exponentiated coefficient estimate
Intercept	7.12	0.23	6.67	7.59	1236.45
s(Area_in_sqm)	−1.60	0.84	−3.21	0.00	0.20
s(Num_of_rooms)	−0.01	0.08	−0.18	0.15	0.99
s(Floor_num × Stories)	−2.13	0.36	−2.82	−1.45	0.12
s(Floor_num × Stories)2	0.79	0.33	0.14	1.43	2.20
s(Floor_num × Elevator[No])	−1.57	47.90	−88.73	113.03	0.21
Elevator[No]	−0.19	0.23	−0.57	0.34	0.83
Not_at_top_floor[No]	−0.02	0.00	−0.02	−0.01	0.98
Registered_in_cadaster [No]	−0.02	0.00	−0.03	−0.01	0.98
Move_in_ready[No]	0.01	0.00	0.00	0.01	1.01
VAT_refund[No]	0.00	0.01	−0.01	0.01	1.00
Belgrade_WF[Yes]	0.28	0.17	−0.05	0.60	1.32
Duplex[Yes]	0.07	0.06	−0.05	0.18	1.07
Belgrade_WF[Yes] × Duplex[Yes]	0.16	0.08	0.00	0.32	1.17
Balcony[No]	−0.02	0.00	−0.02	−0.01	0.98
Recessed_balcony[No]	0.00	0.00	−0.01	0.00	1.00
Phone[No]	0.01	0.00	0.00	0.01	1.01
Intercom[No]	−0.01	0.00	−0.02	0.00	0.99
Air_conditioner[No]	−0.01	0.00	−0.02	−0.01	0.99
Surveillance[No]	−0.01	0.00	-0.02	-0.01	0.99
Internet[No]	0.01	0.00	0.00	0.01	1.01
Cable_TV[No]	0.00	0.00	0.00	0.01	1.00
Garage[No]	−0.05	0.00	−0.06	−0.04	0.95
Parking[No]	0.01	0.00	0.00	0.01	1.01
Basement_storage[No]	−0.01	0.00	−0.01	0.00	0.99
Condition[Lux]	0.01	0.00	0.00	0.01	1.01
Condition[Renovated]	−0.07	0.00	−0.08	−0.06	0.93
Condition[MissingInfo]	0.05	0.00	0.04	0.06	1.05
Type[NewBuilt]	0.01	0.00	0.00	0.02	1.01
Type[UnderConstruction]	−0.08	0.01	−0.09	−0.07	0.92
Type[MissingInfo]	−0.01	0.01	−0.03	0.00	0.99
Heating[DistrictHeating]	0.02	0.00	0.01	0.03	1.02
Heating[ElectricThermalStorage]	−0.06	0.01	−0.07	−0.05	0.94
Heating[MarbleHeaters]	−0.03	0.03	−0.10	0.02	0.97
Heating[MissingInfo]	0.03	0.01	0.01	0.04	1.03
Heating[NaturalGas]	0.02	0.01	0.00	0.04	1.02
Heating[NorwegianPanelHeaters]	−0.05	0.01	−0.08	−0.02	0.95
Heating[TiledStove]	−0.04	0.03	−0.11	0.02	0.96
Heating[UnderfloorHeating]	0.06	0.02	0.02	0.09	1.06
Advertiser[Bank]	−0.02	0.12	−0.26	0.21	0.98
Advertiser[Investor]	−0.02	0.02	−0.06	0.02	0.98
Advertiser[Owner]	0.00	0.01	−0.02	0.02	1.00
AgencyFee[1%]	0.06	0.09	−0.11	0.23	1.06
AgencyFee[1.50%]	0.05	0.15	−0.26	0.35	1.05
AgencyFee[2%]	0.03	0.02	−0.01	0.08	1.03
AgencyFee[2.40%]	0.15	0.10	−0.06	0.36	1.16
AgencyFee[MissingInfo]	0.02	0.02	−0.02	0.07	1.02

References

1. Kang, H.-B., Reichert, A.K.: An empirical analysis of Hedonic regression and grid-adjustment techniques in real estate appraisal. Real Estate Econ. **19**, 70–91 (1991).https://doi.org/10.1111/1540-6229.00541
2. Keskin, B.: Hedonic analysis of price in the Istanbul housing market. Int. J. Strateg. Prop. Manag. **12**, 125–138 (2008). https://doi.org/10.3846/1648-715X.2008.12.125-138
3. Wang, D., Li, V.J.: Mass appraisal models of real estate in the 21st century: a systematic literature review. Sustainability **11**, 7006 (2019). https://doi.org/10.3390/su11247006
4. Court, A.T.: Hedonic price indexes with automotive examples. In: Ross, C.F. (ed.) The Dynamics of Automotive Demand, pp. 99–117. General Motors, New York (1939)
5. Wallace, H.A.: Comparative farm-land values in Iowa. J. L. Public Util. Econ. **2**, 385 (1926). https://doi.org/10.2307/3138610
6. Haas, G.: A statistical analysis of farm sales in Blue Earth County, Minnesota, as a basis for Farm Land Appraisal (1922)
7. Griliches, Z.: Notes on the measurement of price and quality changes. In: Models of Income Determination, pp. 381–418. National Bureau of Economic Research, Inc., Cambridge (1964)
8. Herath, S., Maier, G.: The hedonic price method in real estate and housing market research. A review of the literature, Vienna (2010)
9. Peirson, A., Jovanovic, J.: Belgrade Residential Market I Snapshot Q4 2019, Belgrade (2020)
10. Bajat, B., Kilibarda, M., Pejović, M., Petrović, M.S.: Spatial hedonic modeling of housing prices using auxiliary maps. In: Thill, J.-C. (ed.) Advances in Geographic Information Science, pp. 97–122. Springer, Berlin, Heidelberg (2018). https://doi.org/10.1007/978-3-642-37896-6_5
11. Vidović, M., Radosavljević, I., Mitrović, A., Konjović, Z.: Construction of Training Data for price prediction of a real estate from internet ads. In: Proceedings of the International Scientific Conference - Sinteza 2019, pp. 388–393. Singidunum University, Novi Sad, Serbia (2019). https://doi.org/10.15308/Sinteza-2019-388-393
12. Browne, W.J., Goldstein, H., Rasbash, J.: Multiple membership multiple classification (MMMC) models. Stat. Modell. **1**, 103–124 (2001). https://doi.org/10.1191/147108201128113
13. R Core Team.: R: a language and environment for statistical computing. http://www.r-project.org (2019)
14. Bürkner, P.-C.: brms: an R package for Bayesian multilevel models using Stan. J. Stat. Softw. **80**, 1–28 (2017). https://doi.org/10.18637/jss.v080.i01
15. Gelman, A., Goodrich, B., Gabry, J., Vehtari, A.: R-squared for Bayesian regression models. Am. Stat. **73**, 307–309 (2019). https://doi.org/10.1080/00031305.2018.1549100
16. Valier, A.: Who performs better? AVMs vs hedonic models. J. Prop. Invest. Financ. (2020). https://doi.org/10.1108/JPIF-12-2019-0157
17. Kontrimas, V., Verikas, A.: The mass appraisal of the real estate by computational intelligence. Appl. Soft Comput. **11**, 443–448 (2011). https://doi.org/10.1016/J.ASOC.2009.12.003
18. Peterson, S., Flanagan, A.: Neural network hedonic pricing models in mass real estate appraisal. J. Real Estate Res. **31**, 147–164 (2009). https://doi.org/10.5555/rees.31.2.m3h27210w6411373
19. Antipov, E.A., Pokryshevskaya, E.B.: Mass appraisal of residential apartments: an application of Random forest for valuation and a CART-based approach for model diagnostics. Expert Syst. Appl. **39**, 1772–1778 (2012). https://doi.org/10.1016/j.eswa.2011.08.077
20. Zurada, J., Levitan, A., Guan, J.: A comparison of regression and artificial intelligence methods in a mass appraisal context. J. Real Estate Res. **33**, 349–387 (2011). https://doi.org/10.5555/rees.33.3.q6890722u7375871
21. McCluskey, W.J., McCord, M., Davis, P.T., Haran, M., McIlhatton, D.: Prediction accuracy in mass appraisal: a comparison of modern approaches. J. Prop. Res. **30**, 239–265 (2013). https://doi.org/10.1080/09599916.2013.781204
22. Kuhn, M.: caret: classification and regression training (2017). https://cran.r-project.org/package=caret

23. Fik, T.J., Ling, D.C., Mulligan, G.F.: Modeling spatial variation in housing prices: a variable interaction approach. Real Estate Econ. **31**, 623–646 (2003). https://doi.org/10.1046/j.1080-8620.2003.00079.x
24. Christodoulou, E., Ma, J., Collins, G.S., Steyerberg, E.W., Verbakel, J.Y., Van Calster, B.: A systematic review shows no performance benefit of machine learning over logistic regression for clinical prediction models. J. Clin. Epidemiol. **110**, 12–22 (2019). https://doi.org/10.1016/J.JCLINEPI.2019.02.004
25. Radojevic, T., Stanisic, N., Stanic, N.: The culture of hospitality: From anecdote to evidence. Ann. Tour. Res. **79**, 102789 (2019). https://doi.org/10.1016/j.annals.2019.102789
26. Radojevic, T., Stanisic, N., Stanic, N.: Inside the rating scores: a multilevel analysis of the factors influencing customer satisfaction in the hotel industry. Cornell Hosp. Q. **58** (2017). https://doi.org/10.1177/1938965516686114
27. Radojevic, T., Stanisic, N., Stanic, N.: Ensuring positive feedback: factors that influence customer satisfaction in the contemporary hospitality industry. Tour. Manag. **51**, 13–21 (2015). https://doi.org/10.1016/j.tourman.2015.04.002
28. Radojevic, T., Stanisic, N., Stanic, N., Davidson, R.: The effects of traveling for business on customer satisfaction with hotel services. Tour. Manag. **67**, 326–341 (2018). https://doi.org/10.1016/j.tourman.2018.02.007
29. Selim, H.: Determinants of house prices in Turkey: Hedonic regression versus artificial neural network. Expert Syst. Appl. **36**, 284–852 (2009). https://doi.org/10.1016/J.ESWA.2008.01.044
30. Hui, S.K., Cheung, A., Pang, J.: A hierarchical bayesian approach for residential property valuation: application to Hong Kong housing market. Int. Real Estate Rev. **13**, 1–29 (2010)
31. Brown, K., Uyar, B.: A hierarchical linear model approach for assessing the effects of house and neighborhood characteristics on housing prices. J. Real Estate Pract. Educ. **7**, 15–24 (2004). https://doi.org/10.5555/repe.7.1.f687057161743261
32. Arribas, I., García, F., Guijarro, F., Oliver, J., Tamošiūnienė, R.: Mass appraisal of residential real estate using multilevel modelling. Int. J. Strateg. Prop. Manag. **20**, 77–87 (2016). https://doi.org/10.3846/1648715X.2015.1134702
33. Deboosere, R., Kerrigan, D.J., Wachsmuth, D., El-Geneidy, A.: Location, location and professionalization: a multilevel hedonic analysis of Airbnb listing prices and revenue. Reg. Stud. Reg. Sci. **6**, 143–156 (2019). https://doi.org/10.1080/21681376.2019.1592699
34. Bao, H.X.H., Wan, A.T.K.: On the use of spline smoothing in estimating hedonic housing price models: empirical evidence using Hong Kong data. Real Estate Econ. **32**, 487–507 (2004). https://doi.org/10.1111/j.1080-8620.2004.00100.x
35. Lester, B., Visschers, L., Wolthoff, R.: Competing with asking prices. Theor. Econ. **12**, 731–770 (2017). https://doi.org/10.3982/te1846
36. Regulation (EU): EUR-Lex-32013R0575-EN-EUR-Lex (2013)

The Role of Chatbots in Foreign Language Learning: The Present Situation and the Future Outlook

Jasna Petrović and Mlađan Jovanović

Abstract While crucial for all language learners, conversation practice can be very expensive and challenging to get enough of. Chatbots are computer programs developed to engage in conversations with humans. They are designed as software avatars with limited, but growing, conversational capability. The most natural and potentially powerful application of chatbots is related to their fundamental nature—language practice. However, their role and outcomes within (in)formal language learning are currently tangential at best. Existing research has generally focused on chatbots' comprehensibility and the motivation they inspire in their users. In this paper, we provide an overview of chatbots for learning languages, critically analyze existing approaches, and discuss the major challenges for future research.

Keywords Conversational agent · Chatbot · Language learning · Linguistics · Translation · Dialogue · Artificial intelligence · Dialog management

1 Introduction

Conversational agents (also known as chatbots) are becoming an increasingly significant part of our everyday activities. Amazon Alexa [1], Google Assistant [2], to name just a few, are no longer desired and owned exclusively by the tech-savvier segment of the population. On the contrary, an increasing number of people around the globe are realizing the many benefits they are able to provide. From helping organize your daily to-do lists and chores, to bringing you up to speed with the latest weather reports—digital assistants offer a helping hand when you need one. In addition, all of them employ a common technological metaphor—a personal, user-friendly assistant that provides a service by using a natural language [3]. In addition, the more you use

J. Petrović (✉) · M. Jovanović
Singidunum University, Belgrade, Serbia
e-mail: jpetrovic@singidunum.ac.rs

M. Jovanović
e-mail: mjovanovic@singidunum.ac.rs

E. Pap (ed.), *Artificial Intelligence: Theory and Applications*,
Studies in Computational Intelligence 973,
https://doi.org/10.1007/978-3-030-72711-6_17

the assistant, the more adapted it becomes to your voice, word choices and personal preferences. The Artificial Intelligence (AI) becomes more competent every time the assistant is used. With every interaction, it develops a better understanding of our desires, and hopefully, over time, it will be able to anticipate our needs with even more precision.

In the last decade alone we have witnessed an exponential rise in the number of online platforms and applications aimed at facilitating language learning and acquisition. The proliferation of technologies for learning languages has produced personalized services for gaining and maintaining language proficiency at any time, in any place [4]. Conversational agents are increasingly becoming part of this paradigm shift, primarily as a cost-effective means to deliver such services. And that is not all they have to offer. One thing that all of them have in common is their aspiration to ingrain positive working habits that are the sine qua non of language learning and teaching, i.e. repetition, because—repetitio est mater studiorum, and practice in the form of follow-up exercises outside the classroom [5, 6]. Among the chatbot's-strongest points are their ease of use and accessibility—the conversation metaphor using text- or voice-based interfaces make them more intuitive, available on smartphones in any place and at any time [7, 8].

Recent research involving text-based human-chatbot interactions has indicated their potential benefits. Particularly with regards to the motivation and engagement they inspire in their users [8], both of which represent critical requirements for technology-supported language learning. However, the potential of chatbots as language learning partners is still underexplored. Evidence of the effectiveness and efficiency of chatbot-like technologies for learning languages in terms of measurable outcomes (such as acquired knowledge, skills, and attitudes) is either missing or inconsistent [4–6] and [9–11].

This is what motivated our research; reflecting on what has been achieved using chatbots to facilitate language learning, and, in keeping with the results obtained, formulating implications for their future. Due to their nature, we maintain that the chatbots' appearance and behavior are highly interdependent. Therefore, we look at them from two perspectives: technological (i.e., functional) and speaker (i.e., user experience—UX). In particular, this paper:

- gives an overview of the technology for learning languages,
- critically analyzes chatbot technology against the proposed dimensions, and
- outlines the implications for subsequent research on chatbots for learning languages.

The paper is structured as follows: Sect. 2 situates our work in the context of the related technologies. In Sect. 3, we formulate the dimensions for analyzing the relevant aspects of chatbots for learning languages. Section 4 describes and compares the chatbots against the dimensions. Section 5 articulates the research implications for future work. Section 6 proposes a reference architecture for chatbots for learning foreign languages. Section 7 highlights the limitations of our work and Sect. 8 concludes the paper.

2 Technology for Learning Languages

As we have previously underlined, many applications for learning languages have been developed and used thus far. We have chosen to classify them into Web and mobile technologies and gamified technologies. The classification is made according to the technologies they utilize and their mode of content delivery. In view of their abundance, we do not aim to provide an exhaustive list of technologies. Instead, our intention is to cover some representative examples that demonstrate shared properties.

2.1 Web and Mobile Technologies

Time is one of the most important constraints people face nowadays. The hectic pace of contemporary life, employment commitments and family duties impose an onerous burden on modern man who struggles daily to juggle his family and career. Making time for oneself, let alone for a hobby, is often virtually impossible. Here is where technology can come to our aid. Since COVID-19 was declared a global health emergency in January 2020, we have become acutely aware of our dependence on technology in practically every aspect of life. As part of nationwide closures of businesses and schools, the latter had to shift to an online setting. The transition had to happen practically overnight, and soon afterwards most businesses had begun to exist in the virtual world too. Almost all major retailers had to set up or update their websites to comply with the requirements dictated by the changing circumstances. However, in addition to new shopping experiences for both consumers and retailers alike, virtual shopping assistants, (i.e. software that uses machine learning (ML) and natural language processing (NLP) technology to engage in interaction with online shoppers), have also thrived. This begs the question: if one is able to make purchases using a virtual shopping assistant and have their needs fully met, why would a virtual language tutor not be used for the same purpose?

Inspired by the quick transition to online working and learning, the authors decided to test several popular language learning platforms and applications and see how they had adapted to the newly arisen circumstances. The key aim was to test their success in improving accuracy and fluency in the target language. In addition, as conversational agents are the focal point of this paper, the authors used this as an opportunity to see if any of the applications contained the added feature of a virtual language assistant. The following applications were tested.

HelloTalk [12] is a mobile application whose users learn foreign languages by engaging in conversations with native speakers. The application enables users to search for and chat with learning partners (autonomous learning), or join online group conversations moderated by teachers (collaborative learning). Duolingo [13] follows a more systematic and scaffolded approach, i.e. it targets specific language skills (such as speaking, reading, listening, and writing skills) through a series of exercises. Rosetta Stone [14] is a universal platform for learning languages. To attract users, it implements diverse learning strategies. These include topic-based learning (such as work, home, etc.), immersive experiences with rich multimedia content and high interactivity, skill-specific lessons (i.e., vocabulary, grammar, and others), and even interactive storytelling on wide-ranging topics. Memrise [15] facilitates language acquisition through learning by example (users can watch videos of native speakers conversing about an object they have just seen or thought of), learning by exploration (object identification in real-time so that a user can take a photo of an object and ask for a description from native speakers of the chosen language), spaced repetitions, or instant feedback on pronunciation. WordUp [16] is an application for learning English and improving vocabulary. It aims to increase and maintain the user's vocabulary by prioritizing words based on their usefulness, putting words in familiar contexts, and repetition.

All of the applications share some common features. An introductory test is available in each application, which determines the user's language level, and following the results obtained, offers a course tailored to the learner's needs. It's worth noting that it is not entirely clear how language proficiency is assessed. In addition, the tasks appear too simple for more proficient speakers, which may result in a swift loss of interest.

Initially, we began the task by exploring English language learning opportunities. Given the fact that none of the applications offered Serbian as a language of instruction, we had to rely on our knowledge of either Russian (B1), Croatian (C2) or Italian (C1). To gain a better understanding of how language learning applications work, and focusing on whether they are able to sustain the motivation to learn a foreign language long-term, the authors also tested the French and Italian courses, as well as the Russian course, at A1 and B1 level, respectively. Surprisingly, we concluded that all of the studied applications are largely intended for those with a basic or elementary knowledge of the target language. In addition, if one is to make any substantial progress, relying on this type instruction is simply not sufficient. The tasks generally consist of translating from one language into the other, listening and spelling out what is heard when offered words are clicked, selecting the appropriate form of verbs, nouns or articles and placing them into sentences correctly, followed by reiterations of the same exercises. As the user progresses, the application adapts its content and functions to the users' language level.

Finally, it is important to note that each application contains a brief grammar and/or vocabulary tutorial that serves as the preparation for a new lesson or as a review of previously covered content. Every application insists on maintaining continuity. In doing so, the applications employ the motivational elements of self-reflection (users are continuously monitored and presented with their achievement), competition (comparison with peers), and digital nudging (regular reminders to practice). Other than WordUp, all of the applications support multiple languages. Most applications are free to some extent and charge their users for more advanced features. Perhaps it is worth highlighting that the authors used only those applications that offered their services free of charge.

2.2 *Gamified Technologies*

The term gamification is defined as using game elements in non-gaming contexts [17], such as learning foreign languages. These applications use elements of games to incentivize their users, where learning happens as a side effect of playing a game. The user incentives are internal and include fun, pleasure, satisfaction, challenge, achievement, reputation, and competition, among others [18, 19]. For example, Duolingo [13] employs symbols such as hearts, badges, winning streaks, and leaderboards to motivate their users [20]. Kahoot! [21] is an online educational platform used for gamified language learning. The gamified components are multiple-choice questions and trivia quizzes that can be created and moderated by teachers. Ver-taal [22] follows a similar approach, using memory games and crosswords to facilitate foreign language learning.

Zou et al. [19] provide a systematic review of digital games that follow a scaffolding approach for foreign language vocabulary acquisition. The games' genres vary and include storytelling games, problem-solving games, role-playing games, virtual reality (VR) and body-motion games, adventure games, card games, and board games. While there is partial evidence of the games' positive effects on vocabulary acquisition, the issues of sustained user engagement and broader outreach remain. These issues arise due to the inherent diversity of users' motivations to play games, and the heterogeneity of user groups in terms of needs, preferences, and learning styles.

A similar analysis conducted by Hung et al. [18] shows that digital game-based language learning induces positive emotions and attitudes in students. However, the games' positive effects are observed in a controlled environment (classroom). Furthermore, the games are biased towards the English language and mainly improve vocabulary while having less influence on other language skills (such as reading, writing, speaking, and listening).

3 Dimensions for Analysis

We propose the dimensions for describing and comparing existing chatbots for learning foreign languages (Table 1). The dimensions are extracted as themes from an emerging, inductive coding process on existing solutions. The chatbots were selected after an extensive search of available resources by the authors, including chatbot websites [31–34], public collections and articles [23–27], and peer-reviewed publications [4–6, 9–11, 28].

The dimensions are classified into speaker and technology aspects.

The *speaker aspects* include the skill level, language diversity, domain, and interaction modality. The skill level determines whether the chatbot can adapt to the speaker's language proficiency at single or multiple levels of knowledge. The language dimension indicates whether the chatbot supports learning one or more lan-

Table 1 Analytical dimensions

Aspect	Dimension	Value	Description
Speaker	Skill level	Single	Single or inexplicitly separated levels
		Multiple	Multiple levels explicitly separated
	Language	Single	A single target language
		Multiple	Multiple target languages
	Domain	Single	Single knowledge domain
		Multiple	Multiple knowledge domains
	Modality	Text	Textual communication (with possible visuals)
		Voice	Speech communication
		multimodal	Multimodal communication (e.g., text, voice, media)
Technology	Dialog type	Scripted	Predefined, rule-based dialog management
		Statistical	Flexible, ML-based dialog management
	Knowledge base	Open	Open, known language knowledge base
		Closed	Closed, proprietary language knowledge base
	Availability	Open	Free service
		Commercial	Paid service

guages. The domain shows whether a variety of different themes and topics are offered within a single language. The modality can be a text-based conversation (possibly including visual elements), a speech-based conversation, or can combine the previous two in which case it is classified as multimodal.

Concerning the *technology aspects*, the dimensions include dialog type, knowledge base, and availability. For dialog type, we adopted the existing taxonomy of chatbot dialog management systems as (1) predefined, based on matching a finite set of rules, and (2) statistical, which generates responses dynamically from a larger dataset employing ML algorithms [29]. The knowledge base indicates the openness of the chatbot's data source of the linguistic concepts and terms for the language(s) it supports. Finally, the availability shows whether the chatbot is free to use or charges users for its services (fully or partially).

4 Chatbots for Learning Languages

The authors independently installed and tested the chatbots with respect to the aforesaid dimensions. Individual analyses were compared and reconciled in a series of discussions to reach an agreement on the analytical evaluation shown in Table 2.

Table 2 Analytical evaluation of the chatbots for learning languages

Chatbot	Speaker									Technology					
	Skill level		Language		Domain		Modality			Dialog type		Knowledge base		Availability	
	Single	Multiple	Single	Multiple	Single	Multiple	Text	Voice	Multimodal	Scripted	Statistical	Open	Closed	Open	Commercial
Mondly	-	+	-	+	-	+	+	+	+	+	-	-	+	-	+
Babbel	-	+	-	+	-	+	+	-	-	+	-	-	+	-	+
Andy	-	+	-	+	-	+	+	-	-	+	-	-	+	-	+
Lanny	-	+	+	-	-	+	+	-	-	+	-	-	+	-	+

Looking at the table, one can deduce that current chatbots mainly: offer multiple skill levels and languages to learn, combine knowledge from multiple domains, provide textual conversations, implement scripted user dialogs using proprietary linguistic knowledge bases, and charge their users.

Prior to giving a detailed account of the chatbots, (i.e. how they work, what functions they possess, their strenghts and weakenesses), it is worth mentioning that, even though the reference list of avilable chatbots is fairly long, the number of active chatbots is much smaller than expected. Namely, as seen in the table above, by December 2020, we tested only 4 conversational agents that support real-time dialogue with users, i.e. provide immediate answers to prompts, and ask relevant follow-up questions, very much like in any natural and spontaneous conversation with a real person. The curated chatbots [29, 30] including Edwin, Instant Translator, Translator ChatBot, Grammar Guru, CallMom, Tutor Mike, and Dave English Teacher were not available for testing. We will now describe the impressions produced through interaction with chatbots Mondly, Andy, Lanny and Babbel.

Mondly [31] is a language learning application which, in addition to modern and interactive ways of learning, also offers a voice-enabled chatbot. The application itself is partly free of charge, but in order to get access to all its features and content the user must subscribe. Therefore, we only had a short preview of the first lesson in Croatian. The lesson is called *Bok* (the Croatian translation of the word Hi). Communication starts with a greeting—Good morning and the user may select one of the three available prompts: Hello/Good afternoon/Hi. The chatbot then proceeds to ask a few more questions—*How are you? Where are you from? What is your favourite colour?* etc., before it finishes the conversation and wishes you a good day. However, it does offer the option to repeat the same dialogue a few times, but there are very few changes to questions one gets asked. In fact, the only difference lies in the answers offered (each time the user is given, for example, different colours or sports to choose from). It is worth highlighting that the application employs Speech

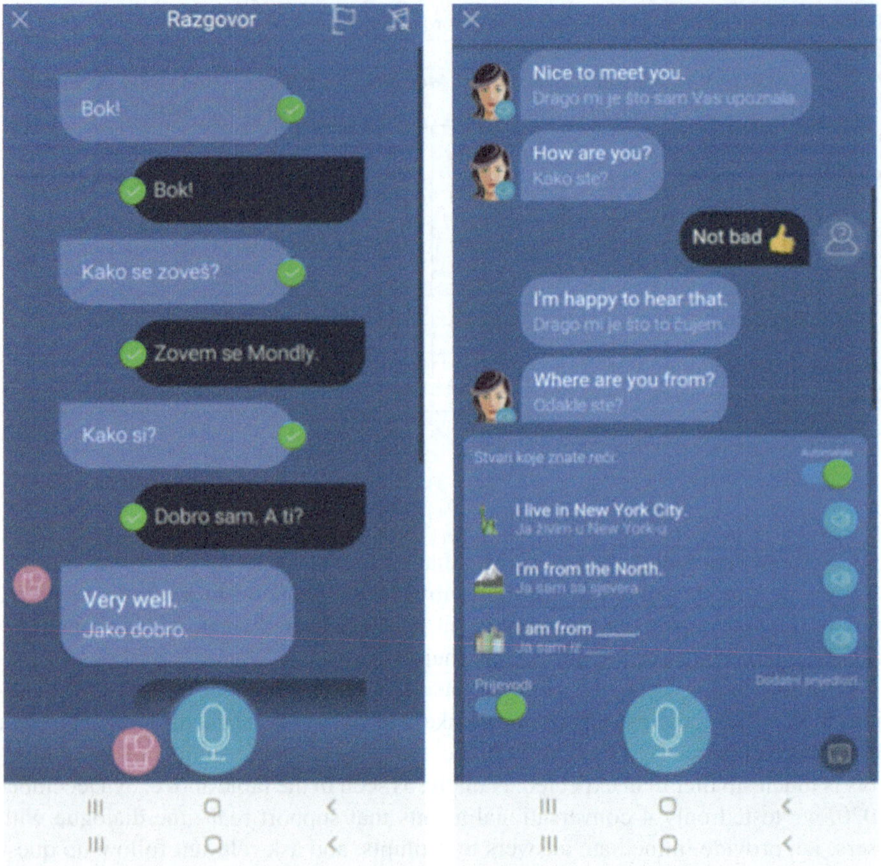

Fig. 1 Mondly—an example of dialogues in the user's native language

Recognition Technology (SRT) which enables the user to record their answers. By recording their answers, the user is able to compare their pronunciation with the native speaker's, which is a very useful feature if you are a beginner. The prompts all seem unnecessary if the level you chose at the start was advanced.

The overall impression is that the conversation neither comes across as very authentic (it is obvious that there is a script it follows very closely), nor does it evolve beyond the simple question-answering sequence. If you are an absolute beginner, or even starting over, Mondly is a great choice as it will teach you a lot of useful vocabulary and phrases for a novice language learner. It also works well with lower intermediate students. However, for a more fluent speaker, Mondly, is not likely to provide an immersive experience. Some of the positives include the fact that it speaks 33 languages and also targets different skills (speaking, listening, reading, writing). The list of others could include: personalization, leaderboards and self-progress reports. The

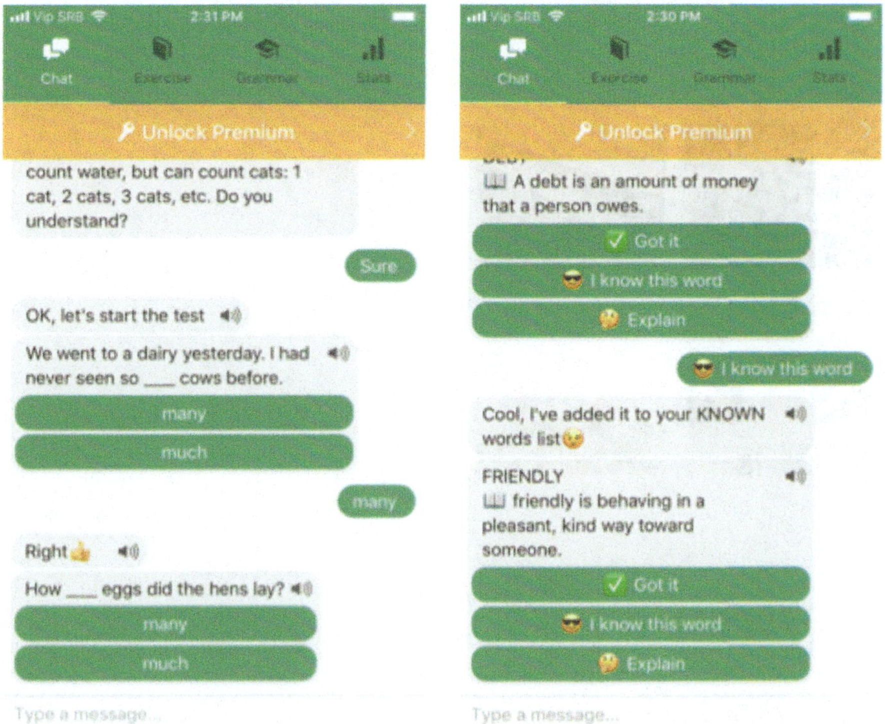

Fig. 2 Andy—a dialog for grammar exercise (left) and a dialog for word (vocabulary) exercise (right)

chatbot also integrates VR technology, with conversations between digital avatars (personas) in artificial environments (such as a train or restaurant) (Fig. 1).

Andy [32] is a phone application aimed at helping users acquire new words and phrases and practice grammar through games and chat. The application is partly free and the trial period lasts for 7 days. If you wish to keep practicing after the trial period expires, it offers you the choice of a paid monthly or annual subscription. The premium subscription enables you to access additional learning features such as grammar lessons and unlimited vocabulary practice. The upside of this application is that the conversation, unlike Mondly, which offers little or no room for a spontaneous chat, is relatively spontaneous and extensive, and conveys the impression that one is conversing with a real person. It is pleasantly engaging and if the reason for using it is to get the feel of talking to a native speaker, Andy is a fairly good option. Level-wise, Andy offers greater diversity. Despite not being able to offer a placement test or any other form of assessment, the overall impression is that the more fluent the speaker is the more dynamic and complex the conversion becomes. However, the major disadvantage lies in the fact that it is lacking a voice-enabled function, which confines communication to text messaging, an activity unlikely to keep one engaged

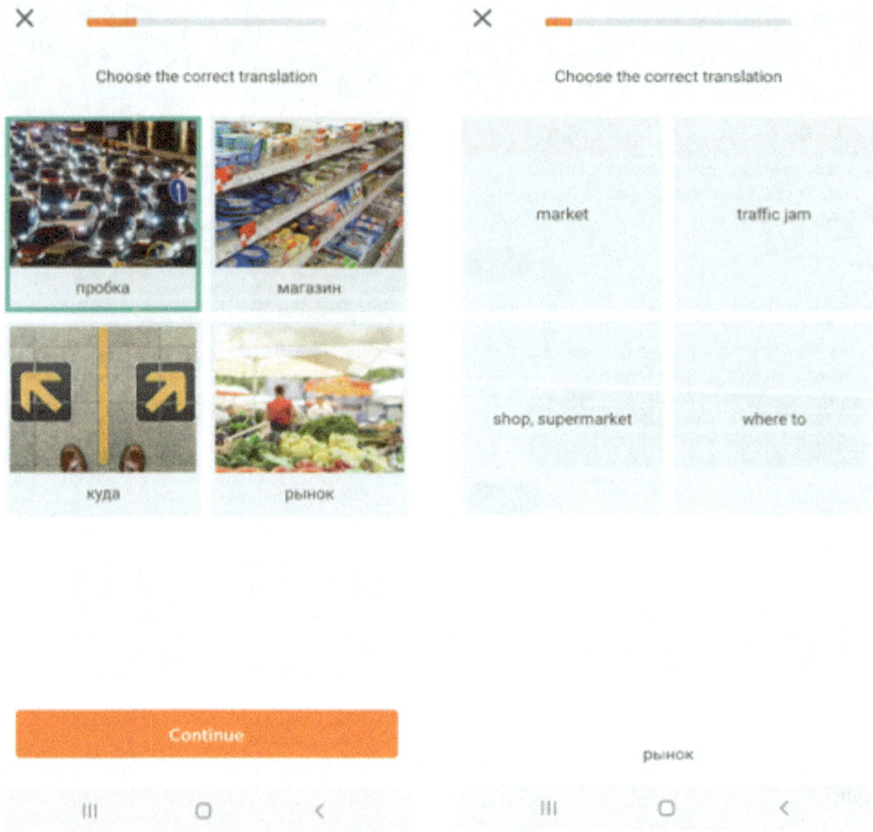

Fig. 3 The Babbel user interface with visual elements for practising translation

for an extended period of time. If you do not subscribe to either the monthly or annual package at the start, the application will ask you to either donate money or subscribe fairly early on in the conversation. It is also worth pointing out, that for the purpose of this paper it was our joint decision not to pay or subscribe to any package deal or services provided in exchange for money (Fig. 2).

Babbel [33] is a virtual assistant that speaks 14 languages. The chatbot offers dynamic and flexible conversation threads. It supports scaffolded learning by monitoring the user's progress and providing level-based lessons. The lessons are contextual, bite-size lectures (10–15 min) covering different topics concerning everyday situations. The assistant targets a variety of skills, including speaking, listening, reading, and writing. However, it is free only for the first month of use and charges its users after that (Fig. 3).

Lanny [34] is a live chat app created for people wishing to master the basics or advance their existing knowledge of Korean. One of the benefits of using this chatbot is that, in addition to conversing in Korean, the chatbot also uses English to

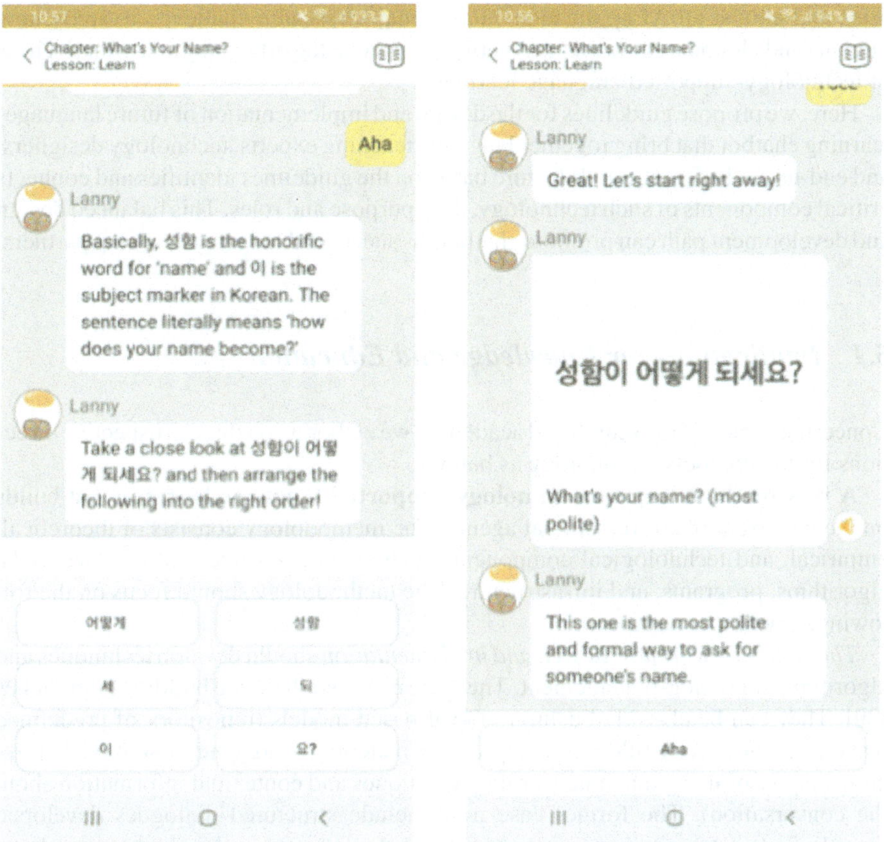

Fig. 4 Lanny—dialogs with the user

answer and ask questions and clarify things. This makes Lanny ideal for people trying to grasp the basics of this language. But Lanny is more than just a conversational partner, it is also a guide to Korean culture and history. The application boasts the use of audio by native Korean speakers that will help learners with pronunciation, reading, writing and speaking Korean. The application is available for Android and iOS users and also has a web version (Fig. 4).

5 Implications for Future Research

Intelligent chatbots for learning languages are meant to work with professionals, not to replace them. Language experts have extensive knowledge and experience that even the most sophisticated AI algorithms of today cannot match. In these applications and almost all AI applications, it is a major research challenge to study how humans and chatbots can work in synergy to improve the efficiency and effectiveness of technology-supported language learning.

Here, we propose guidelines for the design and implementation of future language-learning chatbot that bring together language teaching experts, technology designers, and end-users. Reference architecture based on the guidelines identifies and connects critical components of such technology, their purpose and roles. This balanced design and development path can promote chatbots as successful language learning partners.

5.1 Implications for Knowledge and Education

Concerning scientific research and academia, we elaborate on three prospective directions for the chatbots' technology as below.

A new methodology for technology-supported language learning that builds on the paradigm of conversational agents. The methodology consists of theoretical, empirical, and technological components such as best practice, training materials, algorithms, programs, and infrastructure. The methodology should focus on the following, crucial processes:

The chatbot's workflow design and implementation should develop techniques and algorithms for dialog-management. There are different dialogue building approaches [29]. They can be classified using retrieval-based models (repository of predefined responses selected by rule-based expression matching) and generative models (creating responses using machine learning techniques and contextual information about the conversation). The former case may include structured dialogues developed directly from reference learning materials. In the latter case, the chatbot may learn through conversations, or automatically (given the availability of a training dataset [35]). It is also important to investigate the inclusion of different contextual information such as conversational (e.g., something mentioned earlier in the text), and the user or physical context (e.g., date/time, location, user information) to select the right responses. At the moment, robust, but domain-specific dialog systems using handcrafted rules can be created accurately and straightforwardly using available tools such as Google DialogFlow [30] or ChatScript [36]. ML tools, such as Google TensorFlow [37], can improve the authenticity of such dialogs. However, the accuracy of the tools ultimately depends on the availability and quality of a language-specific training dataset (or model). The existing datasets are biased towards the English language, while other languages are underrepresented.[1] This makes it more challenging to use ML algorithms to learn languages other than English. On the one hand, the creation of these datasets would require manual effort from proficient speakers. On the other hand, it would embrace the socio-cultural characteristics of the specific language speaking region, which is important when learning a foreign language.

[1] There are 7,000 languages in use around the world—this is why they matter: https://www.weforum.org/agenda/2020/02/how-some-endangered-languages-thrive (Retrieved on 28.12.2020).

The chatbot's user interface (UI) design and implementation should concentrate on its appearance and conversational styles (including gamification elements) to elicit positive attitudes (such as increased engagement) and outcomes (such as the quality of engagement) [9]. For instance, mimicking human languages with high message interactivity (conversational cues). It assumes engaging in back-and-forth message exchanges (mainly textual). In addition, the use of human figures (visual cues) leads users to treat chatbots as human and act socially towards them. Finally, human names or identities in the form of labeling a chatbot (identity cue) can be used since humans tend to perceive things by their labels. The form (usability) and function of an interactive product are equally important and cannot be separated. The *operationalization of the chatbot's visual cues, identity cues, and message interactivity as concrete technology design elements* can reveal how interactions among these three factors influence attitudinal and behavioral outcomes (user motivation and language proficiency). For example, how to combine them to compensate for the lack of face-to-face communication. It is also important to investigate the effectiveness of different chatbot input modes (speech and text).

Implementing such a methodology requires cross-cultural, longitudinal qualitative and quantitative studies with end-users. The studies should combine different instruments such as *subjective, self-report* ratings containing measures such as usefulness, usability, emotions, and *objective* metrics such as the level of acquired skills in a foreign language, the amount of conversations, number of errors related to use, and dialogue time. These are all essential predictors of users' acceptance and adoption of the technology [38].

Technology for personalized language learning. A chatbot which keeps track of users' past questions and level of language use, could, over a series of interactions, become familiar to users. The chatbot could reuse past language that had been successfully responded to, thereby enhancing users' self-perceived ability to support interest development. Finally, it could remind the user of the necessity of practice and trying out new words or phrases. A correct interpretation of user requirements, and a match between the chatbot's design and the needs of the language learning task is crucial. The requirements are language- and culture-specific. We envision defining particular scenarios that match specific language learning tasks and domains using co-design techniques [39].

A novel, online conversational service for learning languages at scale (L@S) by creating individual learning experiences, improving learning outcomes, and supporting lecturers. This approach is especially useful for large-scale lecturers and online courses with more than a hundred students per lecture where individual support is not possible due to financial and organizational restrictions. The proposed methodology can deliver a blueprint for the technical infrastructure for developing and deploying such a service.

5.2 Implications for the Economy

The impact of the chatbots on the economy is three-fold, as follows.

Implementation and integration in common platforms (instant messaging tools such as Viber or Facebook Messenger) using open, commercial chatbot technologies (such as Amazon Alexa Skills [1], or Google DialogFlow [30]) and machine learning tools (i.e., Google TensorFlow [37]). These technologies are currently mature enough to develop robust, commercially viable solutions that suit specific domains. Compared to developing from scratch, building a conversational agent with off-the-shelf technologies reduces time-to-market while ensuring the required usefulness and reliability.

Successful combinations of UI elements (usability) and dialog management components (function) can be framed as **specific conversational design patterns (or templates) for the domain of language learning.** Similarly to weather, banking, and other templates used for building commercial, dialogue-based digital assistants. They can improve the scalability and reusability within and across multiple applications.

Hybrid human-machine translation using conversational metaphor. It would be pertinent to examine the efficiency and accuracy of supervised multimodal translation that combines text and speech compared to either text- or speech-based translation. By their very nature, chatbots can offer a holistic approach to translation that can capture some of the frequent cases of lexical diversity between different languages that occur mainly in human speech (i.e., lexical gaps in casual conversations).

5.3 Implications for People and Society

Finally, we show how chatbots can support people in their everyday activities and transform our society.

Chatbots present an **affordable and ubiquitous source of language interaction** for many students learning a Foreign Language. It is, therefore, essential to understand how they can best be put to use both in formal and informal education. As a source of conversation practice and pervasive technology, they enable extra practice during independent study anywhere and anytime.

Chatbots can offer some **language learning mechanisms** that many human language learning partners could not and/or would not. In particular, a broader range of expressions and vocabulary, the scaffolded introduction of new vocabulary, grammar, and expressions, or consistent understandable repetition which a human partner is unlikely to present (except language experts). It can also alleviate the issue of availability of native speakers or expert-speaking users, or even difficulty in finding a skilled and available speaking partner online [29].

An inclusive design and development strategy for language learning chatbots. This strategy is not about replacing lecturers with chatbots but creating a co-dependent and intelligent relationship between teacher and chatbot utilizing their strengths and delivering the best student experience. This strategy should aim for improvements in chatbots' usability by following the principles of conversational user experience design [3].

6 The Reference Architecture

Based on the research implications, we propose a blueprint for the technological architecture for designing and developing chatbots for learning languages (Fig. 5).

The architecture aims to address the issues mentioned above by technology design. The *knowledge layer* contains linguistic concepts and terms necessary for language content generation. Preferably in different languages, but at least two—a learner's native language and a second language. The existing knowledge sources can be exploited [40]. User-related information includes personal information, learning-related information, and dialog-related information. At the *language layer*, the information for learning languages concerning different language domain(s) and purpose(s) is created from the layer below. The layer's content serves as the basis for generating and maintaining conversations with users through either interactive lecturing or examinations. The *dialog layer* creates and manages conversation threads. In particular, it accepts and interprets user input, makes contents of responses with specific semantics and syntax in one or several languages (dialog management), and creates outputs as a natural language. The *UI layer* implements the agent's conversational appearance towards end-users as either text/visual, voice, or a combination of the two.

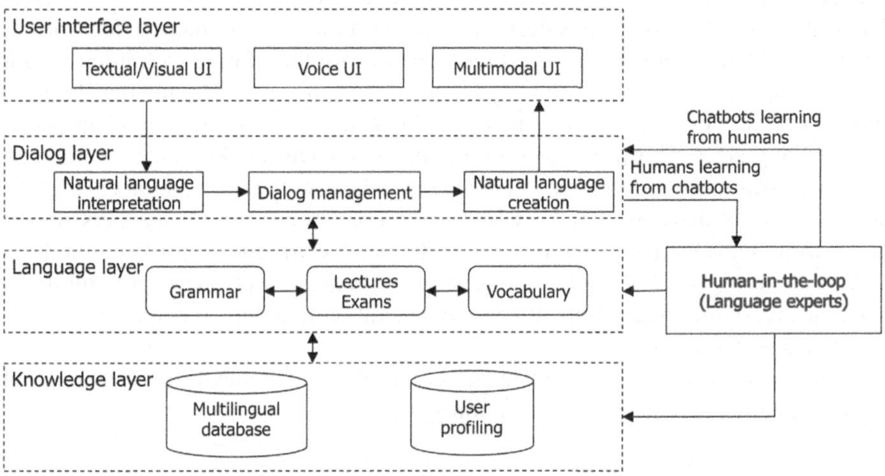

Fig. 5 The reference architecture for chatbots for learning foreign languages

Since automatically generated conversations still cannot match the ones created by people, human intervention is necessary. In this mutual relationship, humans can learn from chatbots. For example, the new meaning of existing words or new words with new meanings (i.e., casual speech input by users from specific speaking regions or specific dialects). Vice-versa, chatbots can learn from humans through expert supervision. Such supervision may include manual interventions in the automated natural language interpretation of the learners' input, creation of responses, and natural language generation. This way, the chatbots can not only improve on their functions but capture the evolution of languages.

7 Limitations

Our goal was to conduct the survey with an analytical (heuristic) evaluation of the chatbots to identify critical aspects, understand their effects, and provide design recommendations for their improvement. This paper does not provide an exhaustive list of chatbots for learning foreign languages. Instead, it extracts common examples that describe the current landscape of language chatbots. In addition, we did not assess the effectiveness of chatbots on learning languages. These are the elements to be covered in future research.

8 Conclusion

To the best of our knowledge, this is the first critical analysis of chatbots for learning languages that explicitly considers and combines end-user and technological aspects. The paper provides a critical reflection on the current state of chatbots for learning languages and indicates some important directions for their future. A limited sample of available chatbots ($N = 4$) shows the complexity of the design and development space. It illustrates the variety of factors influencing the feasibility of chatbots for learning languages. This paper informs end-users about available chatbot technology, communicates their possibilities to language teachers, and guides engineers for their development. Will future chatbots be fully autonomous, or will teachers remain and learn to enhance their methods using chatbots? Currently, we believe the latter to be the case—at least until AI makes the prerequisite advances in natural text and voice recognition, understanding, and generation to equip chatbots with greater effectiveness and autonomy.

In our future research we intend to analyze the role of language chatbots in hybrid, human-machine collaborative translation systems.

References

1. Amazon Alexa voice assistant.: Retrieved from https://developer.amazon.com/en-US/alexa, 28 Dec. 2020
2. Google Assistant services.: Retrieved from https://assistant.google.com/, 28 Dec. 2020
3. Moore, R.J., and Arar, R.: Conversational UX design: a practitioner's guide to the natural conversation framework. Morgan & Claypool, San Rafael (2019)
4. Höhn, S.: Artificial companion for second language conversation. Springer International Publishing, Berlin (2019)
5. Fryer, L.K., et al.: Stimulating and sustaining interest in a language course: an experimental comparison of Chatbot and Human task partners. Comput. Human Behav. **75**, 461–468 (2017)
6. Fryer, L.K., Nakao, K., Thompson, A.: Chatbot learning partners: connecting learning experiences, interest and competence. Comput. Human Behav. **93**, 279–289 (2019)
7. Hwang, I., Lee, Y., Yoo, C., Min, C., Yim, D., Kim, J.: Towards interpersonal assistants: next-generation conversational agents. IEEE Pervasive Comput. **18**(2), 21–31 (2019)
8. Klopfenstein, L.C., Delpriori, S., Malatini, S., and Bogliolo, A.: The rise of bots: a survey of conversational interfaces, patterns, and paradigms. In: Proceedings of the 2017 Conference on Designing Interactive Systems, pp. 555–565. ACM Press, New York (2017)
9. Go, E., Sundar, S.S.: Humanizing chatbots: the effects of visual, identity and conversational cues on humanness perceptions. Comput. Human Behav. **97**, 304–316 (2019)
10. Ruan, S., et al.: BookBuddy: turning digital materials into interactive foreign language lessons through a voice chatbot. In: Proceedings of the 6th ACM Conference on Learning @ Scale, p. 30 (2019)
11. Winkler, R., Söllner, M.: Unleashing the potential of chatbots in education: a state-of-the-art analysis (2018)
12. HelloTalk mobile application for learning foreign languages with native speakers. Retrieved from https://www.hellotalk.com, 28 Dec. 2020
13. Duolingo web and mobile platform for learning foreign languages. Retrieved from https://www.duolingo.com/learn, 28 Dec. 2020
14. Rosetta Stone web and mobile platform for learning foreign languages. Retrieved from https://www.rosettastone.com/, 28 Dec. 2020
15. Memrise web and mobile application for learning foreign languages from native speakers. Retrieved from https://www.memrise.com/, 28 Dec. 2020
16. WordUp mobile application for improving English vocabulary. Retrieved from https://www.wordupapp.co/, 28 Dec. 2020
17. Deterding, S., Dixon, D., Khaled, R., Nacke, L.: From game design elements to gamefulness: defining "gamification". In: Proceedngs of the 15th International Academic MindTrek Conference: Envisioning Future Media Environments, pp. 9–15 (2011)
18. Hung, H.T., Yang, J.C., Hwang, G.J., Chu, H.C., Wang, C.C.: A scoping review of research on digital game-based language learning. Comput. Educat. **126**, 89–104 (2018)
19. Zou, D., Huang, Y., Xie, H.: Digital game-based vocabulary learning: where are we and where are we going? Comput. Assist. Lang. Learn. **34**, 1–27 (2019)
20. Duolingo—translating user research into improved design. Retrieved from, https://uxplanet.org/duolingo-e3e47d2b41ba, 28 Dec. 2020
21. Kahoot! web educational platform. Retrieved from https://kahoot.com/, https://kahoot.it, 28 Dec. 2020
22. Ver-taal language exercises for foreigners. Retrieved from www.ver-taal.com, 28 Dec. 2020
23. Botlist collection of chatbots for learning languages. Retrieved from https://botlist.co/collections/@Taell/express-yourself, 28 Dec. 2020
24. Chatbots for learning languages. Retrieved from https://chatbots.org/industry/learnlookup/, 28 Dec. 2020
25. 7 of the Best Language-Learning Chatbot Apps. Retrieved from https://blog.vsoftconsulting.com/blog/7-of-the-best-language-learning-chatbot-apps, 28 Dec. 2020

26. 5 Resources for Chatbots to Be Your Language Learning BFFs. Retrieved from https://www.fluentu.com/blog/language-learning-chatbot/, 28 Dec. 2020
27. 7 Powerful Chatbot Apps for Learning Languages. Retrieved from https://www.tristatetechnology.com/blog/best-language-learning-chatbot-apps/, 28 Dec. 2020
28. Pham, X.L., Pham, T., Nguyen, Q.M., Nguyen, T.H., Cao, T.T.H.: Chatbot as an intelligent personal assistant for mobile language learning. In: Proceedings of the 2018 2nd International Conference on Education and E-Learning, pp. 16–21 (2018)
29. Harms, J.G., Kucherbaev, P., Bozzon, A., Houben, G.J.: Approaches for dialog management in conversational agents. IEEE Internet Comput. **23**(2), 13–22 (2019)
30. Google DialogFlow platform for chatbot development. Retrieved from https://dialogflow.com/, 28 Dec. 2020
31. Mondly Chatbot for learning foreign languages. Retrieved from https://app.mondly.com/, 28 Dec. 2020
32. Andy Chatbot for learning the English language. Retrieved from https://andychatbot.com/, 28 Dec. 2020
33. Babbel Chatbot for learning foreign languages. Retrieved from https://it.babbel.com/, 28 Dec. 2020
34. Lanny Chatbot for learning the Korean language. Retrieved from https://web.eggbun.net/, 28 Dec. 2020
35. A Trainable Spaced Repetition Model for Language Learning. Retrieved from https://doi.org/10.7910/DVN/N8XJME, 28 Dec. 2020
36. ChatScript—an open source authoring tool for building textual chatbots. https://en.wikipedia.org/wiki/ChatScript, 28 Dec. 2020
37. Google TensorFlow Machine Learning platforms. Retrieved from https://www.tensorflow.org/, 28 Dec. 2020
38. Gursoy, D., et al.: Consumers acceptance of artificially intelligent (AI) device use in service delivery. Int. J. Inf. Manag. **49**, 157–169 (2019)
39. Steen, M.: Co-design as a process of joint inquiry and imagination. Design Issues **29**(2), 16–28. MIT Press (2013)
40. Wordnet—the largest lexical database of English. Retrieved from https://wordnet.princeton.edu/, 28 Dec. 2020

Intelligent Interactive Technologies for Mental Health and Well-Being

Mlađan Jovanović, Aleksandar Jevremović, and Milica Pejović-Milovančević

Abstract Mental healthcare has seen numerous benefits from interactive technologies and artificial intelligence. Various interventions have successfully used intelligent technologies to automate the assessment and evaluation of psychological treatments and mental well-being and functioning. These technologies include different types of robots, video games, and conversational agents. The paper critically analyzes existing solutions with the outlooks for their future. In particular, we: (i) give an overview of the technology for mental health, (ii) critically analyze the technology against the proposed criteria, and (iii) provide the design outlooks for these technologies.

Keywords Human-ai interaction · Intelligent systems · Digital healthcare · Mental health · Mental well-being · Review · Survey · Robotic technologies · Video games · Conversational agents · Chatbots · Affective computing

1 Introduction

Artificial Intelligence (AI) is among the oldest disciplines of computer science that builds systems that resemble humans in learning, thinking, problem-solving, and decision making. The field received significant interest in the previous decade, mainly due to advances in automated machine learning (ML) and deep learning (DL). They learn useful patterns from a large amount of data and keep the acquired knowledge as model structures and parameters that can be further applied to make predictions by interpreting unseen data [1]. These models are either a set of elements or features

M. Jovanović (✉) · A. Jevremović
Singidunum University, Belgrade, Serbia
e-mail: mjovanovic@singidunum.ac.rs

A. Jevremović
e-mail: ajevremovic@singidunum.ac.rs

M. Pejović-Milovančević
Institute of Mental Health and Faculty of Medicine, University of Belgrade, Belgrade, Serbia
e-mail: milica.pejovic@imh.org.rs

E. Pap (ed.), *Artificial Intelligence: Theory and Applications*,
Studies in Computational Intelligence 973,
https://doi.org/10.1007/978-3-030-72711-6_18

that contribute when making decisions (in ML) or are organized into several layers of abstraction (such as neural networks) for general and specific interpretation tasks (in DL).

Healthcare provision and medicine are one of the most significant challenges of AI due to being the pillars for a global society and the necessity of providing higher-quality assistance to the healthcare workforce [2].

An emerging and expanding domain for application of AI is mental health. Readily available and ubiquitous devices and applications enable the provision of flexible mental care—on-demand, at any time, both at healthcare facilities and at home. Societal challenges, such as the ongoing Covid-19 pandemic, can necessitate physical distancing. Consequently, prolonged social isolation brings common risks for mental health and well-being, potentially triggering conditions such as loneliness, anxiety, and depression [3, 4]. The growing demand for telemedicine as contactless, remote, and accessible healthcare services can help users stay connected, visit their doctors remotely, and self-manage their mental functioning during the global Covid-19 lockdown [3].

As they move into the digital era, mental healthcare is seeing substantial benefits from interactive technologies and ML/DL, including treatment both at home and in hospitals. Medical decision-making is by its very nature uncertain, unknown, inconsistent, and lacking data from multi-dimensional spaces. At the same time, it must capture patients' heterogeneity to adjust healthcare decisions, prescriptions, and therapies to the individual. Moreover, healthcare professionals must understand the automated decision-making process to verify it. Prevention is crucial here. It maintains normal mental functioning, avoids the onset of critical conditions, prevents existing illnesses from progressing, reduces healthcare burden, and makes people satisfied and happier [5]. Reducing the gap between health and well-being is something that traditional healthcare may struggle to achieve. These technologies can help by ingraining positive health habits through sustained user engagement [6, 7].

This paper does not focus on the implementation and performance details of the AI models and algorithms, but the consequences they have for the user experience of a specific healthcare technology employing these models. In particular, we concentrate on user-related aspects of these technologies that may influence their acceptance and the effectiveness of mental healthcare provision.

The paper is structured as follows. Section 2 describes a landscape of mental health interventions organized as robotic technologies, video games, and conversational agents that can sense and respond to human emotions. Section 3 introduces the criteria used to analyze the technologies. Section 4 presents the results of the analysis, with the future outlook. Section 5 concludes the paper.

2 Related Work

We describe the intelligent interactive technologies for mental healthcare as they emerged. The systems are grouped as robotic technologies, video games, and conversational agents that can recognize and adapt to users' emotions (being an emerging requirement for digital mental health therapy).

2.1 Robotic Technologies

Social robotics has proved to be a useful method for treating autism spectrum disorder (ASD) in children. Previous research attests to the various benefits of robotic devices targeting specific aspects of ASD behavior. The outcomes range from decreasing repetitive and stereotyped behaviors [8], through enhanced attention and imitation behavior for social learning [9], to the induction of spontaneous linguistic behavior [10–12]. The results of using socially assistive robots to improve mental health are mixed and there is a lack of empirical evidence or utility of using them in mental health interventions [13].

One of the most popular programmable humanoid robots is NAO. It has multiple degrees of freedom and sensors, as well as the integrated speech recognition support. There is a wide range of applications of NAO robot, including interventions for children with ASD [14], diabetes management [15], reduction of apathy and stabilization of cognitive performance in advanced dementia patients [16], supporting the learning of pupils with severe intellectual disabilities [17], rehabilitation training [18], reducing pediatric distress and pain in medical settings [19], and logopedic interventions [20, 21]. Pepper is a humanoid robot that uses the same underlying platform. A study shows there are no differences in expectations between the models related to the physical appearance—before the interaction. However, it changes after the interaction [22].

Pediatric care is a field in which the use of intelligent robotic technology is increasing. Huggable is a smartphone-based social robot designed to address the emotional needs of hospitalized children by engaging them in playful interactions and promoting their socio-emotional well-being [23]. It has 12 degrees of freedom and can move its head, shoulders, elbows, waist, muzzle, and ears. Another example of an Android-based robot platform that leverages smartphones to drive computation and display an animated face is Tega [24]. The platform is designed to support long-term, in-home interactions with children in early-literacy education, from vocabulary to storytelling.

Medi is another example of a social robot used in pediatrics, designed to reduce the level of anxiety and stress when children cope with pain [25]. The robot uses a set of behaviors derived from pain management literature and cognitive behavioral therapy (CBT). The built-in functions include personalized greetings, playing games, storytelling, and dancing. Robots such as the THERAPIST are used in motoric and neurorehabilitation therapies for children through different games activities activities, but require complex cognitive architectures to provide optimal and personalized experiences [26].

CommU is a smaller humanoid robot with 14 degrees of freedom—including directed eye gaze—was used by teenagers with ASD to converse with their teacher

[27] indirectly. The robot was designed to be small and cute to avoid fearfulness among the children. Regardless of the lack of ML/DL, the robot's characteristics make it a compatible for their later implementation. When designing the robot's exterior and appearance, robustness and sanitation ability are important factors for use in the hospital environment. Actroid-F is another robot used as an intermediary for communication with young adults with ASD [28].

Toy robot Pekoppa, shaped like a bilobed plant with a responsive stem and leaves, was used as a listener in communication with neurotypical and children with autism aged 6–7 years [29]. The robot's form is compatible with the finding that children with autism show a preference for minimalist objects to which they can their own mental states or those of others [30].

Other than the previous solutions, open-source robotic technologies are emerging. For example, the Poppy project is a community-centered robotic project that provides free specifications for 3D printing of a humanoid robot, torso, or arm [31]. The printouts can be combined with affordable controllers, including Arduino, RaspberryPi, and others. Other examples are iCub, a humanoid robot designed to support research in embodied AI [32], and Oncilla [33]. Additional sensors could be integrated into the solution to improve the Human-Robot Interaction experience [34].

A large user group that prefers and benefits from interacting with physically embodied robots are older people[35–37]. Social robotics developed for this user group is mainly oriented towards preventing dementia [38] and loneliness [39]. Telenoid [40] and Giraff [41] are examples of robots designed to support telepresence. The Nurse-Bot project is a RaspberryPi based robot system to provide medical assistance in later life [42].

The main concern with robotic technologies is user acceptance, influenced by how likable and trustworthy they are. If a robot looks too human-like but does not match social expectations in terms of behavior, the user might distrust these systems [43]. Social robotics is unlikely to be ubiquitous in the near future due to its higher cost compared to other technologies, while physical presence is not a prerequisite for emotional expression [94].

2.2 *Video Games*

Another stream of research and system development shows the various benefits of video games for the mental health and well-being of different target user groups [44–57].

The most common properties of video games concerning enhanced mental well-being include feedback on progress, points and scoring, rewards and prizes, narratives, personalization, and customization [50]. The related target aspects of mental well-being are anxiety, well-being, alcohol use, and depression. On the other hand, games suffer from methodological issues concerning the validity and generalizability of their findings. Games are used as a black-box solution without understanding their underlying mechanism and how they approach and tackle specific mental health conditions. The concrete health outcomes are described in terms of perceivable user experience elements (as the ones mentioned above), and not in relation to the game mechanics, comprehensive explanations of how they are used, and causal relationships between game tasks/rules and elements of human mental functioning [50]. Resolving the latter challenges is crucial for the correct application of games in mental healthcare.

As we get older, we notice many changes. Our brain may also show signs of ageing. Cognitive decline in thinking, language, memory, understanding, and judgment are often a normal part of ageing. It can also be an early sign of dementia, a group of brain disorders characterized by a more severe, irreversible decline in cognitive functions. Engaging in activities that stimulate the brain throughout life is thought to enhance cognition in later life and reduce the risk of age-related cognitive decline and dementia.

Cognitive training is an intervention that provides structured practice on tasks relevant to different aspects of cognition, such as attention, memory, or executive functions [51]. Training can focus on a single cognitive domain (e.g., memory) or multi-domain training (including attention and processing speed). As such, video games have been used to design the cognitive training.

In cognitive/emotional training, most mental health interventions used action games (i.e., Call of Duty: Modern Warfare3 in [53]), followed by puzzle games (i.e., Angry Birds from [54]) [52]. The prevalent aspects of cognition that are trained include processing speed and reaction times, memory, task-switching/multitasking, and mental spatial rotation [52].

One of the main difficulties in obtaining a clear picture of the effects of video game training is the significant variability of several key variables of the intervention studies—the type of video game used, the cognitive process assessed, how these cognitive processes were evaluated, and the different personal characteristics of the trainees [50, 52, 56]. This variability seems to be the leading cause of the mixed results reported in the literature [52]. The reviewed studies indicate that video game training improves some aspects of cognition but not others [50, 56]. The main limitations are relatively small sample sizes in longitudinal intervention studies, the diversity of motivational factors for playing video games, and the placebo effect resulting from familiarity with the investigators [50, 52, 56].

Concerning specific target groups, video games are increasingly used to help autistic children improve and maintain their social and cognitive skills [45–49]. For example, the ECHOES game [44] uses AI to control a virtual character during a joint attention and communication skills training program. The agent acts as a social partner to autistic children when learning. This technological intervention includes interactive storytelling elements, medium-term goals, and rewards for success. The evaluation revealed that children initiated interactions with the virtual character more often.

The children are motivated to engage with the games, find them easy to use, and learn relevant content over a short period. There are important questions about how effective and how much is learned using video games. Regarding effectiveness, games' clinical validation does not meet evidence-based medicine standards, and there is a lack of connection and compatibility between the game design and the clinical validation [48, 49]. Concerning learning, the results show that it is important provide an educational game context that meets the children's diverse interests and preferences. The majority of existing games have a positive effect on high-functioning individuals [46, 48]. In general, games represent a complex design space requiring research on how different gamification approaches can affect the experiences of participants with inherently varied motivations and interests in gaming [56]. The heterogeneity of the children with autism impacts how the user model of a serious game for different users needs to be constructed.

The success of gamified cognitive training has been demonstrated for both commercial and noncommercial video games [52, 56]. Some studies (dealing with emotion regulation) even suggest using commercial games rather than games developed with specific purposes due to prolonged engagement, which is critical for effective mental health treatment [55, 56]. On the other hand, current gamified technologies lack standardized terminology, development and study protocols, best practices in matching games and mental health tasks, and clear ethical procedures[56].

Similarly, gamification as the practice of using game elements in non-game settings [57] has been recently proposed for digital health interventions [58]. The work described in [58] offers a model of gamification principles for digital healthcare. The principles are articulated as user requirements and cover meaningful purpose, meaningful choice, supporting player archetypes, feedback, and visibility (see [58] for details). They were empirically validated in a user study (N = b113 raters) by assessing user satisfaction against Web and mobile applications (N = 17). The study revealed a significant correlation between three variables (supporting player archetypes, feedback, and visibility) and the test of application quality concerning engagement, functionality, aesthetics, and information quality (known as MARS [59]).

From an AI perspective, video games are a promising technology as many features are integrated into commercial (and available) gaming environments and platforms such as Unity.[1] For example, the Unity games can implement intelligent agents moving in a synthetic game environment, detecting, reacting to collisions, interacting with other agents, games' objects, and avoiding dynamic obstacles.[23] The Unity Perception uses computer vision and augmented reality to detect and interpret real-world and digital game objects and characters.[4] Aside from using AI algorithms, a recent achievement of games is the automated generation and annotation of training data for such algorithms at scale.[5]

---

[1]Unity game engine: https://unity.com/ (Retrieved on Dec 25, 2020).

[2]Exploring new ways to simulate the coronavirus spread: https://bit.ly/37PQWYG (Retrieved on Dec 25, 2020).

[3]Unity Machine Learning Agents: https://bit.ly/34RG2zM (Retrieved on Dec 25, 2020).

[4]Training a performant object detection ML model on synthetic data using Unity Perception tools: https://bit.ly/3hjoLVl (Retrieved on Dec 25, 2020).

[5]Use Unity's perception tools to generate and analyze synthetic data at scale to train your ML models: https://bit.ly/3rwofrt (Retrieved on Dec 25, 2020).

2.3 Conversational Agents

Technology-based health interventions are increasingly using a shared design metaphor—a personal, intelligent assistant (also known as a chatbot) that provides healthcare through natural conversation. The main reason is to make existing services more user-friendly—an agent takes a patient through a turn-taking dialog, exchanging questions and answers to complete a task, similar to how doctors do [60]. Successful examples of such agents in mental health care are Woebot [60] designed to help with depression, and Vincent, aimed at raising self-compassion [61]. The technology is integrable into popular messaging platforms, such as Facebook Messenger, Slack, Telegram, and Skype. Although current chatbots do not possess the same level of physical presence as humans, even unimodal interactions (such as text or voice) can still have behavioral significance, while being less costly to design and deploy [6, 62]. Users display positive sentiments and trust towards virtual agents that provide emotional and functional affordances and dominate the users' experience with these agents [63]. A common finding was that the agents' functionality and social interactivity are equally important and complementary rather than separate [64]. In digital healthcare, chatbots provide low-cost, easy access to medical triage (e.g., Babylon [65]), mental health support and well-being (e.g., Woebot [60]), and health-promoting behavior change (e.g., Florence [66]).

Recently, conversational agents have been used successfully to automate the assessment and evaluation of psychological treatments through social support, supporting mental well-being and promoting psychological functioning [61, 67]. They act as digital avatars and communicate with their users through natural dialogue. In this way, they enable medical professionals to continue mental health counseling and therapy. Some prominent examples are Woebot [60] and Wysa [68]. Woebot [60] is a virtual mental therapist who monitors users' mood and suggests tips and mental activities. Wysa [68] offers emotional support during conversation with its users to improve their mental health and well-being. CBT is an umbrella term for various treatments that aim to enhance or re-establish normal functioning concerning specific mental states and emotions. Agents implement online, remote CBT as a guided dialogue that commences with diagnosing patients' mental conditions. Based on the diagnosis, the agent suggests and assists in specific exercises and monitors them through structured conversations. The measures of a treatment's progress are extracted from patients' responses as free-form text or selected options. The underlying logic analyzes and interprets patients' input using Natural Language Processing (NLP) and ML algorithms to estimate and make successive decisions during therapy. The virtual therapist reduces the gap between health and well-being—facilitating individual resilience to mental disorders (i.e., depression, stress, and anxiety) by developing desirable self-care habits through personalized mental exercise (such as mindfulness, self-awareness, and optimism).

The requirement for emotionally sensitive chatbots assumes multimodal sensing capabilities and expressing emotions in more complex ways. Computer science methods have been applied to visual, audio, and textual data to infer emotion [69]. In many cases, this involves detecting subtle signals in high-dimensional data. While verbal and nonverbal cues both contain rich information about a person's emotional state, researchers have found significant improvement in automated understanding of nonverbal behavior by combining signals from numerous modalities (such as speech, gestures, and language) [69]. Multimodal human behavior analysis requires a sufficient amount of high-quality, diversified datasets annotated manually for facilitating learning [70–75].

In general, the underlying ML/DL algorithms of mental health chatbots can learn about users' behavior from verbal and nonverbal signals.

Verbal signals. Linguistic patterns and word choice could be linked to a user's affective state. The LIWC [76] software package enables the automatic extraction of linguistic style features by capturing the frequency of words from different categories. Matching a person's linguistic style (for example, through word choice) is perhaps one of the simplest ways an assistant can be designed to emotionally bond with a person. Speech signals convey linguistic and paralinguistic information features [77, 78]. Selecting a suitable and robust set of features is a challenging task. The features that have been used in the recently published research are pitch, intensity, formants, Mel-frequency cepstral coefficients (MFCCs), and filter bank energies (FBE) [78]. Tools for syntactic and semantic analysis of discourse allow for natural language understanding and generation to provide meaningful conversations with users based on context analysis [79, 80]. However, speech recognition may be difficult due to the specifics of the patients' spoken language and an insufficient corpus for training speech recognition algorithms in these specific situations [79].

Nonverbal signals. Facial expressions are one of the richest sources of affective information. Automated facial action coding can be performed using highly scalable frameworks [81], allowing analysis of large datasets (for example, millions of individuals). These tools are easily accessible and relatively simple to integrate into other applications. They can even execute on resource-constrained devices enabling mobile applications of facial expression analysis. While expressed affective signals are those that are most used in social interactions, physiology plays a significant role in emotional responses. Computer systems can measure many of these signals in a way that an unaided human could not. The current state of work offers physiological emotion databases combining unimodal signals that can be used for measuring the nervous system's activities connected with emotions. They contain data including electroencephalogram (EEG), functional near-infrared (fNIR), cardiopulmonary parameters (heart and respiration rates), and skin conductance [73, 74]. Despite advances in sensing emotions, specific challenges in objective measurement remain. The sparsity and lack of specificity within unimodal cues (such as facial expression) are key reasons why multimodal affective computing systems have been found to be consistently better than unimodal ones [82]. The most common approach to label emotions is discrete categorization, which uses a set of basic emotions and the associated cognitive, physiological, and behavioral processes [83]. While several categorizations have been proposed, the most used set is the so-called "basic" list of emotions: anger, fear, sadness, disgust, surprise, and joy [83].

Table 1 The dimensions to analyze intelligent interactive technologies for mental well-being.

Dimension	Description
Transparency	Exposing a certain level of accessibility to the system's data and algorithms
Explainability	Clarifying what the system can do, how well the system can do it, and why the system made a decision
Privacy	Protecting user data during collection, analysis and use
Error management	Error prevention and recovery strategies
Context awareness	Recognizing and responding to the environment
Learning and personalization	Learning about user behavior and adapting and evolving accordingly
Empathy and social behavior	Understanding and responding to user emotions, and exposing sociability
Healthcare provision	Healthcare services for prevention, diagnosis, and treatment

3 Analytical Framework

In analyzing intelligent interactive technologies for mental health, we focus on elements of Human-AI interaction. We start from and adapt Human-AI interaction guidelines to explore existing technologies [84]. The original guidelines are instrumental and prescriptive concerning principles or rules to implement in interactive AI technologies [6] In contrast, we provide a descriptive analysis of how end-users perceive and interact with these technologies. Consequently, the dimensions are extracted, combined, and extended to capture relevant user-related aspects of mental health diagnosis, prevention, and treatment services (as summarized in Table 1).

3.1 Transparency and Explainability

Ground truth may not exist in making various medical decisions, including prevention, diagnosis, and treatment. The"black-box" nature of current AI medical systems (meaning that details on how their learning models work are not clear nor visible to non-technical people) can raise concerns with their users [7, 85]. One of the critical aspects of ML systems in medicine is that they do not understand context, being focused on the statistical processing of the data—they can identify phenomena but still do not explain them [86]. For example, the data can reveal that the users who

[6]The interactive cards with practical examples of the guidelines: https://bit.ly/3b3VMUr (Retrieved on Dec 25, 2020).

received a particular treatment demonstrated improvement but cannot tell us why that happened.

There is a tradeoff between increased requirements for ML/DL accuracy and explainability—higher accuracy can be less transparent (i.e., Convolutional Neural Network, CNN), while algorithms offering clear explanations (i.e., Decision Tree classifier) can lack accurate predictions [87]. In general, concerns regarding the lack of control in Human-AI interactions are emerging (such as decision making of the vehicle in autonomous driving). In a situation in which an AI algorithm can assist or be independent in making decisions, it will be necessary for users to understand how the algorithm came to a decision [88]. Consequently, there is an increasing demand for intelligent systems in healthcare that are not only learning about their users well (being personalized and useful) but are transparent, explainable, and understandable for medical professionals and patients [7]. These attributes are also important in building user trust and facilitating acceptance of these technologies. Currently, medical professionals use intelligent technologies as a means of support rather than a replacement. Even if future AI systems possess greater autonomy in healthcare decision making, doctors will still need to understand and follow the process.

Measuring the quality of explanations and communicating the explanations to different stakeholders (including doctors and patients) in a comprehensible and user-friendly way is an emerging requirement for medical AI systems [7, 89]. Some researchers have even proposed explanation interfaces for the underlying AI models and algorithms that can measure explainability as the quality of human-AI interaction [2], similar to usability (describing the ease of use). These interfaces should speak using non-technical language to provide understandable and trustable explanations for their decisions to users. Often, people do not fully understand the effects of certain health behaviors which may negatively influence their motivation. Providing comprehensible explanations can facilitate understanding, raise awareness concerning benefits, and persuade users to adopt a healthy lifestyle [7].

According to the source guidelines [84], transparency and explainability concerning system capabilities should be stated and clarified at the start of interactions with users.

3.2 Privacy

The issue of users' privacy is widely acknowledged in digital healthcare [90]. The main elements include ensuring that user data is always secure, that users can control their data, and that AI systems collect only necessary information about users [91].

ML/DL-based solutions store and use a high amount of data to provide a more advanced and more personalized service. Some systems also engage additional remote resources (usually cloud-based for data storage or processing) to improve performance. Therefore, how personal data is stored and transferred over the network is crucial—namely, what kind of encryption is used and if it is used correctly. This aspect is important through the device/solution's lifespan, including cases where it is stolen, broken, or discarded. Data anonymization functions, if applied correctly, can significantly improve privacy protection by removing personally identifiable information. If the solution uses remote resources or sends statistical data to the provider for further analysis, there is a risk of revealing the user's geographical location and daily patterns.

3.3 Error Management

Since the AI technology in medicine generally does not provide deterministic or predictable outcomes per se, managing errors is important. The occurrence of errors can be avoided by design, and systems can capture, handle, or learn from errors after they occur.

The occurrence of errors in the systems' is highlighted in the guidelines from [84]. They propose efficient support in the event of errors, dismissal and recovery, and error prevention by scoping functions based on the clear understanding of users' goals.

3.4 Context Awareness

Humans act in a material world and spend their time in different environments (including home, work, indoors, and outdoors). Therefore, the ability to sense, interpret and respond to the user's environment is an important factor in learning about their habits and behavior [92]. Combined with AI models that learn and predict specific mental conditions, they can improve automated decision-making accuracy. The capabilities are supported by mobile and wearable technologies for continuous monitoring, diagnosis, and treatment of mental health conditions [92].

The guidelines from [84] stress the necessity of continuous context understanding throughout Human-AI interaction. The system should provide services and information based on the user's current task and environment in a timely manner.

3.5 Learning and Personalization

This dimension's focus is the quality of the ML/DL algorithms' output concerning individual user's needs. The outcome ultimately depends on the algorithm's design, including the design of the model itself, data collection, and training process. However, given the nature of our analysis, we will not analyze the technical details of these algorithms, but their effectiveness as perceived by end-users. Specifically, how well they learn about users over time (perceived as accuracy and suitability of their outputs and decisions as recommendations, instructions, or answers), and what type of control over this process they provide to their users (being customizable concerning their services to match user preferences and expectations).

The guidelines in [84] suggest a Human-AI interaction that evolves so that the system keeps a record of its users' actions, learns about the users' behaviors and adapts its services accordingly, provides timely feedback on user actions and new capabilities and is customizable in that users can communicate their preferences and control the system's behavior.

3.6 Empathy and Social Behavior

Emotions play a crucial role in our well-being and communication with other people. Vice versa, the way we communicate with others influences our feelings. Interactive intelligent computer systems that respond to social and emotional cues can be more engaging and trusted while performing complex tasks in a more socially acceptable manner [93]. The field of affective computing concerns the design and development of computer systems that sense, interpret, adapt to, and potentially respond appropriately to human emotions [69]. Some authors even propose a taxonomy of the perceivable design elements connected with particular emotions [94]. The elements can be visual, verbal, auditory, and invisible (such as tactile sensations).

The starting guidelines [84] indicate that the results should be delivered to respect the user's social context and cultural background and avoid biases and stereotypical behavior towards users.

3.7 Healthcare Provision

We analyze the type of healthcare service that the system offers as prevention, diagnosis, and treatment; the number of target conditions (single or more); and the involvement of different stakeholders (such as patients, family, friends, peers, and medical professionals) [95].

4 Discussion and Outlook

Here, we analyze the state of the art technologies with respect to the dimensions from Table 1. and identify outlooks for the future.

4.1 Discussion

The discussion is structured according to the dimensions described in Chap. 3 (Table 1).

4.2 Transparency and Explainability

Robotic technologies and video games serve two general purposes concerning support for instrumental and hedonic activities. The instrumental activities are focused on fulfilling practical goals, such as activities for daily living and specific mental therapies with robotic technologies [8–12, 14–22, 27–34]. The hedonic activities are engaged in for fun and enjoyment, while the concrete health outcome is a side-effect of the gameplay [50–57].

Concerning explainability, robotic technologies do not generally provide reasons for their decisions, instructions, or suggestions to users [14–22, 27–30, 35–37]. The details of the underlying AI algorithms and data are not accessible to users (which may be expected since the robots have physical appearances.) [14–22].

The video games, by their nature, do not expose the 'serious' parts concerning AI mechanisms and data to their users, nor they explain the rules and decision of the gameplay [44–57].

On the other hand, conversational agents have dialogues with their users, and each talk may require an exchange of explanations for mutual understanding between the collocutors [60–68]. This understanding is critical in conversations between doctors and patients due to healthcare complexity and uncertainty. Mental chatbots, such as Wysa [68] and Woebot [60], clarify their decisions and the reasons for asking the user for specific data to some extent [95]. The chatbots' capabilities need to be set in advance for their users to match expectations regarding mental health provision (stating what they can do).

4.3 Privacy

Observed privacy is mainly manifested with user data collection. By analyzing the existing literature, we found that systems can collect data either explicitly or implicitly. The former assumes that users provide their data explicitly, such as data entry from the system's interface, responding to a questionnaire, or through conversations. The latter concerns using various sensors (such as cameras, eye-tracking devices, and mobile device sensors for measuring position and orientation) that monitor and record user behavior during interaction ors logging user actions for future analysis.

In robotic technologies [8–12, 14–22, 35–41] and video games [44–57], data collection is predominantly implicit. Conversational agents collect data from their users explicitly during conversation sessions [60–68].

Dealing with implicitly collected data requires caution as they can reveal the user's identity from biometric information [101].

4.4 Error Management

The evidence on error prevention and breakdown management is scarce in robotic technologies and video games. More specific strategies and actions concerning error prevention, support, dismissal, and recovery are not clearly described.

Dialog breakdowns are common in conversational agents as their knowledge and capabilities still do not match humans.

A common error prevention technique is scoping the dialog as multiple questions with user choice constraints such as predefined answers [60, 68, 95].

Error repair methods mainly involve preventing users from modifying previous inputs [60, 68, 95]. Although framing conversations with close-ended questions can help chatbots understand their users, user input errors can still occur.

4.5 Context Awareness

Robotic technologies are meant to be used in the physical world and interact with their users in different environments (i.e., homes, schools, public spaces, healthcare facilities, etc.). The prerequisites for a successful interaction include sensing, interpreting, and responding to the user's environment. The majority of the robots are equipped with sensors that collect various data about the physical environment (i.e., light, noise, temperature, objects and structures) and the users (i.e., posture, gait, speech, facial expression and other people) [8–12, 14–22, 27–41].

The collected information is used to supplement previously learned tasks and user behaviors to improve the quality of the robot's services.

Video games run in a virtual environment and the majority of them do not make use of the context information [44–57].

Similarly, chatbots implement natural language understanding and generation solely from the underlying algorithmic data and information collected from dialogs with their users [60–68].

4.6 Learning and Personalization

All technologies can learn about their users and adjust mental healthcare provision for better outcomes. We highlight their specificities. Robotic technologies learn about user activities, habits, preferences and mental health conditions. Over time, they collect data implicitly from monitoring user's activities (using multiple sensors) and recording explicit interactions with them. The collected data are continuously integrated and analyzed. The results of the analyses provide insights on the user's health conditions, based on which the system generates recommendations for activities (in case of diagnosis and prevention) or adapts instructions in ongoing training sessions (in case of therapy or rehabilitation) [8–12, 14–22, 27–41]. Additionally, most robots provide feedback on the user's progress concerning current training and activities, and offer options for users to specify their preferences and needs from the system (enabling customization). [8–12, 14–22, 27–41].

Video games naturally hide healthcare tasks from their users. In this sense, the customization and feedback are implemented, but mainly concerning the gameplay (i.e., fun purpose) not their healthcare role (i.e., serious purpose) [44–57]. The games are designed so that particular gameplay tasks are matched with healthcare tasks. This way, the gameplay represents a facade for prevention, diagnosis, or therapy activities in the background. The data are collected while users are playing the game, and analyzed for the correct assessment of users' health conditions. Based on the user's leaned behaviour, the game adapts to the users' to maintain or improve the health conditions' parameters [44–57]. Some features of the games (such as levels, activities engaging one or multiple human cognitive functions/senses/actuators, and rewards) lend themselves well to mental healthcare. However, it may not always be possible to find or design a suitable game.

Commercial gaming platforms (see Sect. 2.2, last paragraph) are increasingly integrating AI elements into their services. We see this as a future opportunity for developing learning capabilities of digital healthcare applications.

Chatbots learn from the dialogs they conduct with their users on a regular basis (using text, speech, visual elements, or a combination of the three). They analyze user input to extract and understand intentions, create content for responses, and communicate natural language responses. [102]. The described mechanism is used to study about user behavior to deliver mental healthcare in conversation sessions. NLP algorithms are crucial for engaging in a human-like conversation with users. In particular, correct understanding of user utterances and providing appropriate responses are critical due to the complexity and diversity of languages. The chatbots are still not capable of conversing as humans do, such as inferring meaning from the context of a complex text or speech, dealing with dialects, and recognizing metaphors, jokes, or sarcasm, etc.

4.7 Empathy and Social Behavior

Video games expose empathy by communicating messages containing emotional cues to their users during gameplay to incentivize their players (such as motivational tips, greetings, etc.). Overall, they are not capable of recognizing and responding to user emotions during the interactions.

On the other hand, implementing emotionally-aware behavior in robotic technologies and conversational agents has received significant attention in research[62, 82, 93–97]. A related phenomenon regarding robots is the uncanny valley effect [103] describing the relationship between the degree to which a robot resembles a human and the emotional response to such a robot. In case the robot imperfectly resembles a human, they can provoke feelings of uneasiness and repulsion in observers.

At the moment, conversational agents expose emotionally-sensitive behavior mainly from the written dialogs with their users [60, 68]. They extract sentiments from users' utterances and generate appropriate responses as they detect the change in the user's mood during the conversation. In addition, recent research suggests chatbot design elements (textual, visual, and auditory) that can elicit particular emotions from users [94].

4.8 Healthcare Provision

Robotic technologies are used in treating particular mental conditions or facilitating mental well-being as prevention. Some robots address several related aspects of mental functioning, such as in ASD [8–14, 27, 28]. Others tackle different conditions (i.e., NEO robot from [15–20]). Pediatric care robots are mainly used to preserve children's mental functioning [23–30]. Robotic technologies for older adults are also applied primarily during the prevention stage of mental health [35–42]. Stakeholders other than target user groups are not directly involved the use of technologies with respect to their purpose.

Video games are mainly present in the prevention period to alleviate certain conditions (such as ASD) [44–49] and maintain cognitive functioning in old age [50–57]. Overall, gameplay does not include stakejolders from their users' social circles (i.e., doctors, formal/informal carers, and family members).

Mental health chatbots encompass all aspects of healthcare: prevention, diagnosis, and therapy [60–68]. Interactions are conducted with patients. They offer sustained conversations with users in their homes. In light of the current pandemic, telemedicine is becoming more ubiquitous [104, 105] as a remote and contactless healthcare provision. Chatbots represent a fundamental part of this change' gear of that shift.

4.9 The Outlook

We articulate the outlook for designing intelligent interactive technologies for mental health along three main lines.

Emotionally-aware systems. The technology can be described from the user perspective, in terms of the desirable hedonic and pragmatic qualities. Hedonic or experiential attributes refer to the affective and mental affordances it should provide (such as trust, emotions, personalization, and learning from interactions). Pragmatic or instrumental qualities concern the functional affordances it should enable (such as usability or ease of use, and usefulness).

One of the main criticisms of existing technologies for mental healthcare is that they are not capable of empathy, namely recognizing users' emotional states and tailoring responses that reflect these emotions [62, 93, 96, 97]. The lack of empathy negatively affects engagement with the technology. Sustained and regular user engagement is an prerequisite for effective health intervention and crucial for creating training data to improve algorithms for learning about users. A sucessful affective system would be one that could learn about a person's nuanced expressions and responses, and adapt to different situations and contexts.

These models have common requirements. Firstly, the models should be machine-readable and computable. Secondly, they should combine verbal and nonverbal cues from different modalities (such as speech, text, and vision) to improve emotion recognition accuracy.

Scalable and scaffolded learning about user behavior. Cognitive psychology proposes bottom-up information processing (created from our senses) and top-down inference (based on the existing knowledge about the things from the world such as words, concepts, objects, actions, and relations) [98].

Similarly, hybrid learning that combines top-down models (such as knowledge graphs, rule-based systems, and symbolic representations) and bottom-up approaches (such as ML and DL) can yield higher quality automatic predictions and decisions [99], or make (healthcare) AI systems more comprehensible to their users [100]. Although algorithms use approximations of real-world data and simulate human abilities, knowledge about users can unfold over time while making the system readily understandable. The technological framework can integrate different modalities (e.g., facial expressions, text and speech content and linguistic patterns, and physiological signals) to optimize accuracy, performance, and computational cost.

Concerning the interpretation of emotions, different models of emotions can be employed, such as a circular, two-dimensional space in which points close to one another are highly correlated [83]. This approach may provide a continuous and fine-grained interpretation of multiple emotions at once. For example, in some cases, none of the "basic" emotion labels may apply to a specific, observed emotional response, but that response can lie somewhere within the dimensional space. The deployed framework can generate domain-specific training data (or model) for emotion recognition applications that embrace the target user group's cognitive and socio-demographic characteristics.

Continuous and affordable mental healthcare provision. A multi-stage intervention that combines in-hospital treatment and out-of-hospital care. Current healthcare systems aim for a balance among improving the health of specific populations, delivering quality care to individuals, and cost control. Computerized, personalized, and online assistance and mental health treatment can reduce the burden on healthcare providers and services while increasing their effectiveness. This approach can also support informal caregiving (at places other than hospitals, such as in the home) in mental therapy to improve its overall effectiveness. This methodology brings three significant benefits.

Such an approach can enable customized and adaptive online treatment (including a mobile version) that can combine different design elements and assess their effectiveness in improving mental health and well-being with decreased data acquisition and evaluation costs.

Subsequently, advanced evaluation and quality assurance of mental health treatments in terms of existing medical standards, user attitudes, usability, and treatment effectiveness. The possibility to engage a larger number and a broader range of participants with different levels of cognitive functioning or home bound users.

Finally, the formation of a multidisciplinary community for human-friendly AI for healthcare that brings together experts from related fields (such as medicine, artificial intelligence, human-computer interaction, and information systems). The community can raise the profile of this research within these communities.

5 Conclusion and Limitations

This paper provides an overview of existing interactive AI technologies for mental health and well-being. The technologies were analyzed from different aspects of Human-AI interaction. The findings can be informative for various stakeholders, including medical professionals, researchers, engineers, and end-users. In line with the existing initiatives for more inclusive and accessible healthcare [104, 105], we consider this paper a step towards this global goal.

Being a literature/research review, the present study does not claim to be comprehensive. Instead, we summarize the existing literature and research at the time of writing and focus on the shared elements of different technologies from the Human-AI interaction perspective. We identify and describe the aspects concerning the analytical framework and derive the outlook for future research. Our analysis is based on papers and not on real systems due to their lack of availability (this excludes publicly available chatbots such as Woebot [60] and Wysa [68]). It may influence overall findings. Further analysis of the specific technologies is required.

References

1. Du, M., Liu, N., Hu, X.: Techniques for interpretable machine learning. Commun. ACM **63**(1), 68–77 (2019)
2. Holzinger, A., Langs, G., Denk, H., Zatloukal, K., Müller, H.: Causability and explainability of artificial intelligence in medicine. Wiley Interdiscipl. Rev.: Data Min. Knowl. Discov. **9**(4), (2019)
3. Killgore, W.D., Cloonen, S.A., Taylor, E.C., Dailey, N.S.: Loneliness: a signature mental health concern in the era of COVID-19. Psych. Res. **113117**, (2020)
4. Rossi, A., Panzeri, A., Pietrabissa, G., Manzoni, G.M., Castelnuovo, G., Mannarini, S.: The anxiety-buffer hypothesis in the time of COVID-19: when self-esteem protects from the impact of loneliness and fear on anxiety and depression. Front. Psychol. **11** (2020)
5. Ebert, D.D., Cuijpers, P., Munoz, R.F., Baumeister, H.: Prevention of mental health disorders using internet-and mobile-based interventions: a narrative review and recommendations for future research. Front. Psych. **8**, 116 (2017)
6. Fitzpatrick, K.K., Darcy, A., Vierhile, M.: Delivering cognitive behavior therapy to young adults with symptoms of depression and anxiety using a fully automated conversational agent (Woebot): a randomized controlled trial. JMIR Ment. Heal. **4**(2) (2017)
7. Dragoni, M., Donadello, I., Eccher, C.: Explainable AI meets persuasiveness: translating reasoning results into behavioral change advice. Artif. Intell. Med. **101840** (2020)
8. Wainer, J., et al.: A pilot study with a novel setup for collaborative play of the humanoid robot KASPAR with children with autism. Int. J. Soc. Robot. **6**(1), 45–65 (2014)
9. Zheng, Z., et al.: Robot-mediated imitation skill training for children with autism. IEEE Trans. Neural Syst. Rehabil. Eng. **24**(6), 682–691 (2015)
10. Kim, E.S., et al.: Social robots as embedded reinforcers of social behavior in children with autism. J. Aut. Dev. Disord. **43**(5), 1038–1049 (2013)
11. DE-Enigma: Playfully empowering autistic children. EC-funded research project. http://de-enigma.eu/. Cited 25 Dec 2020
12. DREAM: Development of Robot-Enhanced therapy for children with AutisM spectrum disorders. EC-funded research project. https://www.dream2020.eu. Cited 25 Dec 2020
13. Scoglio, A.A., Reilly, E.D., Gorman, J.A., Drebing, C.E.: Use of social robots in mental health and well-being research: Systematic review. J. Med. Internet Res. **21**(7) (2019). https://doi.org/10.2196/13322
14. Desideri, L., Negrini, M., Malavasi, M., Tanzini, D., Rouame, A., Cutrone, M.C., Bonifacci, P., Hoogerwerf, E.-J.: Using a humanoid robot as a complement to interventions for children with autism spectrum disorder: A pilot study. Adv. Neurodev. Disord. **2**(3), 273–285 (2018). https://doi.org/10.1007/s41252-018-0066-4
15. Robinson N.L., Connolly J., Hides L., Kavanagh D.J.: A Social Robot to Deliver an 8-Week Intervention for Diabetes Management: Initial Test of Feasibility in a Hospital Clinic. In: Wagner A.R. et al. (eds) Social Robotics. ICSR 2020. Lecture Notes in Computer Science, vol. 12483. Springer, Cham (2020). https://doi.org/10.1007/978-3-030-62056-1_52
16. Sánchez, S. M., Mora-Simon, S., Herrera-Santos, J., Roncero, A. O., Corchado, J. M.: Intelligent Dolls and robots for the treatment of elderly people with dementia. ADCAIJ: Adv. Distrib. Computi. Artif. Intell. J. **9**(1), 99–112 (2020)
17. Aslam, S., Standen, P. J., Shopland, N., Burton, A., Brown, D.: A comparison of humanoid and non-humanoid robots in supporting the learning of pupils with severe intellectual disabilities (2016)
18. Assad-Uz-Zaman, M., Rasedul Islam, M., Miah, S., Rahman, M.H.: NAO robot for cooperative rehabilitation training. J. Rehabil. Assist. Technol. Eng. **6**, 2055668319862151 (2019)
19. Trost, M.J., Ford, A.R., Kysh, L., Gold, J.I., Matarić, M.: Socially assistive robots for helping pediatric distress and pain. Clin. J. Pain **35**(5), 451–458 (2019)
20. Egido-García, V., Estévez, D., Corrales-Paredes, A., Terrón-López, M.-J., Velasco-Quintana, P.-J.: Integration of a social robot in a pedagogical and logopedic intervention with children: a case study. Sensors **20**(22), 6483 (2020). https://doi.org/10.3390/s20226483

21. Polycarpou, Panayiota Andreeva, Anna Ioannou, Andri Zaphiris, Panayiotis.: Don't Read My Lips: Assessing Listening and Speaking Skills Through Play with a Humanoid Robot. HCI International 2016 – Posters' Extended Abstracts, 618. 255-260. (2016) https://10.1007/978-3-319-40542-1_41
22. Manzi, F., Massaro, D., Di Lernia, D., Maggioni, M.A., Riva, G., Marchetti, A.: Robots Are Not All the Same: Young Adults' Expectations, Attitudes, and Mental Attribution to Two Humanoid Social Robots. Behavior, and Social Networking, Cyberpsychology (2020)
23. Jeong, S., Breazeal, C., Logan, D., Weinstock, P.: Huggable: The impact of embodiment on promoting socio-emotional interactions for young pediatric inpatients. In: Proceedings of the 2018 CHI Conference on Human Factors in Computing Systems, pp. 1–13 (2018). https://doi.org/10.1145/3173574.3174069
24. Westlund, J. K. et al.,: Tega: A social robot. In: 11th ACM/IEEE International Conference on Human-Robot Interaction (HRI), pp. 561–561 (2016). https://doi.org/10.1109/HRI.2016.7451856
25. Aghel Manesh, S., Beran, T., Sharlin, E., Greenberg, S.: Medi, human robot interaction in pediatric health. In: CHI -14 Extended Abstracts on Human Factors in Computing Systems, pp. 153–154 (2014). https://doi.org/10.1145/2559206.2579529
26. THERAPIST: Towards an autonomous socially interactive robot for motor and neurorehabilitation therapies for children. JMIR Rehabil. Assist. Technol. (2014)
27. Shimaya, J., Yoshikawa, Y., Matsumoto, Y., Kumazaki, H., Ishiguro, H., Mimura, M., Miyao, M.: Advantages of indirect conversation via a desktop humanoid robot: case study on daily life guidance for adolescents with autism spectrum disorders (2016). https://doi.org/10.1109/roman.2016.7745215
28. Kumazaki, H., Warren, Z., Corbett, B.A., Yoshikawa, Y., Matsumoto, Y., Higashida, H., Kikuchi, M.: Android robot-mediated mock job interview sessions for young adults with autism spectrum disorder: a pilot study. Front. Psych. **8**, 169 (2017)
29. Giannopulu, I., Montreynaud, V., Watanabe, T.: Minimalistic toy robot to analyze a scenery of speaker-listener condition in autism. Cognit. Process. **17**(2), 195–203 (2016). https://doi.org/10.1007/s10339-016-0752-y
30. Giannopulu, I., Montreynaud, V., Watanabe, T.: Neurotypical and autistic children aged 6 to 7 years in a speaker-listener situation with a human or a minimalist InterActor robot. In: Proceedings of IEEE RO-MAN, pp. 942–947 (2014)
31. Poppy Project - Open source robotic platform. https://www.poppy-project.org/. Cited 25 Dec 2020
32. Metta, G., Sandini, G., Vernon, D., Natale, L., Nori, F.: The iCub humanoid robot: an open platform for research in embodied cognition (2008)
33. Sproewitz, A., Kuechler, L., Tuleu, A., Ajallooeian, M., D'Haene, M., Moeckel, R., Ijspeert, A.: Oncilla robot: a light-weight bio-inspired quadruped robot for fast locomotion in rough terrain (2011)
34. Zhao, N., Zhang, Z., Wang, Y., Wang, J., Li, B., et al.: See your mental state from your walk: Recognizing anxiety and depression through Kinect-recorded gait data (2019). https://doi.org/10.1371/journal.pone.0216591
35. Fan, J., Beuscher, L., Newhouse, P. A., Mion, L. C., Sarkar, N.: A robotic coach architecture for multi-user human-robot interaction (RAMU) with the elderly and cognitively impaired (2016). https://doi.org/10.1109/roman.2016.7745157
36. Tapus, A.: The role of the physical embodiment of a music therapist robot for individuals with cognitive impairments: Longitudinal study. In: Virtual Rehabilitation International Conference, Haifa (2009)
37. Cavallo, F., Esposito, R., Limosani, R., Manzi, A., Bevilacqua, R., Felici, E., Di Nuovo, A., Cangelosi, A., Lattanzio, F., Dario, P.: Robotic services acceptance in smart environments with older adults: user satisfaction and acceptability study. J. Med. Internet Res. **20**(9), e264 (2018)
38. Kang, H.S., Makimoto, K., Konno, R., Koh, I.S.: Review of outcome measures in PARO robot intervention studies for dementia care. Geriat. Nurs. **41**(3), 207–214 (2020). https://doi.org/10.1016/j.gerinurse.2019.09.003

39. Low, J.T.S.: LECABot, mini robot companion for elderly (2020)
40. Kuwamura, K., Nishio, S., Sato, S.: Can we talk through a robot as if face-to-face? Long-term fieldwork using teleoperated robot for seniors with Alzheimer's disease. Front. Psychol. **7**, 1066 (2016)
41. Orlandini, A., Kri'ffersson, A., Almquist, L., Björkman, P., Cesta, A., Cortellessa, G., ... Loutfi, A.: Excite project: A review of forty-two months of robotic telepresence technology evolution. Presence: Teleoper. Virtual Environ. **25**(3), 204–221 (2016)
42. Alvarez, J., Campos, G., Enríquez, V., Miranda, A., Rodriguez, F., Ponce, H.: Nurse-bot: a robot system applied to medical assistance. In: 2018 International Conference on Mechatronics, Electronics and Automotive Engineering (ICMEAE), pp. 56–59. IEEE (2018)
43. Hortensius, R., Hekele, F., Cross, E.S.: The perception of emotion in artificial agents. IEEE Trans. Cognit. Dev. Syst. **10**(4), 852–864 (2018)
44. Porayska-Pomsta, K., et al.: Blending human and artificial intelligence to support autistic children's social communication skills. ACM Trans. Comput.-Hum. Interact. (TOCHI) **25**(6), 35 (2018)
45. Tsikinas, S., Xinogalos, S.: Studying the effects of computer serious games on people with intellectual disabilities or autism spectrum disorder: a systematic literature review. J. Comput. Assist. Learn. **35**(1), 61–73 (2019)
46. Malinverni, L., et al.: An inclusive design approach for developing video games for children with autism spectrum disorder. Comput. Hum. Behav. **71**, 535–549 (2017)
47. Ip, H.H., et al.: Enhance emotional and social adaptation skills for children with autism spectrum disorder: a virtual reality enabled approach. Comput. Educ. **117**, 1–15 (2018)
48. Grossard, C., Grynspan, O., Serret, S., Jouen, A. L., Bailly, K., Cohen, D.: Serious games to teach social interactions and emotions to individuals with autism spectrum disorders. ASD Computers & Education, 113, pp. (2017)
49. Chen, J., et al.: A pilot study on evaluating children with autism spectrum disorder using computer games. Comput. Hum. Behav. **90**, 204–214 (2019)
50. Cheng, V.W.S., Davenport, T., Johnson, D., Vella, K., Hickie, I.B.: Gamification in apps and technologies for improving mental health and well-being: systematic review. JMIR Ment. Heal. **6**(6) (2019)
51. Ballesteros, S., Kraft, E., Santana, S., Tziraki, C.: Maintaining older brain functionality: a targeted review. Neurosci. Biobehav. Rev. **55**, 453–477 (2015)
52. Pallavicini, F., Ferrari, A., Mantovani, F.: Video games for well-being: a systematic review on the application of computer games for cognitive and emotional training in the adult population. Front. Psychol. **9**, 2127 (2018)
53. Hutchinson, C.V., Barrett, D.J., Nitka, A., Raynes, K.: Action video game training reduces the Simon Effect. Psychon. Bull. Rev. **23**(2), 587–592 (2016)
54. Clemenson, G.D., Stark, C.E.: Virtual environmental enrichment through video games improves hippocampal-associated memory. J. Neurosci. **35**(49), 16116–16125 (2015)
55. Villani, D., Carissoli, C., Triberti, S., Marchetti, A., Gilli, G., Riva, G.: Video games for emotion regulation: a systematic review. Games for Heal. J. **7**(2), 85–99 (2018)
56. Colder Carras, M., Van Rooij, A.J., Spruijt-Metz, D., Kvedar, J., Griffiths, M.D., Carabas, Y., Labrique, A.: Commercial video games as therapy: a new research agenda to unlock the potential of a global pastime. Front. Psych. **8**, 300 (2018)
57. Deterding, S., Dixon, D., Khaled, R., Nacke, L.: From game design elements to gamefulness: defining "gamification". In: Proceedings of the 15th International Academic MindTrek Conference: Envisioning Future Media Environments, pp. 9–15 (2011)
58. Floryan, M., Chow, P.I., Schueller, S.M., Ritterband, L.M.: The model of gamification principles for digital health interventions: evaluation of validity and potential utility. J. Med. Internet Res. **22**(6) (2020)
59. Stoyanov, S.R., Hides, L., Kavanagh, D.J., Zelenko, O., Tjondronegoro, D., Mani, M.: Mobile app rating scale: a new tool for assessing the quality of health mobile apps. JMIR Health Health **3**(1) (2015)

60. Woebot - a text-based, virtual companion who advises people about their mental well-being. https://woebot.io/. Cited 25 Dec 2020
61. Lee, M., et al.: Caring for Vincent: A Chatbot for Self-Compassion. In Proceedings of the 2019 CHI Conference on Human Factors in Computing System, p. 702. ACM (2019)
62. Yang, X., Aurisicchio, M., Baxter, W.: Understanding Affective Experiences With Conversational Agents. In Proceedings of the 2019 CHI Conference on Human Factors in Computing Systems, p. 542, ACM. (2019)
63. Zhou, M.X., et al.: Trusting virtual agents: the effect of personality. ACM Trans. Interact. Intell. Syst. (TiiS) **9**(2–3), 10 (2019)
64. Jain, M., et al.: Evaluating and informing the design of chatbots. In Proceedings of the 2018 Designing Interactive Systems Conference, pp. 895–906. ACM (2018)
65. Babylon - accessible and affordable health service. https://www.babylonhealth.com/. Cited 25 Dec 2020
66. Florence - personal health assistant. https://florence.chat/. Cited 25 Dec 2020
67. Miner, A., et al.: Conversational agents and mental health: Theory-informed assessment of language and affect. In: Proceedings of the 4th ACM Intern. Conf. Human Agent Interaction, pp. 123–130 (2016)
68. Inkster, B., Sarda, S., Subramanian, V.: An empathy-driven, conversational artificial intelligence agent (Wysa) for digital mental well-being: real-world data evaluation mixed-methods study. JMIR mHealth and uHealth **6**(11) (2018)
69. Poria, S., et al.: A review of affective computing: from unimodal analysis to multimodal fusion. Inf. Fusion **37**, 98–125 (2017)
70. Jackson, P., Haq, S.: Surrey audio-visual expressed emotion (SAVEE) database. University of Surrey, Guildford (2014)
71. Martin, O., Kotsia, I., Macq, B., Pitas, I.: The eNTERFACE'05 audio-visual emotion database. In: 22nd International Conference on Data Engineering Workshops (ICDEW'06), pp. 8–8. IEEE (2006)
72. Wang, Y., Guan, L.: Recognizing human emotional state from audiovisual signals. IEEE Trans. Multimed. **10**(5), 936–946 (2008)
73. Correa, J.A.M. et al.: Amigos: a dataset for affect, personality and mood research on individuals and groups (2018)
74. Song, T., et al.: MPED: a multi-modal physiological emotion database for discrete emotion recognition. IEEE Access **7**, 12177–12191 (2019)
75. Ong, D., Wu, Z., Tan, Z. X., Reddan, M., Kahhale, I., Mattek, A., Zaki, J.: Modeling emotion in complex stories: the Stanford Emotional Narratives Dataset (2019)
76. LIWC (Linguistic Inquiry and Word Count) http://liwc.wpengine.com/. Retrieved on Dec 25, 2020
77. Schuller, B., et al.: Paralinguistics in speech and language-State-of-the-art and the challenge. Comput. Speech Lang. **27**(1), 4–39 (2013)
78. Kamińska, D., Sapiński, T., Anbarjafari, G.: Efficiency of chosen speech descriptors in relation to emotion recognition. EURASIP J. Audio Speech Music Process. **3** (2017)
79. Hwang, I., Lee, Y., Yoo, C., Min, C., Yim, D., Kim, J.: Towards interpersonal assistants: next-generation conversational agents. IEEE Pervas. Comput. **18**(2), 21–31 (2019)
80. Abdul-Kader, S.A., Woods, J.C.: Survey on chatbot design techniques in speech conversation systems. Int. J. Adv. Comput. Sci. Appl. **6**(7), 72–80 (2015)
81. McDuff, D., Girard, J.M., El Kaliouby, R.: Large-scale observational evidence of cross-cultural differences in facial behavior. J. Nonverbal Behav. **41**(1), 1–19 (2017)
82. McDuff, D. et al.: A Multimodal Emotion Sensing Platform for Building Emotion-Aware Applications (2019)
83. Marsella, S., Gratch, J., Petta, P.: Computational models of emotion. Blueprint for Affective Computing - A Sourcebook and Manual, pp. 21–46. Oxford University Press (2010)
84. Amershi, S. et al.: Guidelines for Human-AI Interaction (2019)

85. The IEEE Global Initiative on Ethics of Autonomous and Intelligent Systems,: Ethically aligned design: A vision for prioritizing human well-being with autonomous and intelligent systems. Affective Computing, pp. 90–109. First Edition. IEEE (2019). Retrieved on Dec 25, 2020 from https://bit.ly/3pTNtzv
86. Pearl, J., Mackenzie, D.: The book of why: the new science of cause and effect (2018)
87. Bologna, G., Hayashi, Y.: Characterization of symbolic rules embedded in deep DIMLP networks: a challenge to transparency of deep learning. J. Artif. Intell. Soft Comput. Res. **7**(4), 265–286 (2017)
88. Adadi, A., Berrada, M.: Peeking inside the black-box: a survey on explainable artificial intelligence. XAI IEEE Access **6**, 52138–52160 (2018)
89. Miller, T.: Explanation in artificial intelligence: insights from the social sciences. Artif. Intell. **267**, 1–38 (2019)
90. Nurgalieva, L., O'Callaghan, D., Doherty, G.: Security and privacy of mHealth applications: a scoping review. IEEE Access **8**, 104247–104268 (2020)
91. Broadbent, E.: Interactions with robots: the truths we reveal about ourselves. Ann. Rev. Psychol. **68**, 627–652 (2017)
92. Bardram, J.E., Matic, A.: A decade of ubiquitous computing research in mental health. IEEE Pervas. Comput. **19**(1), 62–72 (2020)
93. McDuff, D., Czerwinski, M.: Designing emotionally sentient agents. Commun. ACM **61**(12), 74–83 (2018)
94. Feine, J., Gnewuch, U., Morana, S., Maedche, A.: A taxonomy of social cues for conversational agents. Int. J. Hum.-Comput. Stud. **132**, 138–161 (2019)
95. Jovanovic, M., Baez, M., Casati, F.: Chatbots as conversational healthcare services. IEEE Internet Computing (2020)
96. Nadarzynski, T., Miles, O., Cowie, A., Ridge, D.: Acceptability of artificial intelligence (AI)-led chatbot services in healthcare: a mixed-methods study. Digit. Heal. **5** (2019)
97. Grudin, J., Jacques, R.: Chatbots, Humbots, and the quest for artificial general intelligence. In: Proceedings of the 2019 CHI Conference on Human Factors in Computing Systems, p. 209. ACM (2019)
98. Delorme, A., Rousselet, G.A., Macé, M.J.M., Fabre-Thorpe, M.: Interaction of top-down and bottom-up processing in the fast visual analysis of natural scenes. Cognit. Brain Res. **19**(2), 103–113 (2004)
99. Mao, J., Gan, C., Kohli, P., Tenenbaum, J. B., Wu, J.: The neuro-symbolic concept learner: Interpreting scenes, words, and sentences from natural supervision (2019)
100. Gaur, M., Faldu, K., Sheth, A.: Semantics of the Black-Box: can knowledge graphs help make deep learning systems more interpretable and explainable? (2020)
101. Neal, T., Sundararajan, K., Woodard, D.: Exploiting linguistic style as a cognitive biometric for continuous verification. In: 2018 International Conference on Biometrics (ICB), pp. 270–276 (2018). https://doi.org/10.1109/ICB2018.2018.00048
102. Harms, J.G., Kucherbaev, P., Bozzon, A., Houben, G.J.: Approaches for dialog management in conversational agents. IEEE Internet Comput. **23**(2), 13–22 (2019)
103. Mori, M., MacDorman, K.F., Kageki, N.: The uncanny valley [from the field]. IEEE Robot. Autom. Mag. **19**(2), 98–100 (2012). https://doi.org/10.1109/MRA.2012.2192811
104. Khemapech, I.: Telemedicine-Meaning, Challenges and Opportunities (2019)
105. Craig, J., Patterson, V.: Introduction to the practice of telemedicine. J. Telemed. Telecare **11**(1), 3–9 (2005). https://doi.org/10.1258/1357633053430494

The manufacturer's authorised representative in the EU is Springer
Nature Customer Service Centre GmbH, Europaplatz 3, 69115 Heidelberg,
Germany. If you have any concerns regarding our products, please
contact ProductSafety@springernature.com

Printed and bound by CPI Group (UK) Ltd, Croydon, CR0 4YY
28/09/2026
02224090-0018